Strangers in the Night

DAVID E. FISHER

AND MARSHALL JON FISHER

Strangers in the Night

A Brief History of Life on Other Worlds

A CORNELIA & MICHAEL BESSIE BOOK

COUNTERPOINT
WASHINGTON, D.C.

Library of Congress Cataloging-in-Publication Data
Fisher, David E., 1932–
    Strangers in the night : a brief history of life on other worlds / David E. Fisher and Marshall Jon Fisher.
        p.   cm.
        "A Cornelia and Michael Bessie book."
        Includes bibliographical references and index.
    ISBN 1–887178–87–2 (alk. paper)
    1. Life—Origin.   2. Exobiology.   3. Life on other planets.
    I. Fisher, Marshall.   II. Title.
    QH325.F47   1998
    576.8'.39—dc21   98–35513

Book design and electronic production by David Bullen Design

Printed in the United States of America on acid-free paper that meets the American National Standards Institute z39-48 Standard

☛ A Cornelia & Michael Bessie book
COUNTERPOINT
P. O. Box 65793
Washington, D. C. 20035-5793

Counterpoint is a member of the Perseus Books Group.

10   9   8   7   6   5   4   3   2   1

FIRST PRINTING

For Ron, Liz, and Dave

Contents

Acknowledgments

We are grateful to the following for conversations and/or correspondence: Ron Fisher; Everett Gibson, Jr.; Chris Romanek; Steve Squyres; Harry McSween, Jr.; Don Bogard; Bruce Hapke; John Kerridge; William Cochran; Gerald Gulley; Paul Horowitz; to librarians Kay Hale and Helen Albertson and various other members of the Rosenstiel School of Marine and Atmospheric Science in Miami and the Lunar and Planetary Institute in Houston. Since the opinions of several of these people are diametrically opposed, it should be clear that we alone bear the responsibility for interpretations and opinions expressed in this book. We want to express our appreciation to members of the the geology department at Southern Illinois University for their hospitality during preparation of the manuscript, and to the Sloan Foundation for financial support.

Notes

Further information on many of the topics under discussion is provided in the Notes section.

Preface

The question of whether life exists anywhere in the universe besides our own Earth is one of the most important religious, philosophical, and scientific questions that can be asked. Although Copernicus and Galileo convinced us almost four hundred years ago that we are not situated at the center of the physical universe, we have continued to argue over whether or not we are the center of the metaphysical one. Do we live in a universe with a nearly infinite number of islands of life scattered throughout it, or are we unique? Is life a natural phenomenon, to be found wherever some rather simple physical and chemical conditions are in place, or is life a miracle, found nowhere but here on this lonely, lovely planet?

The scientific consensus is strong that miracles do not happen. By definition they are a violation of physical law, and we choose to believe that we live in an orderly—if chaotic—universe in which physical principles are inviolate. It therefore follows that life should spring forth as part of the natural evolution of the universe, in at least a statistically significant portion of those cases where the right conditions are present. Since those conditions, so far as we now know, are rather simple and ordinary, and as the possible locations for them are so numerous throughout the universe, it follows that life should be a nearly ubiquitous phenomenon; at the very least, life here on Earth should not be a unique phenomenon.

But until we actually find life somewhere else we should admit that we know nothing about the probability of its occurrence. Life on Earth could be unique; until we *know* otherwise, we are arguing faith instead of science.

In this book we take a look at the history of our ideas about life on other worlds, from antiquity to the present. Our focus on these ideas has been stimulated by the very exciting scientific breakthroughs made in the past couple of years: the discoveries of possible fossil organisms in a meteorite from Mars; of planets around a variety of different stars; of a possible ocean on one of Jupiter's moons, combined with the presence of life deep in our own oceans, hidden from sunlight; and the vast improvements in computer technology and radio astronomy which have made possible the first really optimistic searches for signals from extraterrestrial civilizations.

We stand on the brink of an unimaginable glimpse into a previously unfathomable universe. With this book we plant our feet firmly on the precipice, and carefully peer over the edge.

ONE

Closest Neighbors

*Foolish, philosophically absurd,
and formally heretical.*

POPE URBAN VIII,
on the discoveries of Galileo (the man)

Wow!

POPE JOHN PAUL II,
on the discoveries of *Galileo* (the spacecraft)

1 | Lowell of Mars

*An ounce of imagination is worth
any ten facts in the world.*
RAY BRADBURY

1.

In his favorite portrait photograph, showing him as he wished
to be seen, Percival Lowell is a mature gentleman of obvious
means. His eyes are brilliant and clear as he looks out at the
camera, leaning casually with one hand on a cane, with the
other holding his hat at his hip. A cavalier attitude with his
body, a firm and commanding stare with his eyes—altogether
a man to be reckoned with, a man of destiny.

The first Percival Lowell was an English merchant and already
rich when at the age of sixty-seven he decided that English politics in 1638—with the growing violence between Parliament
and king—were too rambunctious to be borne. Notwithstanding his age, he packed up his belongings and his family
and set sail across the Atlantic to the Massachusetts Bay Colony, where he lived and prospered for another twenty-six years

before dying peacefully at the head of a large and dominant New England family. For the next two hundred years the Lowells were fruitful, establishing themselves as the prototype of the new American nobility in Boston—"the home of the bean and the cod, where the Lowells talk to the Cabots, and the Cabots talk only to God."

Young Percival was born in 1855, educated at Miss Fette's school in Boston and, after the age of nine, at boarding schools in France. He was fluent in Latin even as a child, writing a tragedy about the wreck of his toy sailboat in hexameter verse. After graduating Phi Beta Kappa from Harvard in 1876 he set out for the mandatory Grand Tour of Europe. He had a lovely time, nearly enlisted in the Serbo-Turkish War, and returned home to work in the family businesses for five years, after which he retired from commercial life and began to devote himself to more intellectual pursuits. He decided to become an Asiatic scholar, but a Lowell didn't pursue such ambitions by the means afforded to ordinary people: he didn't simply enroll in graduate school and study the subject for several years, carry out a research project under the tutelage of a professor, earn a Ph.D., and finally begin his life's work.

Instead of all that plebeian nonsense he sailed to Japan, where his family connections opened wide the proper doors. He became a Korean envoy, met everyone who counted, and during the course of the next half-dozen years wrote four books on oriental culture before becoming bored with the whole mise-en-scène.

What to do next? Percival Lowell returned to Boston. He had never been fond of the city; it was too full of Lowells. He tried polo, which was fun; but he wanted more than fun. He wanted desperately to *do* something, to *be* somebody. His younger brother, Abbott Lawrence Lowell, was already a successful and influential Boston lawyer and would soon be president of Harvard; his uncle, James Russell Lowell, was America's most distinguished man of letters; his little sister Amy was to be a renowned poet. What was *he* to be?

He looked up, and found his answer among the stars.

Or, to be more precise, among those "wandering stars," the planets. From childhood, he had always been interested in astronomy. He had lugged a decent-sized telescope across the Pacific and kept it with him during his journeys through Asia. And then, in the late 1880s just as today in the 1990s, astronomy burst through onto the front pages of the newspapers. Suddenly, everyone was talking about Mars.

The speed with which the planets whirl around the sun decreases with their distance from the sun, and so Mars moves more slowly than Earth. It also follows an orbit around the sun that is a shade more elliptical than that of the Earth. Its orbit is still very nearly a circle, but it does swing a bit far out at its farthest point and in closer at its nearest. Though the ellipticity is small, the distances involved are so very great that this makes a real difference in how close Mars comes to Earth. Every two years and two months Mars and Earth come into opposition, when the planets are closest together; at this time they come to within about 50 million miles of each other, while at conjunction, when it is farthest away, Mars is nearly 250 million miles from us. Every fifteen years the ellipticities of the two planets' orbits are such that an extremely close opposition occurs, and Mars can approach to within 35 million miles.

This proximity makes a significant difference in how clearly Mars can be seen from Earth, and so at these times various astronomers around the world take their best shot at viewing the planet. During the opposition of 1862, when Lowell was still wearing knee-breeches at Miss Fette's school, Father Secchi in Italy had seen something no one had ever seen before: there seemed to be thin scraggly lines drawn across the surface of the planet. Writing to the astronomical community about his observations, he called them *canali*, or channels. This inspired Giovanni Schiaparelli, director of the Brera Observatory in Milan, to take a closer look during the next close opposition. In 1877, which would be one of the closest approaches of the century, Schiaparelli saw these *canali* quite clearly. This time some American newspapers picked up the story, and someone translated the Italian word as "canals," which made all the difference in the world.

A "channel" could mean anything, but the Suez Canal had been dug in 1869, and by 1877 planning for the Panama Canal was under way. These achievements were known throughout the world as the ne plus ultra of mankind's engineering prowess. A canal was the epitome of civilization's success at making a planet habitable. It was the work of industrious, intelligent beings.

Could it be different on Mars?

2.

The science of *exobiology*—the study of life on other worlds— has often been ridiculed as being, like theology, a subject without an object. But the idea that not merely the lowest forms of life but indeed intelligent, humanlike creatures exist on other worlds can be traced back as far as human history goes. By about the fifth or sixth century B.C., Democritus and Leucippus had some thoughts which Epicurus (ca. 300 B.C.) later refined into the following syllogism:

1. There is no God, and
2. Nature is the end result of an evolutionary process, and
3. All matter is created of atoms, and, finally,
4. Since these atoms are indestructible and infinite, they are borne far out into space, not being used up on one world, and so

Life exists elsewhere in the universe.

In an epistle, written in the form of a letter to Herodotus,

Epicurus states plainly: "There are infinite worlds both like and unlike this world of ours. . . . In all worlds there are living creatures and plants and other things we see in this world."

As Lucretius (99–55 B.C.) put it, given that empty space extends without limit in all directions and atoms innumerable are rushing helter-skelter through the universe, "it is in the highest degree unlikely that this Earth and sky is the only one to have been created and that all those particles are accomplishing nothing. . . . Nature is free and uncontrolled by proud masters," he continued. "[She] runs the universe by herself without the aid of gods."

By 180 A.D. we had our first science fiction novel, the Greek poet Lucian's *Vera Historia,* in which a whirlwind carries our hero's ship to the moon and to a series of adventures there. But Lucian was quick to distance himself from the real world: "I write of things which I have neither seen nor suffered nor learned from another, things which are not and never could have been, and therefore my readers should by no means believe them."

Others thought differently. The Pythagoreans taught that though Lucian had neither "seen nor suffered nor learned from another," the scenes he visualized certainly "could have been." They confidently declaimed that "The moon is inhabited as our Earth is, and contains animals of a large size . . . [which] . . . in their virtue and energy are fifteen degrees superior to ours, and emit nothing excrementitious."

The existence of extraterrestrials, excrementitious or not, was fiercely argued by the scholars of early Christianity. For though the Bible is explicitly silent on the question, implicitly it is quite clear. Its first words are: "In the beginning God created the heavens and the Earth." Plural heavens, but only one Earth. The Bible goes on to say that God created the sun to give light to the Earth by day, and the moon to light it by night; in other words, the heavens were created for the express benefit of the Earth alone.

All very clear. But in this argument Thomas Aquinas and the scholars of the Middle Ages got themselves hung up on the horns of a dilemma: If God is omnipotent, He could have created as many worlds as He liked. If He created only one, what limited Him? His imagination? But if there is any limit placed on God, then He is not omnipotent. They settled the argument by simple fiat: Aquinas wrote that God *is* omnipotent, and he *didn't* create other worlds, and that is simply that. Nicole Oresme, bishop of Lisieux (1325–1382), explained the reasoning like this: "God can and could in His omnipotence make another world besides this one or several like or unlike it. But of course there has never been nor will there be more than one corporeal world." So that settled it. (The argument reminds us of Ring Lardner's famous line in the *You Know Me, Al* series: "'Shut up,' he explained.")

The question was important at the time because of the concept of the Jewish Messiah, which Christians saw embodied in the person of Jesus Christ. It seemed ridiculous to suppose that God would have been cruel enough to have had His Son die over and over again in an infinite number of worlds, and totally abhorrent to suppose that beings on other worlds could attain eternal life without Him, and mean-spirited to think that God would create these other beings only to damn them to eternal death. Therefore these other worlds simply could not exist. The argument was well put by Philipp Melanchthon in the sixteenth century: "The Son of God is One; our master Jesus Christ was born, died, and was resurrected in this world. Nor does He manifest Himself elsewhere, nor elsewhere has He died or resurrected. Therefore it must not be imagined that Christ died and was resurrected more often, nor must it be thought that in any other world without the knowledge of the Son of God, that men would be restored to eternal life." Therefore there were no other populated worlds, and we would just have to live with the idea that God decided things were best that way, and if we didn't understand His reasons—

why, that was just too bad. For He moves in mysterious ways, His wonders to reveal.

But then Science reared its ugly head in the person of Copernicus, and with him came the concept of our world as merely one of a group of similar planets, all of them orbiting a central sun. And if those planets were so similar, why couldn't they be inhabited? The idea liberated the spirit of Giordano Bruno, and doomed his body.

Born in Nola, Italy, in 1548 to a military family, Bruno began his intellectual life at the age of fifteen in the great Dominican convent in Naples, where Thomas Aquinas had once taught. He was a pious friar and excellent student until he came under the influence of a famous magician, Giambattista della Porta, who had established a rival academy in Naples for the investigation of nature by means of mysticism and magic.

Often spoken of as a medieval scientist, Bruno was instead a deeply religious and poetic mystic. He came to believe that Christianity was a usurping, weakened religion which had overthrown the one true religion—that of the ancient Egyptians—and which was about to collapse of its own weight. Bruno saw the Copernican idea (which had been privately published and circulated in 1514 for fear of the Church's reaction, and was not generally released until Copernicus's death in 1543) as a magic sign of this coming revolution. The Earth, he believed, was alive; and so the fact of its moving, impossible to understand by the Catholic teachings of the day, was clear to him. In fact, the entire universe was alive: the stars consisted of an infinite number of living worlds prowling through space like Blake's Tyger, burning bright in the forests of the night.

And yet Bruno remained a Christian, and indeed a Catholic. He seems to have believed that Catholicism would be transformed but not destroyed; Christ was not a Son of God but a "benevolent magus," as indeed Bruno himself was.

It all seems a bit heretical even today. To the Catholic authorities of the time it was despicable. He left the convent in Naples one step ahead of the Inquisition, fleeing to Rome. Out of the frying pan, into the. . . . He fled Rome for Geneva, from there went to Toulouse, and then to Paris where his public lectures on the role of magic gained him an audience with King Henry III, who sent him on to England with a royal letter of introduction. Suitably impressed, the French ambassador invited him to live rent-free in the embassy, where he laid up a store of future trouble by describing Queen Elizabeth as "divine": the Inquisition would later rudely ask him how any human could be divine, let alone a woman.

But for the moment he lived happily in England, writing and lecturing about the plurality of life in different worlds, and indeed in different living universes; he was not advocating science instead of religion as the way to search for truth, he was simply advocating "a new religion based on love, art, magic, and mathesis." For Bruno, Copernicanism was not the beginnings of a scientific investigation into nature, but was a background for his magical view of the universe: the Earth moves, therefore it is alive.

It was not a good time to hold such views. In France the Catholic League was dominating politics and soon forced the recall of the ambassador to England, who was known as a liberal. With him came Bruno, forced out of his rent-free residence in London. He came back to Paris, but he didn't shut up, and soon he was forced to flee to Germany, which he soon left for Prague. He published a book there that brought him to the attention of a Venetian nobleman, Zuan Mocenigo, who invited him to come to Venice and teach him the mysteries of the universe. What he taught was, however, too much for Mocenigo, who betrayed him to the Inquisition.

After a lengthy trial, Bruno regretted and recanted all his beliefs, the result of which was only that he was sent to Rome for a further trial which lasted for eight years. Instead of beating him down, however, the imprisonment and continued interrogations so disgusted him that he recanted his recantation, regretted his regrets, and obstinately proclaimed his mystical beliefs without further wavering.

After eight years of this, the Inquisition decided that enough was enough. On Saturday, February 19, 1600, they made a final effort to save his soul. He was carried naked and gagged from his cell, tied to a stake, and paraded through Rome to the Square of Flowers in front of the Theater of Pompey. The Master Inquisitor stood in front of him with a lit torch in one hand, a picture of Christ in the other, and demanded repentance. Bruno turned his face away, and the inquisitor touched the torch to the kindling at his feet, and Giordano Bruno was burned alive.

He is revered by some as a martyr to truth and science, but he is none of that. He was an enemy of the Church certainly, and a friend to all those who value the freedom of an unfettered imagination; but with all the good will in the world he cannot be considered a friend or martyr to science. He was no scientist—he never sought to prove his ideas with experiment nor to limit his speculations to the testable; he was instead an exponent of science fiction. He would have loved Ray Bradbury's dogma that an ounce of imagination is worth any ten facts in the world. The Inquisition was horribly wrong in killing him, but no more horrible and no more wrong that it was in burning alive the thousands of Jews, Protestants, Moors, Gypsies, intellectuals, and atheists, who all died at the stake to save their souls from the sin of heresy (i.e., individual thought). His primary sin, in the eyes of his murderers, had nothing to do with a scientific view of the universe; it was simply that in proclaiming the abundance of life throughout the universe he was in their eyes necessarily denying the uniqueness and divinity of Christ. He was a martyr certainly, but not to science; a martyr instead to the demand that our minds and imaginations be allowed to roam free, unfettered by Church dogma or McCarthyism or even sense and sensibility. And we can't think any the worse of him for that.

3.

Science began with the Danish nobleman Tyge Brahe, in the late sixteenth century—not, as many people think, with Copernicus. The idea that the Earth revolved around the sun had been around since antiquity (at least since the days of Aristarchus of Samos, ca. 270 B.C.). It had always been clear to men of sense that the observed passage of the sun, moon, and stars across the sky could be explained as easily by a revolving Earth as by the movement of all the heavenly objects. The idea never found general favor because in those times you could always *feel* movement. If you were riding on a horse or falling off a cliff you could tell even with your eyes closed that you were in motion, but standing still on the Earth you felt no such sensation. (Today, with the experience of flying in a jet at five hundred miles an hour and feeling no sensation of motion, we are not so bound by our "common sense," that most misleading of mental mechanisms.)

Still, even in those days there were second thoughts. Riding on a horse, you felt the wind in your face; but sailing *with* the wind, that natural sensation disappeared. Perhaps all sensation of motion disappeared with a revolving Earth. Copernicus had organized this idea into a simple and comprehensive scheme but, like Bruno, he can't really be called a scientist. Merely being right is no criterion; scientists are wrong as often as they're right. The sole advantage of science is that when you're wrong, eventually you find out by testing your ideas with observations, with experiments. Copernicus did no such thing, nor did he suggest that others do so. He presented his arguments on a philosophical basis, which is not much different from the mystical arguments of Bruno.

Tycho Brahe was born in 1546. (Everyone from those days has had their names latinized for us; thus Tyge became Tycho, just as Niklas Koppernigk became you-know-who.) He was obsessed with astronomy and its intellectual retard, astrology;

he was convinced that our lives and souls are influenced by the stars and, particularly, by the planets. What makes him important today is that, in consequence, he sought to measure accurately the positions of these heavenly objects. He produced his first important work in his mid-twenties, when a bright star (a supernova) appeared suddenly in the sky and his careful measurements of its position proved that it was indeed a star, far from the Earth – moon system. Yet Aristotle had taught that the stars were eternal and unchanging. Five years later he combined careful observations with simple geometry to prove that a comet seen that year (1577) was at least three times further from the Earth than the moon is. Thus, since comets were seen to travel through space, it could no longer be supposed that, as Aristotle had taught, the heavens were solid; so Aristotle was wrong once again. Could the Aristotelian concept of a geocentric universe also be wrong?

Perhaps. After much thought, Brahe accepted the Copernican idea that the planets revolved around the sun—Aristotle was wrong—but he then came to the conclusion that Copernicus too was wrong: the sun and the planets together had to revolve around the Earth. He argued that the Earth couldn't move because it was so heavy, because there was no sensation of motion, and finally because of the absence of stellar parallax.

This last point was a good one. Imagine, for example, that you observe a star directly above your head in January. If the Earth is really moving around the sun, the star should no longer be directly overhead six months later, at the opposite point in the Earth's supposed orbit:

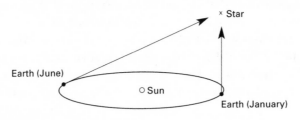

But this apparent change in stellar positions was never observed by Brahe, and therefore he quite reasonably concluded that the Earth could not move.

So Tycho faced three competing theories: the old Greek idea that everything revolved around the Earth, the Copernican theory, and his own. We say that science began with him because instead of appealing to the Bible or to philosophy or to magic to settle the argument, he suggested that we find the truth by firm observations.

At the time, each of the theories predicted positions of the planets that agreed with the observations which had been recorded since ancient times. Tycho thought that if he made a new series of observations, more precise than the old ones, he would find that only one of the theories could predict the correct, precise positions. So he embarked on a lifelong quest for ever more precise observations. King Frederic II of Denmark offered him the island of Hven, near Copenhagen, and support for the rest of his life. He built an observatory there in 1576, and lived and worked there for twelve years until King Frederic died, and it turned out that when the king had said Brahe would be supported "for the rest of his life," Frederic had meant his life, not Tycho's. The new king had no interest in astronomy, and Brahe was dispossessed.

Luckily he found another patron in the Emperor Rudolf II, and he continued his work in Prague. But there is more to science than observations. His experiments had to be compared to the predictions of the three theories, and he found that the mathematics involved was beyond him. So he hired the world's first theoretician, Johannes Kepler.

Kepler was a "small, frail man, near-sighted, plagued by fevers and stomach ailments," a philosopher, theologian, astrologer and mathematician, a young man in his twenties who had just published a book giving a mathematical but misguided argument in favor of the Copernican system when Brahe hired him to carry out the necessary computations.

Kepler's Copernican theory was built on geometrical figures inserted between the planetary orbits, and was based on a mystical feeling about the mathematical structure of the universe. He now jumped at the chance to get his hands on some hard data with which to prove his point. He was also happy to get a steady job, which was none too easy to come by for a scholar in the early 17th Century. Especially a job unencumbered by the demands of the Church.

He had been born in Weil der Stadt, Germany, in 1571 to a father he described as "criminally inclined, quarrelsome, liable to a bad end," and a mother who would later be arrested and tried as a witch. At the age of twenty he received a master's degree from the University at Tübingen, and reluctantly gave up the life of a clergyman when he received an offer to serve as teacher of mathematics in Graz. In 1595 he published an astrological calendar that predicted a bitterly cold winter, peasant uprisings, and an invasion by the Turks, all of which duly took place and which established him as an eminent seer, although he later wrote that "If astrologers do sometimes tell the truth, it ought to be attributed to luck." Nevertheless, throughout his life he continued to publish such prognostications; as he put it, "writing astrological calendars is better than begging."

He married a rich wife—"fat, confused, and simple-minded," as he described her—only to find her fortune tied up in legal and church intricacies, the latter related to the Keplers' status as Protestants in a Catholic Europe. In 1598, in fact, on the 28th of September, he and all Protestant teachers in Graz were ordered to leave town before sunset. After a couple of years he learned that Brahe had established a new observatory in Prague, and he went there to apply for a job.

The two men established an argumentative but fruitful collaboration, and when Brahe's chief assistant resigned, Kepler was hired. Brahe died the following year, before the results were in. Kepler stayed on as his successor, and found that Brahe's theory did not match his observations. Neither did the old Greek geocentric ideas. Surprisingly, to his astonishment and regret, neither did the Copernican theory. Finally, however, after considerable fiddling and head-scratching, Kepler found that the precise positions of the planets, as observed by Brahe, were matched by the predictions of the Copernican theory if it was modified so that the planets moved in ellipses. With that discovery, the question should have been settled. But it wasn't, because just as few people in the world would have been able to carry out Kepler's calculations, few people were able to understand them. The answer was there, but it was written in mathematics, a language almost no one understood. Something else was needed.

4.

The marriage of science to technology began with the telescope at the turn of the seventeenth century, when an anonymous young apprentice to Hans Lippershey, a Dutch maker of spectacles, invented the "looker," or, as it was later called, the "optick stick." Fooling around in Lippershey's shop after he had swept it out and completed his other chores, the apprentice was looking with first one eye and then the other through the variety of lenses on the shelves. With no particular thought in mind, he tried looking through two lenses at once: instead of one in front of each eye, he put two back to back and squinted through them, making the world pass in and out of focus and having a grand old time. And then, by chance, he juxtaposed one concave and one convex lens and, as he moved his hands, changing the distance between them, he caught a glimpse of a suddenly magnified world which vanished as quickly as it had come.

That was weird, he must have thought. He looked down at the two lenses in his hands, then tried it again. At first everything was blurry, but as he carefully moved them back and forth he found one particular distance at which the view sharpened into focus, and brought everything nearer. He went to the open door and looked down the cobbled street, and laughed with excitement.

When Lippershey came back to the shop he found his apprentice bubbling over with the joy of his new game. Lippershey himself tried the lenses, and—being a sensible businessman in a commercially minded country—immediately his interests turned to thoughts of war. Holland was in the midst of a war with Spain that was being fought on the high seas. It was obvious how such a discovery could be used in those days of wind and sail, for if one ship could see its opponent while still far enough away to be invisible, it could dictate the conditions of encounter or escape.

Lippershey put together a tube with one lens at each end, of a length designed to bring distant objects into focus, and took it to the Dutch naval bureaucracy. Like all bureaucracies before and since, it was more interested in not making a fool of itself than in obtaining something new, and so of course it made a fool of itself. It bought the "lookers," but did nothing with them, and the war was over before their usefulness in military affairs could be realized.

In that same year, 1609, a provincial professor of mathematics at the University of Padua heard about the discovery and decided to build a telescope for himself. Instead of warfare, he thought about the sky, and with that thought he doomed himself and freed the rest of us.

Galileo Galilei looked through his telescope at the stars and saw nothing new. To the naked eye they are spots of light, and through the telescope they were nothing more. But when he looked at the planets, those spots of light thawed and resolved themselves into solid, clear disks. The planets were not stars, they were spherical worlds, perhaps like the Earth. He saw four moons revolving around Jupiter, so it was obvious that not everything in the heavens revolved around the Earth. Most important of all, he saw that the planet Venus showed phases like the moon, an observation that could not be explained by any geocentric theory, but that followed naturally from the Copernican model.

So it was now clear that the Earth revolved around the sun, but for his insistence in proclaiming this Galileo was arrested and imprisoned by the Inquisition. He saved himself from Bruno's fate by publicly recanting; he spent the rest of his life under house arrest, but his proof escaped beyond those walls. Anyone with a telescope could see the phases of Venus: the Earth no longer stood still.

And as the telescopes improved in quality, people began to see more and more. In 1646 Francesco Fontana of Naples discovered markings on Mars, showing that the planet was not just an amorphous disk, but had features on its surface, perhaps mountains or oceans like Earth. Actually, the first marking he saw, a black spot, turned out to be a flaw in the glass of his telescope. But when he told the world about what he thought he saw, he stimulated others to look, with better telescopes, and soon a variety of real markings was seen. Christian Huygens, for example, was a Dutch lawyer turned mathematician who, stimulated by Fontana, taught himself to grind lenses and make telescopes. By 1659 he had discovered the rings and moons of Saturn, and had seen surface markings on Mars clearly enough to watch as they revolved with the planet's rotation, enabling him to estimate the Martian day as remarkably similar to that of the Earth (today we know it is 24 hours, 37 minutes, and a few seconds). This similarity led him to hypothesize about the inhabitants of Mars and what they might look like, although his book, *Cosmotheoros*, was not published until 1798.

In 1666 Giovanni Domenico Cassini discovered the polar caps of Mars, again emphasizing its similarity to Earth. Twenty years later Bernard de Fontenelle was writing that "The Earth swarms with inhabitants. Why then should nature, which is fruitful to an excess here, be so very barren in the rest of the planets?" In 1749 Ben Franklin, in his *Poor Richard's Almanac*, stated that "It is the opinion of all the modern philosophers and mathematicians, that the planets are habitable worlds." By the 1770s Johann Elert Bode had rediscovered a rule formulated in 1766 by Johann Daniel Titius about the relative distances of the planets from the sun (which has come to be known as the Titius-Bode law): each planet is roughly twice as far from the sun as the previous one. He went further than Titius, however, and argued that in consequence the inhabitants of each planet are twice as spiritual as those of the next interior planet. And not only the planets were inhabited: the sun, too, was inhabited by "solarians . . . rational inhabitants . . . [who] are illuminated by an unceasing light."

We had gone from a geocentric universe where the heavens were created for the express purpose of illuminating the Earth to a universe in which the Earth was only one of many equal worlds, not only equally populated but perhaps with superior beings. The existence of these other creatures was no longer the center of the argument; we were now arguing over what sort of creatures they were.

Modern astronomy is often said to have begun with William Herschel, a court organist and amateur astronomer of Bath. Born in Hannover in 1738 of a musical family, he came to England when he was offered a musical living by King George III, but he brought with him a keen interest in the sciences of all kinds. At the age of thirty-four he bought his first telescope, found that it wasn't very good, and began to grind his own lenses. In a few years he was making better telescopes than anyone else, and so he set out to solve Tycho Brahe's problem: If the Earth was moving around the sun, why didn't we see the parallax of the stars?

He never did solve that problem, but in 1781 he "perceived [a star] that appeared visibly larger than the rest; being struck with its uncommon magnitude . . . [I] suspected it to be a comet." It wasn't; it turned out instead to be a turning point in his life. It was a planet, the first new planet to be discovered since ancient times, and it brought him instant fame and financial security. He named it *Georgium Sidus*, George's Star, and King George rewarded him with an annual stipend of two hundred pounds (which was graciously continued even after the community of astronomers insisted on calling the planet Uranus). He resigned his musical post and, supplementing his salary by selling a few telescopes, lived the rest of his life as a professional astronomer, eventually earning a fellowship in the Royal Society and a knighthood.

He is regarded in most books as the very model of the modern professional astronomer: interested not so much in theories as in facts, spending his life making better and better telescopes and more and more precise observations of the stars. But he was guided, even driven, by one overwhelming theory—he wanted to find proof of life among the stars. And indeed he did—or thought he did.

He was influenced in this regard primarily by James Ferguson, a Scottish sheepherder some thirty years older. With only three months' education in a grammar school, Ferguson became a maker of scientific instruments and an astronomer so accomplished that King George III settled fifty pounds a year on him for life. (George, in addition to losing the American colonies and dying in madness, was an intellectual who supported the arts and sciences well beyond the normal fashion of kings and governments of the time.)

In 1768 Ferguson wrote an *Easy Introduction to Astronomy for Young Gentlemen and Ladies*. One of the most popular books of the time, it was reprinted a dozen times. Instead of writing in a didactic style, Ferguson has a series of characters talking to each other, discussing astronomy in a casual way. One of them, Eudosia, remarks that she "cannot imagine the inhabitants of our Earth to be better than those of other plan-

ets. On the contrary, I would fain hope that they have not
acted so absurdly with respect to [God] as we have done."

Which illustrates the general feeling among the gentry of
the time that of course the other worlds were populated, just as
they were learning every day that the furthest reaches of the
Nile and the distant islands of the South Seas harbored their
own peculiar forms of life, human, pseudohuman, and otherwise. Herschel was greatly taken with the simplicity and rea-
sonableness of the argument, and set out to prove it. In 1779
Mars came to a particularly close opposition to Earth, approaching to nearly 35 million miles. Herschel was ready for it.
Night after night he observed the planet, and discovered clouds
drifting across its surface, from which he deduced "a considerable but moderate atmosphere."* The Martians "probably
enjoy a situation in many respects similar to our own," was his
conclusion, put in a manner as casual as that of Ferguson but
carrying even more the solid weight of authority.

Not only Mars was inhabited. His observations of lunar
craters led him to "the great probability, not to say almost absolute certainty, of her being inhabited." The craters, you see,
were not volcanic eruptions or meteorite impacts but lunar
cities, constructed in a circular pattern because of the thinness
of the atmosphere which induces the inhabitants to. . . . Well,
the argument gets a bit hazy. So too were his observations, for
he thought he saw "growing substances" on the moon, gigantic forests of towering "trees at least 4, 5 or 6 times the height of
ours." He also looked at the sun and observed the sunspots
and measured their tremendous size; echoing Bode's ideas, he
thought they indicated cooler regions of the sun, within which
solar creatures lived and prospered.

In the early years of the nineteenth century interest passed
beyond merely passively observing these Martians, Solarians,
and other extraterrestrial creatures wherever they were. Plans

*These were probably dust clouds; the Martian atmosphere is too thin for
him to have been able to see Earthlike clouds

were advocated for growing a wheat field in Siberia in the form of a right triangle, bordered with pine trees, big enough to be seen by the Martians, to demonstrate both our existence and our knowledge of mathematics. Joseph Johann von Littrow, of the Vienna Observatory, suggested digging miles-long ditches in the Sahara Desert to outline various geometric figures, filling them with kerosene, and setting them alight. A French inventor proposed that a giant mirror be constructed so that it would focus on Mars, and by concentrating the sunlight it could be used to burn messages to the Martians in the soil of their own deserts, as children use magnifying glasses to burn grass.

These people were not all crackpots. The person who has sometimes been called the greatest mathematician of all time, Karl Friedrich Gauss, joined them. In 1818, being hired to help a large surveying operation in Germany, he invented the heliotrope which enabled sunlight to be focused and reflected over great distances, a sort of primitive laser. In 1822 he proposed building a giant heliotrope, "with 100 separate mirrors, each of 16 square feet. Used conjointly, one would be able to send good heliotrope-light to the moon. . . . This would be a discovery even greater than that of America, if we could get in touch with our neighbors on the moon."

These proposals weren't acted on, and no communication was established with our lunar neighbors, but in 1835 the next best thing happened: the newspapers reported a firm, scientific, personal observation of living creatures on the surface of the moon.

In August of that year a series of articles appeared in the *New York Sun* by a staff reporter named Richard Adams Locke, describing the results of new telescopic observations of the moon from South Africa. Sir John Herschel, son of the great William, had published articles in the *Edinburgh Journal of Science*, Locke wrote, releasing data from a new telescopic observatory in the Cape of Good Hope. With his telescope, superior to any previously operating anywhere in the world, he had been able to see great temples on the moon, broad avenues, and the lunar people themselves huddling around their campfires . . .

Eventually the story was revealed as total fiction, nothing but a hoax. Well, not exactly a hoax, Locke explained; it was a satire on the silliness of astronomers and those who believed them. He had hoped that the effect of his satire would be to wake people to the ludicrous claims of those who believed other worlds to be inhabited, but as so often happens the sensational articles were read by many more people than those who subsequently read the nonsensational revelation that they were a hoax; for many years afterward people all over the world—but particularly in America—believed that living humans (Locke named them Selenites) had really been seen on the moon.

Those who believed in the plurality of inhabited worlds without direct evidence, meanwhile, continued to argue on speculative and philosophical grounds. Scientific people reasoned that the innumerable stars were suns, so why should they not have their own planets? That these would be too small and too far away and too hidden in the glare of their star to be seen from Earth was no reason to doubt their existence. That is, absence of evidence is not evidence of absence. And if these planets existed, why should there not be life on them?

Because God created life only here on Earth, the religious right replied. To say otherwise is to deny the Bible. But even within religious groups the argument raged: to deny the power of God to flower the universe with a diversity of life was itself blasphemy, was it not?

Then in 1859 Darwin published his *Origin of Species,* and it became clear to those of a scientific persuasion that life was to be described by a natural series of events rather than as something magical. And what could be more natural than that it should pervade the universe of which the Earth was only a small, insignificant part? The forces of religion united to face this new attack, which in the mouths of many (though not Darwin himself) claimed that God was irrelevant, that life was a force unto itself.

This scientific argument gained support from two new quarters in the nineteenth century. Chemistry entered into it in 1828 when a twenty-eight-year-old German chemist named Friedrich Wöhler synthesized urea from ammonium cyanate. The significance of this achievement cannot be overstated—it is one of the defining moments of human intellectual achievement, along with *Hamlet* and Beethoven's Fifth. Chemists had already known how to synthesize the compound ammonium cyanate from its constituent elements nitrogen, oxygen, carbon, and hydrogen. But urea was a compound of biological origin: we synthesize it in our bodies and excrete it in our urine. Until 1828 it had been thought that the chemistry of living things was distinct from nonliving; men could make *inorganic* chemical compounds in their laboratories but only God could breathe life into dust, and the origin of all things biological lay in this miracle. But now Wöhler showed that we could create in the laboratory the same chemicals created in the body,* and further research showed that there was no difference between the compounds formed by these two methods. The chemistry of life was the same as all other chemistry; by extension, life *was* chemistry, and nothing else.

In the years that followed, Wöhler's example was followed by hordes of organic chemists who synthesized all manner of "biological" compounds. (It was, in fact, too much of a good thing. Wöhler himself, bewildered by the complexity of all the new compounds, complained, "Organic chemistry drives me mad! It's like a primeval tropical forest full of the most remark-

---

*In a letter to the master chemist Berzelius on February 22, 1828, you can hear Wöhler chuckling: "I can no longer hold back my chemical urea, and I have to let it out that I can make it without a kidney."

able stuff—a dreadful endless jungle into which you'd better not enter because there seems to be no way out!" (And those of us who took organic chemistry in college know exactly how he felt.)

The second line of scientific evidence came from the emerging field of meteoritics, a field that was itself thought to be nonsensical until the nineteenth century. The notion that huge hunks of stone and iron could fall out of the sky seemed to deny common sense. For if it is true that what goes up must fall down, it seemed equally true that what falls down must first go up—and how could stones and irons be levitated like birds into the atmosphere? When in 1807 two professors from Yale University published a report favoring the meteoritic origin of an unusual stone and the newspapers of the day carried the story, Thomas Jefferson is reputed to have commented "I would more easily believe that two Yankee professors would lie, than that stones could fall out of the sky."

But a few years earlier, in 1803, while the French Academy of Sciences was meeting to debate this very issue, a shower of meteorites fell on a small village, L'Aigle, and an investigating team reported back to the Academy that in fact these stones were not terrestrial. There were long arguments on where they came from—the leading contenders were a disrupted planet, perhaps the parent of the asteroids which by then were known to orbit between Mars and Jupiter, or from volcanic eruptions on the moon—but one could no longer doubt they came from somewhere in space. So when, in 1859, Wöhler analyzed a meteorite and found it contained "a carbonaceous substance which can have no other than an organic origin, the proposition that life existed beyond the confines of Earth found a powerful argument."

Of course, the word *organic* is not quite synonymous with *biological*, as Wöhler himself had already proven: urea is organic, and can be biological in origin, but can also be synthesized in the laboratory. However, in 1864 another carbonaceous meteorite fell in Orgueil, France, and several chemists found in it organic substances similar to those found in petroleum (thought to be the decay products of ancient life). So the origin seemed to be biological.*

The pendulum was swinging far to the left. The previous speculations about life "out there" were now buttressed by firm scientific evidence, so much so that when Sir William Thomson (later Lord Kelvin) gave the 1871 presidential address to the British Association, he not only favored the evidence but also suggested that life on Earth itself originated from organic contaminants that fell in meteorites (a view still held by some today):

> We all confidently believe that there are at present, and have been from time immemorial, many worlds of life besides our own [and] we must regard it as probable in the highest degree that there are countless seed-bearing meteoric stones moving about through space . . . [and] that life originated on this Earth through moss-grown fragments from the ruins of another world. [This hypothesis] may seem wild and visionary; all I maintain is that it is not unscientific.

This, from the president of the British Association, arguably the greatest and most prestigious scientist of the age. So when a few years later Father Secchi saw vague *canali* on Mars, and when these were seen more clearly in 1877 by Giovanni Schiaparelli, who estimated that they were seventy-five miles wide and three thousand miles long, the suggestion that they were evidence not only of life on Mars but also of intelligent life, seemed not only logical but overwhelmingly obvious.

What, precisely, did the Martian canals signify? What was their purpose? The answer was clear to earthly Marsologists, given

*In 1996 a similar but more sophisticated argument was presented as the basis for a new claim to have discovered extraterrestrial life. This will be discussed in Chapter 4.

what we "knew" of Mars in particular and of the planets in general. In the latter half of the eighteenth century Immanuel Kant and Pierre Laplace had put together what seemed to be a reasonable theory of planetary formation. (Kant, who was later to become famous for his philosphical works such as the *Critique of Pure Reason*, was the first to suggest that the solar system formed as a nebular disk around the sun. His idea was later put into stricter form by Laplace, who had become a professor of mathematics at the age of nineteen and would later flourish scientifically and politically under both Napoleon and the Bourbon restoration.) They speculated that the early sun had been surrounded by a spinning disk, and that this disk spun off particles of matter which coagulated into the planets. This speculation—for such it is and no more; they had no evidence to back up their conclusions—is surprisingly similar to the model we have today, with one notable exception: in that model they saw the disk as spinning off material for one planet at a time, starting with the farthest away. So Uranus was formed first (Neptune and Pluto had not yet been discovered), then Saturn, Jupiter, Mars, Earth, and so on inward toward the sun. Kant had argued that "most of the planets are certainly inhabited. . . . The intelligence of these inhabitants becomes more perfect the further their planets are from the sun." His reasoning was once again based along rather mystical lines, but by 1877—now that we had learned what Darwin taught—the same conclusion could be based on a scientific basis: If Mars was older than the Earth, then—in accordance with evolutionary theory—Mars and the Martians were more evolved, more advanced than Earth and the Earthlings.

In this view Mars was beginning to suffer the fate in store for the Earth's distant future: it was losing its atmosphere and its supply of water by evaporation into space. But not to worry: if we had the capability to build the Suez Canal, surely the more advanced Martians had the capability to engineer their entire dying planet. It was clear, according to this reasoning, that they had built a system of planetwide irrigation canals, and these are what Schiaparelli had seen through his telescope. The problem facing this theory was that Schiaparelli was the only Earthling who was seeing them.

Since his announcement every telescope on this planet had been aimed at Mars, and no one else had seen anything like the canals. By 1882 Camille Flammarion in France, who produced an annual astronomy book, was the only reputable astronomer to accept the canals (though he himself never observed them). But it is not surprising that he was ready to accept as true what he had never seen: he would later spend more and more of his time delving into psychic phenomena. Then in 1886 someone else—J. Perrotin at the Nice observatory—finally saw them. The canal believers argued that the Milan and Nice observatories had the world's best telescopes; the reason no one else saw the canals was that their equipment wasn't good enough. Perrotin admitted that they were difficult to see, yet really quite clear if one had good eyesight and a good telescope. But Edward E. Barnard, with a telescope at Mount Hamilton in California which was clearly superior to the European equipment, said, "To save my soul, I can't believe in the canals as Schiaparelli draws them. I see details where some of his canals are but they are not straight lines at all. . . . I verily believe, for all the verifications, that the canals as depicted by Schiaparelli are a fallacy."

The controversy raged on. Well, perhaps "raged" is not quite the right word. Most of the world's astronomers ignored the canals, while those who believed in them continued to publish mostly in the popular press. A few other astronomers began to see them, although not all accepted the hypothesis that they were indicative of intelligent life. Schiaparelli himself, who began cautiously enough ("I am careful not to combat this suggestion [of intelligent Martians], which contains nothing impossible") had shifted ground by 1882: "There is nothing analogous [to the canals] in terrestrial geography. Everything indicates that here there is an organization special to the planet Mars . . ." And then in 1888 bright flashes on the planet, extending over the terminator (the line separating the sunlit area from the dark) were clearly seen by several observers; unlike the canals, there was no doubt as to their reality and in the popular press they were excitedly ascribed to signals to Earth from living, breathing Martians.

Again, most astronomers ignored this interpretation (the flashes were later understood as natural reflections of sunlight from mountain peaks), but that same year America's most distinguished scientific journal, *Science*, suggested that the Martian maria were not seas, but vegetation. Vegetation was a far cry from canal-building, light-signaling Martians, but it was at least life, and *Science* was as respectable an authority as you could get. Camille Flammarion followed this up in 1892 by going a bit further. In a book which received a wide readership, *La Planète Mars et Ses Conditions d'Habitabilité*, he pronounced that "The present inhabitation of Mars by a race superior to ours is very probable." Mars is older than Earth, he explained, and so "we may hope that mankind there will be more advanced and wiser."

At the next close approach to Earth, more projections across the terminator were seen by the Harvard astronomer William Pickering, who also saw dark, round "lakes" at the intersection of canals, reinforcing the interpretation—and with the authority of Harvard—that the canals not only existed but were in fact irrigation devices. Things were heating up.

To Percival Lowell, looking up at Mars and trying to decide what to do with his life that would prove suitable for a Lowell of Boston, it must have been as if he heard that hero of Victorian literature, Sherlock Holmes, calling urgently, "Hurry, Watson! The game's afoot!"

He would be an astronomer. He would map out the canals of Mars more clearly than anyone had yet done, and he would prove the existence of the race of intelligent Martians. But he would not do this in the normal course of events, he would not follow the path usually necessary for those who aspired to a research career: he would not study for a Ph.D. any more than he had done when his ambitions lay in the direction of Asian scholarship. No, what he would do is, he would build his own observatory. And not just any old observatory, he would build the world's best. And he would build it in time for the next opposition of Mars, due in just two years' time, in 1894.

He had no time to waste. He contacted William H. Pickering for advice, and was told that the "seeing" varied greatly from place to place. The Harvard Observatory in Cambridge was about as bad a location as one could choose; dust and grime from the city streets, together with smoke from the industrial smokestacks and residential chimneys clouded the night skies. And even when the occasional wind blew the gunk away, the night lights of the city provided an inescapable optical pollution that was growing worse year by year. This was why he, Pickering, had taken a leave of absence from Harvard and was now head of the new Arequipa Observatory high in the Peruvian Andes.

Do the same thing, he advised Lowell: Find a site far away from the cities. Get away from the crowded East Coast. The atmosphere too, invisible to ordinary eyes, is a thick ocean each telescope must penetrate, so get as much above it as possible. A mountaintop is ideal; but not too high, or the breathing will be difficult. And although you must get away from the lights of civilization, you need its railways; you can't haul a world-class telescope across the country on a mule train.

Lowell hired a team of surveyors and sent them out west, where they found the town of Flagstaff, in the Arizona Territory. Seven thousand feet high, with a railway right there in town, it was little more than a few buildings with a population of less than one thousand, a center for the surrounding area's ranchers and the lumber trade. The night sky would be clear and unpolluted by smoke or light, the atmosphere was thin but breathable, and the railroad would deliver the necessary equipment practically right to the door. (The townspeople, eager for the fame a scientific establishment would bring them, had agreed to build a wagon trail direct from the railway stop to the proposed observatory.) The only problem was that you couldn't possibly build an observatory in two years and Mars was fast approaching. Even a Lowell couldn't slow it down.

But a Lowell could do damn near anything else. He could hire the best people, and as many of them as he needed, and he could belt them and flay them (financially speaking) until they did the impossible. When Mars came into opposition in April of 1894, the building at the Flagstaff observatory wasn't yet complete, all the modern conveniences weren't yet in place, but the telescope was there and so was Percival, fresh from presenting his first paper about Mars to the Boston Scientific Society, hanging onto his perch in the cold, clear nights, catching his first unobstructed views as the planet grew larger and redder, night by night.

He saw it all: the polar caps, the surface markings, and above all, the canals. One hundred and eighty-four of them, more than twice as many as Schiaparelli had seen on his best night.

2 | The Gardens of Mars

*One does not have to search far to hear the
opinion that somewhere on Mars there is a
Garden of Eden—a wet, warm place where
Martian life is flourishing. This is a daydream.*
NORMAN HOROWITZ

1.

Like William Herschel before him, Percival Lowell's sin lay in
being sure of what he was going to find before he searched for
it. Even before he went to Flagstaff to begin observations of
Mars in his new observatory he had begun writing his first
book on the subject, and his ideas were fixed firmly in his
mind. The basic starting point was the canals of Schiaparelli,
whose existence Lowell accepted despite the failure of many to
see them. If they existed, as he was sure they did, they stretched
in straight lines for thousands of miles: no geologic process
could produce such an array. Therefore they had been built by
intelligent creatures.

For what purpose? It was clear to him: Mars was older than
Earth, and smaller. It was harder for its gravity to hold on to its

atmosphere and water, and the planet was beginning to dry up. But its civilized races, being evolutionarily further developed than Earth's, had organized themselves into a world government and had built planetwide canals to irrigate the drying regions with water from the ice caps on the poles. The canals themselves were not wide enough to be seen from Earth, but the vegetation that sprouted along them, like the vegetation growing along the banks of the Nile, spread out far enough to make the tracks visible.

As for the Martians themselves, Lowell was open-minded enough to accept them for what they were: not human in form, not like us at all, but intelligent and resourceful. Could a planet like Mars, with little atmosphere to breathe, support such creatures? Of course, he chided us gently; we could see that if we would only divest ourselves of our preconceived and chauvinistic notions of what intelligent life must be like. "That beings constituted physically as we are would find [Mars] a most uncomfortable habitat is pretty certain. But there is nothing in the world or beyond it to prevent, so far as we know, a being with gills, for example, from being a most superior person. A fish doubtless imagines life out of water to be impossible; and similarly to argue that life of an order as high as our own, or higher, is impossible because of less air to breathe . . . is to argue, not as a philosopher but as a fish."

That's a typical piece of Lowellian prose: forceful, dramatic, and imaginative but offered without a shred of evidence. The evidence is what he proposed to find with his new observatory. In the meantime, his pen went marching on. Before leaving for Flagstaff he read a paper to the Boston Scientific Society and told them exactly what his work would show. In fact, he never for the rest of his life deviated from the ideas he laid out in 1894; whatever he saw through his telescope served only to prove what he had already known. This is not science; it is either precognition or obsession.

He told the assembled group at the Boston Scientific Society that his observatory would be dedicated to "an investigation into the conditions of life in other worlds." In particular, he was going to concentrate on Mars, because the canals seen there by Schiaparelli were undoubtedly "the work of some sort of intelligent beings." He was an enthusiastic and optimistic speaker, insisting that "there is strong reason to believe we are on the eve of pretty definite discovery in the matter."

The reception to all this was mixed. Some of the audience were thrilled to be sitting in on the "eve of discovery," and indeed the discovery of the century. Nonsense, said several others. "Misleading and unfortunate," said the director of the Lick Observatory, while Charles Eliot, the president of Harvard, would soon write that Mr. Lowell was "undoubtedly an intensely egoistic and unreasonable person . . . In my opinion his frame of mind towards the [Flagstaff] Observatory is a hopeless one. Fortunately," he added, "he is generally regarded in Boston among his contemporaries as a man without good judgement." Edward Charles Pickering, the director of the Harvard Observatory and the older brother of the man who had been advising Lowell in regard to the observatory, recommended that Harvard withdraw all support, and Eliot agreed. The younger brother, however, William H. Pickering, had long been a vociferous proponent of life on other worlds; in fact, he had recently been recalled in academic disgrace from the observatory in Arequipa, Peru, because of "martian mania and sensational pronouncements." (His brother had admonished him that when he viewed dark spots on Mars he should call them exactly that, rather than referring to them as "lakes," and suggested he not send telegrams to the New York newspapers with such "discoveries" until he was more sure of the facts; but he was incurably adamant.) Naturally enough he became a strong supporter of Lowell's ideas, although he preferred to think of them as his own. When Lowell asked him to come to Flagstaff, Harvard was glad to give him unpaid leave and thus to be rid of him, at least temporarily.

After the close opposition of 1894, in which both Lowell and Pickering, working in the half-completed observatory, saw with clarity a network of Martian canals more complete and more intricate than anyone before or since, Lowell surged into print. His audience was not only the cadre of the world's professional astronomers, but the great mass of the people. His book, titled simply *Mars,* was published in 1895; it is a clear and lucid explanation, albeit in flowery and dramatic prose, of his ideas and explorations. Astronomic photography was in its infancy and not yet to be trusted, but the book was full of drawings of Mars that showed the canals unmistakably. Lowell wrote more than a dozen articles in magazines ranging from *Popular Astronomy* to the *Atlantic Monthly,* went back to Boston to give four popular lectures, and then traveled to Europe to meet with Schiaparelli himself.

Thus began the golden Martian era of our civilization. Most astronomers aligned themselves solidly against him: *Science* magazine warned that "The [professional] astronomer cannot afford to waste his energies on hopeless speculation about matters of which he cannot learn anything . . . ," and the director of the Lick Observatory described Lowell's writings as "misleading and unfortunate . . . half-truths." But the popular press leaped on his words, and he was praised for "his humor, his ready wit, his complete knowledge of the subject," as well as "an enduring enthusiasm, a proper regard for facts, and a clear literary style." The battle lines were clearly drawn, and those of philosophical or scientific pretensions came down solidly on one side or the other.

Actually, there were three sides. The canals either did or did not exist, providing two sides to the argument. The third side consisted of people who were convinced of the canals' existence, but not that they indicated intelligent creatures as their creators. This was the main battle line. Most people, including many astronomers, were persuaded that the canals did exist despite the many observations that didn't see them. The reasoning was that if something is difficult to see, it's easy to miss. Therefore the many people who looked and didn't see the canals could easily be mistaken. But seeing is believing: if just one reputable person actually sees something, then that something surely exists. And by now several people besides Schiaparelli had seen the canals, chief among whom were Lowell and Pickering.

This argument took a beating from an English astronomer, Edward Walter Maunder, a stern-faced man, slightly balding on top, with an impressive and aggressive Victorian beard. In his mid-forties when Lowell exploded from his amateur background onto the professional scene, Maunder was a sound professional who had graduated from King's College, London, before taking up duties at the Greenwich Observatory. Five years previously he had founded the British Astronomical Association, and in 1894 he became its president. In that same year he published a paper claiming that the canals did not exist at all: they were merely an optical illusion. The human mind has an intrinsic desire for order, he explained, and so if a random collection of barely visible spots is shown, the brain subconsciously assembles them into a recognizable pattern. Small, random craters on Mars would be seen as a collection of lines, straight or wavy depending on the subconscious meanderings of the observer's mind. Ten years later Maunder would carry out an experiment to prove this: At the front of a classroom filled with a group of schoolboys from the Royal Hospital School at Greenwich he hung a diagram on which was drawn a random series of small circles, and asked them to reproduce what they saw. The boys at the front of the room saw the circles clearly, and drew them accurately. The boys at the rear couldn't see them at all, and turned in blank papers. Some of the boys in the middle rows turned in drawings covered with thin lines: they had "seen" canals on Mars.

But not many people, at least among the reading public, accepted this argument. First, because it was published in the

Monthly Notices of the Royal Astronomical Society rather than in the *New York Times*, the *Atlantic Monthly*, or *Popular Astronomy*, and so none but professionals read it; secondly, because people who do not make their living squinting through telescopes and trying to describe what they see cannot appreciate the difficulty of astronomical seeing. For most people the canals continued to exist, and the interesting question was, "Who—or what—built them?"

Lowell knew the answer, without a doubt in his mind. To settle the question, and to convince those who still doubted the canals or their origin in intelligent engineering, he bought a larger telescope and had it installed in Mexico, where he expected the skies to be even clearer than on the Flagstaff mesa, in time for the 1896 opposition of Mars. But, as everyone knew, Mars would not come as close that year or ride as high in the sky, and so it was hopeless to expect greater clarity of vision than two years previously.

And so it turned out. Mars came and went without bringing any more convincing evidence of its canals or its makers than Lowell had claimed in 1894. But he did see something else that was to cause him a good deal of trouble. He saw lines on Venus, lines that no one else ever saw or would see. Without hesitation he labeled them canals, and began to speculate about the reason for their occurrence on a planet known to be so hot that liquid water could not exist on its surface. The opposition to these speculations was so widespread that eventually he thought better about it and never again referred to the lines as canals; but still, he insisted, those lines were there.

No one believed him. A psychiatrist, Charles K. Hofling, suggested that Lowell was influenced by "incompletely sublimated voyeuristic impulses due to unresolved oedipal conflicts." Not many people were willing to go that far, but Lowell certainly was a sick man. Bouts of neurasthenia, which had plagued him all his adult life, now became so severe that for the next four years he was unable to make any observations. Andrew Ellicott Douglass, who had been Pickering's assistant in Arequipa and had come with him to Flagstaff, was put in charge of the observatory, and continued the Mars observations. At first a firm supporter of Lowell, he slowly came to conclude that at least many of the canals were in fact optical illusions. In a letter to Lowell's brother Abbott, he regretted "Mr. Lowell's indifference to take up the psychological question . . . [since] some of the well known planetary appearances could . . . be regarded as very doubtful . . . [and his] work is not credited among astronomers . . ." Douglass had tried to talk to Lowell about the problem and about the need to forgo wild speculations and sensational pronouncements to the press, and instead to opt for the calmer, less indulgent scientific method of patient observations, withholding final judgment until all the evidence was in, but "I fear it will not be possible to turn him into a scientific man." As a last resort, trying to salvage the reputation of the Flagstaff Observatory, Douglass was writing to Abbott in the hopes that he might influence his brother.

Instead, Abbott showed Percival the letter, and Percival reacted without hesitation. He fired Douglass, and never looked back. His scientific reputation failed to rise but his popularity among the general public was great. He wrote book after book, magazine article after article, and all found a loving response. The *Astrophysical Journal* might refuse to publish his work, but prestigious—if scientifically naive—magazines such as *Harper's Weekly* and the *Atlantic Monthly* were proud and happy to do so, and he had no trouble finding book publishers.

The public lapped it up. They couldn't get enough to read about the Martians. This, remember, was right in the middle of the great age of earthly exploration. We were going to the North Pole by dogsled and balloon, we were trekking to the South Pole, we were discovering the source of the Nile and the existence of Stone Age tribes in strange Pacific Islands; every day the newspapers were searching for some more exotic, more mysterious unexplored regions to penetrate. The ultimate destination was Mars, and Percival Lowell was not shy about leading the way.

Others were quick to hop on the bandwagon. In Germany an obscure con man named Spiridion Gopcevic, using the pseudonym Leo Brenner, began making fake announcements, claiming to have discovered numerous new canals. He published four books and numerous journal articles; when the professional journals began to reject his increasingly extravagant claims he founded and edited his own, *Astronomische Rundschau* (*Astronomical Review*), which flourished for eleven years. In the end he simply faded away, back into the obscure mists from which he had sprung.

Nor were the professional astronomers united in turning up their noses. In 1898 Samuel Phelps Leland, professor of astronomy at Charles City College in Iowa, wrote *World Making*, published by the Iowa Women's Temperance Publishing Company. In it he looked forward to the forty-inch telescope under construction at the Yerkes Observatory in Wisconsin, with which he expected the astronomers "to see cities on Mars, to detect navies in its harbors, and the smoke of great manufacturing cities and towns."

During the opposition of 1896 many astronomers saw the canals, but no one saw them very clearly and so the issue was still not settled. A typical comment was that of Eugène Michael Antoniadi, who in 1896 was director of the British Astronomical Association's Mars Section, and who was one of the truly professional astronomers who "verified" Lowell's observations: "Had it not been for the . . . foreknowledge that 'the canals are there,' I would have missed at least three-quarters of those seen." In other words, the other side of the seeing-is-believing coin turns out to be believing-is-seeing, whether the believed object exists or not: "If I hadn't believed it I never would have seen it."

Never mind. The tabloids began to pick up messages more interesting by far than those silly yesterday's-news canals. The *New York Herald* reported that the words "The Almighty" were seen on Mars—in Hebrew, no less. *Popular Astronomy*, long a source for the latest news on the canals, drew the line here: "It is a burning shame," they reported, "that such nonsense finds places in our best and greatest daily papers." (It would be a mistake to look back on such nonsense as indicative of an age more ignorant than our own. In 1976 our own best and greatest daily papers reported that a human face was seen carved into a mountain of Mars. The more things change . . .)

In 1897 Kurt Lasswitz, a German professor of literature, was inspired by Leo Brenner's pseudoscience and Jules Verne's book *From the Earth to the Moon* to write *On Two Planets*, the first science fiction novel about Martians invading Earth. Landing on the North Pole, they tried to take over the Earth, but Earthlings soon learned the secrets of their technology, and a truce resulted. This was followed in 1898 by H. G. Wells's *War of the Worlds*, in which Lowell's concept of Mars was totally realized. The dying planet, its water unremittingly drying up, is unable to extend its life any longer by building more canals. The desperate Martians invade Earth, and their superior military technology overwhelms us, but in the end they are defeated by earthly microorganisms to which they have developed no immunity, much as the Native Americans were laid low by the smallpox and measles microbes carried to their shores by the first European invaders.

Sometimes it seemed as if the scientists were striving to keep up with the novelists and the newspapers, instead of the other way around. Sightings of the canals proliferated, and the original proponents had to hustle to stay ahead of the pack. By 1905 Lowell had returned to the observatory, and startled the world once again by announcing that the art of astronomical photography had advanced to the point where his staff had been able to photograph the canals. Until now, planetary observers had been forced to draw their own pictures of what they were seeing, or claimed to be seeing. But with this announcement those culpable days were gone: the camera doesn't lie, he proclaimed, and this would be the end of all doubts.

Unfortunately, when he produced his photographs in 1907, the canals proved to be still elusive. The pictures were tiny—less than a quarter of an inch in diameter—and could not be blown up or reproduced without losing all detail. The quality of the prints was so bad and the proposed lines so tenuous that the situation remained the same: those who saw them, saw them; those who didn't, didn't. Among those who saw them were the editors of the *Wall Street Journal*. In their year-end summary of the past twelve months, they picked as the "most extraordinary event" not the financial crash that panicked most of their investors but Lowell's photographs: "The most extraordinary event of the year . . . is . . . the proof afforded by astronomical observations . . . that conscious, intelligent human life exists upon the planet Mars."

Among those who did not see anything that resembled canals in the photographs was Alfred Russell Wallace, the British scientist who, independently of Darwin, had formulated a theory of evolution. Asked to write a review of Lowell's new book, *Mars and Its Canals*, Wallace rumbled on so angrily and at such length that his review turned into a full-length book of his own. Titled *Is Mars Habitable?* it answered itself emphatically, "No."

For starters, he argued, any intelligent Martians would be too intelligent to build a planetwide system of irrigation canals that would necessarily be doomed to failure. The Lowellian argument was that the canals were used to irrigate the planet, to bring water thousands of miles from the polar caps through desert regions to oases where the people lived. But on a planet which we know to be dry, arid, and of extremely low atmos-

pheric pressure, liquid water in those canals would spontaneously evaporate and be lost to space. There was no possible
way the water would remain liquid, flowing in those supposed
canals over those great distances. Martians intelligent enough
to build such immense engineering projects would certainly
be intelligent enough to realize their futility; and even if they
were not, they would soon enough realize their mistake when
the water evaporated behind them as they dug. There is no
way you can argue with the laws of physics, Wallace insisted:
the canals cannot hold water, and neither does the case for intelligent Martian engineers.

He went further. "Not only is Mars not inhabited by intelligent beings as Mr. Lowell postulates, but it is absolutely UN-
INHABITABLE." He was willing to accept that enough astronomers had seen the canals to provide a strong argument
for their existence, but those supposed engineering structures
could easily be natural cracks formed by the cooling crust,
having nothing to do with either water or Martians. He calculated that due to Mars's distance from the sun it had a temperature of $-35°$ Fahrenheit, so that any water on the planet would
be frozen. If the planet occasionally warmed, which he did not
expect but could not rule out, the unfrozen water could not be
channeled into canals but instead would quickly vaporize and
escape the planet's small gravity. Nowhere on the planet, he argued, would there be liquid water. The ice caps were frozen
carbon dioxide (CO_2), the stuff we on Earth call "dry ice."

Wallace's arguments were well regarded among most scientists. The water situation seemed to doom any chance of
Martian life. Still, there were a few who asserted that no linear
cracks of such magnitude existed in the crust of the Earth or
the moon, so why should they have broken the surface of
Mars? Lowell remained undimmed in his enthusiasm, and
Schiaparelli allowed himself to muse that "The regular and
geometric lines (the existence of which is still denied by many

persons) do not yet teach us anything about the existence of intelligent beings on this planet. But I think it worthwhile if somebody collected everything . . . that can reasonably be said in favor of their existence." This was not quite as strong as his earlier statement that the suggestion of intelligent Martians "contains nothing impossible," but he still found it prudent now to add, in an aside to explain his thinking, "Once a year it is permissible to act like a madman." The rest of the world, or at least the reading public, tended to be swept away by Lowell's enthusiasm and powerful literary style, and by the many newspaper and magazine articles that echoed his faith.

Truth to tell, it was hard to be sure of anything that might occur on a strange world hundreds of millions of miles away. Most scientists were becoming vaguely aware of a sentiment not put into words for several more decades, when the English biologist and essayist J. B. S. Haldane would point out to us his suspicion that "Not only is the universe queerer than we suppose, it is queerer than we *can* suppose." In the first two decades of the twentieth century the settled physical laws of the Victorian universe were crumbling into unrecognizable equations. A Viennese physicist, Wilhelm Roentgen, had closed out the previous century with the discovery of mysterious rays that passed through solid matter. In Germany a man named Planck seemed to show that energy traveled in discrete clumps, while others were beginning to argue that material objects like electrons were really nothing but immaterial waves. And out of Switzerland were coming vague rumors of a patent clerk who wanted to change the nature of time itself. So when William Randolph Hearst, never one to avoid a good story, telegraphed to an eminent scientist, "Is there life on Mars? Please cable one thousand words," the reply he received was, "Nobody knows," repeated five hundred times.

2.

Those millions of miles separating us from Mars were tantalizing because they were so empty, so transparent. We could *see* the planet; sometimes it almost seemed as if we could reach out and touch it. If there were people—or creatures—there, was there no way to contact them? Surely, there must be. In the last days of the last year of the 19th century the French Academy of Sciences announced the establishment of the Prix Pierre Guzman, funded by the widow Guzman: 100,000 francs would go to the first person to establish communication with another world. Madame Guzman didn't want to limit anyone's imagination, but she thought that communication with Mars should be excluded because it would be too easy. (The French Academy decided to hedge on that: communication with any extraterrestrial world would be acceptable.)

William Pickering, who had by now left the Flagstaff observatory because of increasing friction with Lowell, each claiming for himself the honor of being the foremost Marsologist on Earth, was one of the first to get in line, proposing in 1909 a $10 million set of mirrors that could send signals to Mars. The apparatus was never built, but the proposal sparked a series of articles in *Scientific American* and other magazines in which many and varied were the responses, pro and con.

Radio had entered the picture as early as 1901 (just six years after Guglielmo Marconi's first tentative transmissions) when Nikola Tesla, the engineering genius who devised the system of alternating current technology that is the basis for all our electrical power systems, received signals from Mars. Scientists scoffed, but Tesla insisted that he wasn't at all surprised that the Martians would be capable of sending them. "The force of gravity on Mars being only two-fifths of that on Earth, all mechnical problems must be much easier of solution," he explained. "The planet being much smaller, the contact between individuals and the mutual exchange of ideas must have been much quicker," he went on, and this is why "intellectual life on that planet should have been phenomenal in its evolutions."

There were no further signals for nearly twenty years, but then in 1919 Marconi himself said he had received strong signals out of the ether which seemed to come from someplace outside the Earth and which might conceivably have proceeded from the stars. By 1921 he was more definite: MARCONI SURE MARS FLASHES MESSAGES, the *New York Times* reported on September 2. "Marconi is now convinced that he has intercepted wireless messages from Mars," the article went on. "What convinced Signor Marconi was . . . the wave length . . . and their regularity . . . and the letter V of the international code, continued time after time, much after the manner of station calls or test signals sent out from radio stations."

Reasonable men could reasonably think that Pickering and Lowell were crackpots, but Marconi and Tesla were respected engineers, and engineers are not usually noted as having runaway imaginations. The United States Navy was impressed enough with their claims to shut down radio communication with the entire Pacific fleet in an unprecedented effort to establish contact with Mars. On August 22, 1924, every naval station in the Pacific was ordered to stop all transmissions so that we could listen for Martian signals for three days. William F. Friedman, chief of the code section of the Army Signal Corps and the military's outstanding codebreaker (who would later break the Japanese radio code a year and a half before Pearl Harbor), was waiting to decode whatever messages would be received.

Unfortunately, his services were not required. The Martians maintained radio silence with better restraint than the Japanese did seventeen years later. Nothing was heard, and in three days the Navy went back on line. Still, that didn't mean no one was up there on Mars, despite the opinion of most radio experts that the Marconi-Tesla signals were nothing more than atmospheric disturbances and overactive imaginations. Absence of evidence is not evidence of absence, the French astronomer Camille Flammarion responded. Mars, after all, is the home of an ancient civilization: "Perhaps the Martians tried before, in the epoch of the iguanodon and dinosaur, and got tired," he suggested.

If so, they were permanently tired. They never tried again, and communication with them rested solely with our novelists and nuts. H. G. Wells's *War of the Worlds* was followed by a host of others. In particular, Edgar Rice Burroughs, the creator of *Tarzan*, utilized Lowell's vision of an ancient, dying planet, but replaced his civilized world, which had a planetary government and the willing cooperation of all its citizens, with a myriad conglomeration of warring Martians who, in a long series of novels, welcomed the intrepid planetary explorer John Carter, gentleman of Virginia, as their ally or foe, depending largely on the color of their skin and the humanoid characteristics they shared.

Lowell died in 1916, and the legacy he had intended to leave to science was swept up by science fiction. When Philip Nowland's *Buck Rogers* woke up in 1928—well, actually in 2419 since he had inhaled a gas just after the end of the First World War and fallen asleep for five hundred years—he found a solar system inhabited on every planet and moon, providing innumerable opportunities for villainy and adventure. He expanded into comic strips, but his movie career was cut short in 1941 when Earth was under attack from galactic creatures, and Buck flew to Saturn to recruit allies. When the Saturnians arrived to save Earth they turned out to be yellow, and when the Japanese attacked us later that year it suddenly became politically incorrect to have yellow allies. The movie serial was pulled from distribution and never released again (although it is currently available on videocassette under the title *Destination Saturn*).

In 1938 the Martians attacked in earnest, when Orson Welles broadcast Wells's *War of the Worlds* on radio. The panic that ensued, with thousands of people calling the police, fleeing out of town or huddling terrified in their cellars, is a tribute to the power of the medium, to the reality of Lowell's vision as understood by the general population, and to a general smattering of scientific ignorance. (The broadcast announced sightings of flashes on Mars, and nearly simultaneously the Martian war vessels landed on Earth; not even Martians could have managed that.)

The ignorance, hysteria, and paranoia were touched off anew on June 24, 1947, when Ken Arnold, piloting his small plane over Washington, was startled by a formation of nine curious objects flying in a tight vee formation and zooming past him at a tremendous speed. "Flat, they were," he breathlessly reported. "Like pie tins." His drawings looked more like Batman's Batarang, but the newspapers picked up on his verbal description, and the "flying saucer" craze was begun.

It began as crazily as it was to continue. Arnold estimated the speed of the saucers as twelve hundred miles per hour, incredibly faster than any airplane of the time. Yet it is literally impossible to estimate the speed of an object if you have no idea how far away it is: the sun and moon appear to move across the sky at a leisurely pace compared to a jet plane, yet their apparent motion ("apparent" because it's due to the rotation of the Earth) is much greater than that of the jet; they seem to move slower because they are farther away. A bumblebee flitting past your nose moves *fast*, but if it were seen a mile away it would appear to be almost motionless. The point is that if you see something you have never seen before, and have nothing to relate it to so that you have no idea what size it is or how far away it is, you can't possibly evaluate how fast it's moving: it could be moving as fast as the sun or as slow as a bumblebee, and you just wouldn't know.

But twelve hundred miles per hour was such a nice number. The year was 1947, the Second World War was just over, and airplanes that had been pulled along behind their propellers at two and three hundred miles an hour were suddenly jet-propelling themselves across the skies at five and six hundred miles an hour. Twelve hundred miles per hour was slow enough to be believable and fast enough to be just beyond humanity's reach. Never mind that the estimate had absolutely no basis in fact; it was *sexy*, and the tabloids pounced on it.

A rash of reports followed. Fourteen days after Arnold's sighting, Roswell, New Mexico, leaped into the headlines. The Army Air Force's 509th bomb group was stationed there, and on July 8 the nation's radios brought out the news: "The Army Air Force has announced that a flying disk has been found and is now in possession of the army. . . . It has been sent to Wright Field for further inspection."

General Roger Ramey, the officer in charge, said it was a weather balloon. But who wanted to believe that? Everyone knew—everyone *knows*—how the government lies to us. And if you asked what possible reason they could have for lying about flying saucers, you were just naive.*

A rancher found a strewn field of metal shards, obviously from outer space. Someone else found I-beams with strange symbols engraved in them. Another person found crumpled

---

*The Roswell apparition was a Project Mogul balloon. Mogul was a top secret research project aimed at perfecting high-altitude balloons in an effort to detect by stratospheric sound waves the first Russian atomic bomb test, which knowledgeable people in the military knew was coming but of which the public was innocently unaware. Since the project was top secret, the military responses to the Roswell sighting were of necessity a lie. The tangled web they wove led many people to decide there was a government conspiracy to cover up the truth, and indeed there was. But the truth was a military operation aimed at a very terrestrial enemy, and had nothing to do with extraterrestrial invasions.

up bits of stuff, like tinfoil which was "all wadded up; and when it hit the table it spread out like it was water." You couldn't cut it or burn it. "When I was holding it in my hand I couldn't feel it in my hand," a twelve-year-old girl recalled fifty years later. She also remembered that her family was threatened by an Air Force officer to be quiet: "This is a big desert out here. No one will ever find your bodies."

The Roswell incident is a classic in flying saucer paranoiadom. No explanation was offered to the public beyond the idea of a weather balloon. If you want to go over the evidence today, you can't. There is no evidence. No one actually has any of that strange metal. Which, if you have a peculiar turn of mind, only proves the point. Of *course* there's nothing left; *they* have gotten rid of it all. It's a big desert out there. . . .

Other incidents were reported. A plane crashed, and the injured passengers were strange little people, one alive and two dead. They were taken to a nearby airfield which had a hospital, and then never were heard from again. There were too many reports for the Air Force to ignore, and a series of official studies was initiated. But how can you study something that isn't there? Project Sign evolved into Project Grudge and then into Project Blue Book, which introduced the term unidentified flying object, or UFO. Eventually most of the objects were identified. They were seagulls, jet fighters, reflections of ceiling lights on photos taken through windows, or outright hoaxes (some were crude: cutouts pasted on a window or small objects swung from a fishing pole held out of the line of sight; some were more subtle, involving trick photography and double images). Some remained unidentified, but none gave any indication of an extraterrestrial origin.

And right here we have the crux of the UFO debate, which closely resembled the Martian canal controversy. If someone claims to see something, do you believe it? Some people believe anything anyone says, unless it's proven false. Other people believe nothing until it's proven true. Some people believe things if they appear to be reasonable. Other people believe that "the universe is queerer than we can suppose," so *reasonable* is an irrelevant term, and anything is possible.

Most scientists believed that, given the scale of the universe, any intelligent life would try to contact its neighbors first through the transmission of electromagnetic signals, probably in the radio part of the spectrum (which will be discussed in Chapter 6). Given the absence of such signals, the concept of flying saucers, implying humanoid-or-otherwise visitors from other worlds, was an unreasonable one. There were enough claimed eyewitness reports to make it a reasonable subject for further investigation, but when the years stretched on and the investigations turned up absolutely no physical evidence, nearly everyone agreed it was time to pull the plug. The Air Force terminated its investigation in 1969.

But public interest continues. Raymond Palmer, the publisher of *Fate* magazine (later called *Flying Saucers* magazine), carried all kinds of stories, usually featuring people who were forcibly taken aboard a flying saucer and subjected to all kinds of interesting indignities. As we write this, in March of 1997, thirty-nine people in California killed themselves in order to eliminate their earthly trappings and allow their souls to flee to the UFO that is hiding in the tail of Comet Hale-Bopp. There is no shortage of such people, who either believe themselves to be in contact with Venusians or Martians or God or whatever, or who simply want to get their pictures in the papers. If you believe them, they constitute all the evidence you need that not only are we being visited by aliens who behave like lunatics, but also that every government in the world is in collusion to lie to us. And so the flying saucers continue in popularity, and are not likely to disappear anytime soon. But meanwhile, science has been catching up to science fiction.

3.

In 1919 an obscure professor at an obscure New England college published a suitably obscure paper suggesting that liquid-fueled rockets might be used for space exploration. The *New York Times* dutifully carried the story but informed its readers it really wasn't news that was fit to print: rockets would never work in the vacuum of outer space, it explained, because there wasn't any air up there for the rocket gases to push against.

Robert H. Goddard, a young professor at tiny Clark University in Worcester, Massachusetts, knew better, as had nearly every scientist since the seventeenth century (when Newton published his laws of physics, including the one that drives the rockets: every action has an equal and opposite reaction, regardless of whether or not there is any air to push against). Goddard must have been irritated by the *New York Times*'s reaction, but not surprised. Everyone thought he was crazy; why should the *Times* be different?

But actually, the idea of rockets to travel through space was not the aspect that was original with Goddard. Novelists such as Jules Verne had been using rockets for more than fifty years, and there was nothing in the laws of physics that said a rocket similar to the ones the Chinese had first fired in 1232 (in defense of Kaifeng-fu, against the invading Mongols) couldn't penetrate and navigate the emptiness of space. What was original with Goddard was his idea of using liquid fuel to power them—since liquids, although harder to handle, would burn faster and more efficiently—coupled with his intense and patient mechanical work to try to get them to fly straight for ever greater and greater distances, instead of tumbling out of control and falling back disastrously to the Earth which chained them, and us, to its bosom.

Goddard worked largely on his own in the great tradition of American pioneers. In the 1920s there was little federal money available, and what little there was did not go to visionaries who talked of traveling to the stars someday in the indeterminate future. Until the exigencies and opportunities of the Second World War, when he left his dream to work with weapons, he was underpaid and perpetually in need of money, working with few assistants, patching his rockets together with strings and sealing wax, trying to make pigs fly in a country where few people believed they could and still fewer cared. He was the sort of dreamer who kept his eyes on the screwdriver in his hand. Though his ultimate dream was to build a rocket capable of leaving the Earth behind, he fell asleep each night planning what screw he would tighten in the morning, what design change he could make, what little detail he might correct to make the rocket fly a little straighter for a little greater distance than the previous one, before it too failed.

Building on each failure, by 1926 he had put together the first liquid-fueled rocket, which reached a height of forty-one feet over his Aunt Effie's Massachusetts farm, attaining a speed of sixty miles an hour and traveling 184 feet. Throughout each succeeding long Massachusetts winter he would build, bit by bit, one rocket each year. When his summer vacation came he would cart his rocket to the uninhabited deserts in the western part of the country, where the weather was better and the skies clearer, and where a rocket might fly for much further than forty-one feet without crashing into a neighbor's barn, and he would fire off the results of that long winter's work. The rocket would rise twenty feet into the air, eighty feet, perhaps one hundred feet, and then it would begin to wobble, and then the wobble would grow into a gyration and the gyration into a tumble, and the rocket would spin out of control and plummet to the desert floor, and shatter. Robert Goddard would pick up the pieces, each and every one of them, and cart them back to Massachusetts and analyze why the rocket had failed, and then he would build another one.

And he would write reports which few people read, and the

next summer he would take another rocket into the desert and try again. Year after year, one rocket at a time, sometimes improving on the previous year's work, more often not; sometimes finding a technical detail that could be improved upon, more often following a dead trail that ended in a shattering crash on the desert floor. And he would pick up the pieces and start over again the following fall.

There was one person who was interested in everything he did, though Goddard never knew it. In Germany Werner von Braun read everything Goddard wrote, learned from every mistake Goddard made, added his own contributions, and finally convinced the German general staff that he could build a rocket capable of reaching not the moon, but London. And the money flowed in from Berlin, and the slaves arrived from the east, and the world's first military/scientific industrial complex devoted to rocketry began to take shape on an isolated island in the Baltic.

Goddard too began to profit from the war fever racking the world. He was never able to preach the vision of giant rockets reaching across the oceans, but the military came around to ask if he could build rocket shells that might penetrate a tank's armor, or might fly a few hundred yards with greater accuracy and less weight than artillery shells, and soon instead of working by himself to build one rocket a year he was turning out hundreds. By the end of the war, Goddard's rockets had become a useful military weapon; and von Braun's had changed the political shape of the world.

From the time the United States had become a nation until the Second World War, its foreign policy was the result of a continuing conflict between two philosophies: the first urged the nation to take its place as one of the powerful nations leading the world; the other urged us to take advantage of our two oceans, to retreat behind those impenetrable watery barriers and prosper unaided by and unthreatened by the rest of the world.

After the Second World War we no longer had that choice, for on von Braun's drawing board was the A-10 rocket that could leave from anyplace in Europe and reach the cities of America. There was no defense against it, and no refuge left beyond its reach. Combining it with the atomic bomb meant that the United States had been brought into world politics whether we liked it or not.

Irresistibly, the rocket became the anchor and fulcrum of our national defense and the symbol of our military offensive might. A curious phenomenon flourished in these years: the military argued that the nuclear-bomb-carrying rocket was the cheapest form of military insurance against aggression, providing the "biggest bang for the buck" in the then-current phrase. We bought that argument, and the result was—instead of a reduced military budget—the largest military spending spree in history, growing from millions into billions.

And then, in the fall of 1957, a curious new light appeared in the night sky. All over the world people looked up to see it, and tuned in their radios to hear its soft high-pitched beeping, and repeated to each other that strange-sounding word. Sputnik. And the world would never be the same again.

Events moved quickly, spurred on by that insistent light that traveled the sky every night to remind us of what had happened, and what could be. At this time, in the 1950s, we still knew virtually nothing about Mars, and so everything was possible. In 1954 Fred Whipple, the chairman of the astronomy department at Harvard, was writing, "Nobody knows what the canals really are—or if they even exist." And in 1956 von Braun, writing with the rocket expert Willy Ley, insisted that "the canali have been seen by very many reputable observers on very many occasions."

But the age of speculation was about to end. In 1960, just three years after the launch of *Sputnik*, Russia sent off the world's first rocket ships aimed at Mars. They sent two, just in case one of them failed. In the event, both of them failed; they dropped back to Earth and were lost in the ocean. But by the next year the first Russian cosmonaut had traveled around the Earth in orbit, and an American scientist had found firm evidence for life on Mars.

William M. Sinton, using the two-hundred-inch telescope at Mt. Palomar and sending the received infrared light into a spectroscope, found absorption lines corresponding to chlorophyll. Previous work had shown that certain areas on Mars (Pickering's "lakes") underwent seasonal darkening that could conceivably be caused by the growth of vegetation. So Sinton aimed his telescope at these darkening regions, and it was there that he found the evidence of chlorophyll. A careful scientist, he then carried out the perfect control experiment: he pointed the telescope at bright desert regions on the planet, where there should be no life, and the infrared lines disappeared. So clearly they were indigenous to the dark areas, and provided the first firm evidence that life existed on Mars. (Six years later another group would suggest that the lines, which Sinton had accepted as a fingerprint of chlorophyll, were actually due to hydrogen deuterium oxide (HDO), or heavy water, a compound that has no basis in living things. Sinton and his coworkers followed this up and found it to be true; they also found that the HDO was not even on Mars, but was a contaminant in the Earth's atmosphere. His original work was as careful as it possibly could have been, but it was wrong. To this day no one knows why the control experiment failed to show the HDO lines.)

At any rate, in 1961 the Sinton experiment looked conclusive, and the Space Science Board of the National Academy of Sciences convened at the National Radio Astronomy Observatory in Green Bank, West Virginia, to discuss "intelligent extraterrestrial life." They concluded that "The evidence taken as a whole is suggestive of life on Mars." They still hedged their bets on intelligence: "The limited evidence we have is directly relevant only to the presence of microorganisms; there are no valid data for or against the existence of larger organisms and motile animals."

In the fall of 1962 two more Russian Mars rockets failed to leave the Earth. The next one achieved orbit successfully and then took off for Mars. It traveled for months, and then suddenly, 66 million miles from Earth, nearly at the red planet, all contact was lost and it was never heard from again.

By 1963 Earth-based observations showed that the atmospheric pressure on Mars was much lower than had been thought, too low to allow for the presence of liquid water. The situation can best be explained by a "phase diagram":

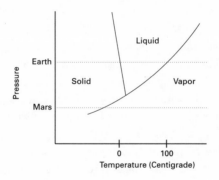

We are plotting temperature against pressure, and the solid lines are the boundaries between the phases solid, liquid, and vapor. The dotted lines correspond to the atmospheric pressures on the surfaces of Earth and Mars. So, for example, on Earth, at a temperature of 0° Centigrade we are on the solid-liquid boundary, which means we are at the melting point of water. At 100°C we are at the boiling point, the boundary between liquid and vapor. At the atmospheric pressure ambient on Mars, we are below the lowest pressure at which liquid water can exist. Depending on the temperature, water there must be either solid or vapor.

(We have to remember that this is an *equilibrium* diagram. So, for example, if you were to land a spaceship on Mars and take out a bucket of water, for a short time you would have liquid water on the surface of Mars. But it would quickly either freeze or evaporate. There can be no permanent liquid water on the surface of Mars and therefore, according to all we know about life, no life there. This point will be discussed further in Chapter 3.)

So by 1963 the pendulum of opinion had swung far to one side, indicating virtually no possibility of life on Mars. The issue of life was dead, so far as scientific consensus was concerned, but still the possibility of exploring other worlds drew us on, and the National Aeronautics and Space Administration (NASA) conceived and launched a series of *Mariner* spaceships to the planets. *Mariners 1* and *2* were planned for Venus: *Mariner 1* crashed into the Atlantic, while *Mariner 2* flew past Venus but returned no pictures. The remaining spaceships were aimed at Mars.

In the initial planning stages there was a proposal for a life detection experiment, designated Gulliver, designed by Norman Horowitz and Gilbert Levin. The idea was to put the spacecraft in orbit around Mars, then parachute a package down to the surface. Sticky coiled springs would spill out and then retract, dragging themselves along the surface and collecting like flypaper whatever microbes might live there. Inside the spacecraft, the now-dirty coils would be soaked in "chicken soup," a nutritious broth which the microbes would feed on, not knowing that the essential carbon atoms in the broth were radioactive carbon-14 isotopes. The microbes would metabolize the carbon and expel it as radioactive carbon dioxide gas, which would then be detected by Geiger counters in the chamber. The data obtained would be radioed back to Earth, and life on Mars would be detected (or denied, if the Geiger counting rate was essentially zero).

It was an elegant idea, but it was canceled before any of the spacecraft got off the ground because no one really believed that there might be any life there to be detected, and space aboard the craft was badly needed for other experiments more likely to succeed.

The first U. S. Mars mission, *Mariner 3*, was lost when its atmospheric shield failed to detach in space, canceling all communications. *Mariner 4* got to within about seven thousand miles of Mars on July 14, 1965. It sent back only twenty-two closeup pictures, but these were enough to settle the hundred-year controversy about Martian canals. Although by now virtually no one thought there could be life on Mars, the specter of Lowell's canals remained alive and well. A NASA *Sourcebook on Space Sciences*, published shortly before *Mariner 4* arrived, concluded that "most astronomers would probably agree that there are apparently linear markings . . . of considerable length on the surface of Mars."

Now, finally, we would know for sure. Naturally enough, there was suspense and drama built in. The twenty-two pictures were taken in only twenty-five minutes, but it took 8½ hours to transmit each photo to Earth. The first few weren't clear enough to be definitive, and the earthbound scientists waiting minute by minute and hour by hour back at the Jet Propulsion Lab were growing increasingly frustrated, until finally photo no. 7 came in loud and clear, and it destroyed Lowell: there were no canals, there was nothing even remotely resembling them. There was nothing down there (or up there) but craters. It was a planet as lifeless as the moon. There was no water and no visible erosion that might be ascribed to past water; there was virtually no atmosphere, and no active tectonics. The surface they looked at, in detail incredibly clear, was that of a cold, dead planet.

Despite the previous evidence indicating that the planet would be lifeless, the disappointment was intense. J. N. James, one of the staff, expressed it well: "When we were first permitted into the secure planetology science room and shown a small photograph of Mars's moonlike surface, we agreed that we had a two-minute jump on the rest of the laboratory staff to get into another line of work. We knew that a large segment of the public still expected to see canals at least, if not green people."

The photo-interpretation team, headed by R. B. Leighton of Cal Tech, reported that it was now "difficult to believe that free water in quantities sufficient to form streams or to fill oceans could have existed anywhere on Mars since that . . . surface was formed." Still, after all this time, Lowell wasn't quite dead. The team found it necessary to add a palliative addendum, stating that their data "neither demonstrate nor preclude the possible existence of life on Mars."

4.

Life, as we all know, is tenacious. It exists under the sands of the Sahara, beneath the frozen surface of lakes in Antarctica; it hung on in Auschwitz and in the Gulag; it miraculously reappears from under layers of smoldering volcanic ash. And life on Mars was no different. The next few years found evidence for it waxing and waning. After two more Soviet missions to Mars failed, *Mariners 6* and *7* found some bright areas that were sparsely cratered, indicating that they were young; that is, that they had been altered.

The idea behind this interpretation is that early in the solar system's history there was an immense amount of interplanetary debris floating around. In fact, the planets themselves were formed by the accumulation of such debris. As the forming planets swirled around the sun, this debris fell on them and dug numerous craters on the surface. Our moon shows the evidence of this early bombardment, since it is a lifeless, inactive body. No volcanos later erupted to cover the craters with magma, no water flowed to erode them. They have remained unchanged for billions of years in mute testimony of that chaotic time.

The early photos of Mars showed a similarly cratered, ancient terrain. But when *Mariners 6* and *7* found a previously unphotographed region, Hellas, which had many fewer craters, it indicated that the surface had *not* remained unchanged since its formation. What could have changed it? What could have erased the craters that must have been there in the beginning? Perhaps, the suspicion arose, there was some erosion after all. But if there had ever been sufficient water to erode the craters, where had it come from and where did it disappear to?

Other regions photographed on these same missions showed evidence of chaotic terrains, in the form of jumbled piles of rubble. Perhaps, the idea arose, there were pockets of frozen water under the surface. And perhaps that water occasionally melted. If it did, it would shrink in volume and the ground above it might collapse, opening a way for the now liquid water to flow up onto the surface, and leaving behind the piles of chaotic rubble the *Mariner* photos showed.

So at some time in the past there might have been water on Mars, reviving the idea of life there, an idea that took a tremendous jump when the data came in from *Mariner 7*'s infrared detectors. The spacecraft flew over the south polar ice cap, and George Pimentel, of the University of California at Berkeley, found that temperatures at the very edge of the ice cap were too high for it to be composed of frozen CO_2. What else could that white stuff be but water? His group also found evidence for methane and ammonia in the atmosphere. These are reduced gases, but the surface of Mars is highly oxidized. The chemical disequilibrium of reduced gases on an oxidized planet indicates that something must be continually producing them. What else could it be but biological activity? On August 7, 1969, the *New York Times* reported on NASA's press briefing. "Two Gases Associated with Life Found on Mars Near Polar Cap" was the headline, and the article went on to say that "Two gases intimately associated with the origin and existence of life—methane and ammonia—have been detected in the Martian atmosphere . . . [The] findings produced a general gasp among scientists and newsmen."

Pimentel and his associate tried to rein in their excitement, but it burst through in a personal statement: "One cannot restrain the speculation that [these gases] might be of biological origin."

A few weeks later, however, laboratory experiments showed that the absorption bands they had attributed to methane and ammonia were also found in solid CO_2. When it was further suggested that the higher temperatures they had found at the edge of the ice sheet could be due to solid ground showing through as the edge of the CO_2 ice cap volatilized, all evidence for life down there in the polar regions vanished. With the morning dew, one might say.

Norman Horowitz (who had helped design the first experiment aimed at finding Martian life), reflecting back on those days, reminisced that "The prospects for life on Mars seemed so dim by 1970 that there seemed little good reason to emphasize biological questions in planning the spacecraft that would land there in 1976."

The spacecraft he was talking about was *Viking*. But there were several others that flew before *Viking* could be ready. In 1971 *Mariner 8* failed and dropped into the Atlantic; the Russians tried again with *Cosmos 419* and an unnamed rocket; both failed. They succeeded in placing *Mars 2* in orbit around the planet, but it found a planetwide dust storm going on. The spacecraft had been preprogrammed to drop a lander to the surface—the first probe sent out from Earth to do this—and there was no way to countermand this command. So, despite the dust storm, it launched its lander, which vanished into the dust and was never heard from.

Mariner 9 also made it into orbit around Mars, and also found a dust storm raging below. But it was a more versatile craft, not preprogrammed but accepting commands from Earth. The scientists directed it to spend its time photographing the moons of Mars, and it hung around long enough for the storm below to dissipate. When it turned its attention back to the Martian surface it found the most exciting things yet: canyons, lava flows, volcanos; all of which was evidence for a tectonically active and therefore more interesting planet than they had thought. And they found more, much more: there were a few wandering, straggling things down there that looked like nothing so much as dry river beds.

As more and clearer photographs came in, the excitement mounted. There were carvings in the surface of that planet that simply could not have been formed by anything but flowing water. Attention turned back to the chaotic terrain found by *Mariner 7*. It now looked extremely likely that subsurface frozen water had, sometime in the past, melted and come flowing out onto the surface. And not just little bitty dribs and drabs of water: the carved-out channels indicated raging rivers, floods, perhaps even lakes and—dared we say it?—oceans. And where you have an ocean, can life be far behind?

Further evidence was surely needed, but it was hard to come by. In 1973 Russia sent out a fleet of spacecraft that reached the planet the following year, but with not much result. *Mars 3* orbited and sent out its lander, but it conked out after two minutes, and sent back no photographs. *Mars 4* reached Mars but failed to orbit. *Mars 5* achieved orbit and sent back sixty pictures, but they added nothing of interest to the question of life. *Mars 6* landed on the planet but stopped signaling seconds before it reached the surface. *Mars 7* missed the planet entirely.

In 1975 we launched *Viking*.

3 | The Vikings of Mars

I am afraid that the experiments you quote,
M. Pasteur, will turn against you. The world into
which you wish to take us is really too fantastic.
LA PRESSE, 1860

The best laid plans o' mice and men
 Gang aft' agley,
And leave us naught but grief an' pain
 For promised joy.
ROBERT BURNS, 1785

1.

In 1970 the prospects for life on Mars were so low that Norman Horowitz saw little use in preparing a biology experiment to be sent to the planet. One year later, when *Mariner* found evidence that water might have existed on the surface, that expectation of life was totally reversed and a dozen scientists began racing to perfect experiments that would land on Mars and actively search for life.

Why? Why is water so important? What else is important

for life? How and where does life form? These are all questions about which we had nothing but speculation for hundreds of years, but by the latter half of the twentieth century we were putting together a pretty good idea of what the answers might be.

The path to those answers was sinuous and twisting, with lots of wrong turns along the way. The basic conundrum was similar to that faced by physicists who were arguing during this same time period about the question of the creation of the universe. The first law of thermodynamics was the basic law of physics: matter and energy can neither be created nor destroyed. But then where did all the material and energy of the universe come from? Either they always existed or they were created; there is no other alternative. Unwilling to face a universe that always existed, the physicists simply finessed the point: they relegated the creation of matter and energy to a time "before physics began." In this they were probably influenced by their Judeo-Christian religious upbringing, which stated that "In the beginning God created the heavens and the Earth." After that, physics took over.

The biologists had the same problem, but curiously enough they came up with a different answer. Life, like the universe, either has always existed or it is created. Religion taught that the breath of God was necessary for life, but the biologists took another tack. They said that life is always being created, it takes place every day in every way.

Well, why not? There was plenty of evidence that this is true. Hundreds of years ago, when the world wasn't as sanitized as it is now for most of us, everyone was familiar with maggots. Leave a dead animal, or person, lying around in the gutter for a few days, and you'll find maggots crawling in and out of its eyes and nose and any open wounds; indeed, in and out of any openings at all into its body. The maggots weren't there when the creature was alive, but now they are.

The conclusion was obvious: dead, rotting flesh produces maggots. And worms. And flies, and higher forms of animals as well. Today the main argument against the natural, or spontaneous, generation of life is that despite all our scientific ingenuity and knowledge we haven't been able to create life in the laboratory. But this was not always true. In the 1600s a respected Belgian physiologist named Jean Baptiste van Helmont wrote down the results of his experiments in producing life—his *successful* experiments:

> If one stuffs a dirty shirt into the orifice of a vessel containing some grains of wheat, the fermentation exuded by the dirty shirt, modified by the odor of the grain, after approximately twenty-one days brings about the transformation of the wheat into mice.

Which brings us to a basic misconception about science. Most people think that science works by testing theoretical ideas with experiment or observation, and the experiment either proves or disproves the theory. But it's more complicated than that. For example, the first experiment done to test Einstein's theory of relativity came up with the result that the theory was wrong. But the theory wasn't wrong; it has since been tested time and again, and has been "verified to unprecedented levels of accuracy," according to the journal *Physics Today*; it was that first experiment which was wrong. "It is easy to do experiments, but hard to do them flawlessly," Louis Pasteur wrote, commenting on the experiment by van Helmont (1579–1644) that produced mice from grains of wheat and an old undershirt.

Pasteur himself did experiments flawlessly. In the 1860s he killed the idea of spontaneous generation by two series of experiments. The spontaneous generation theory was based on the idea that there is a mysterious force in our atmosphere which produces living creatures out of nothing (or out of dirty shirts and wheat). Pasteur responded that all life comes from previous life of the same kind; mice come only from mice, people come only from people. Life cannot be created, only reproduced. To prove this by experimentation, he turned his attention to the simplest situation: whenever a nutritious broth (such as yeast and sugar) is left exposed to air, bacterial mold soon forms. This was proof to the spontaneous generationists that life forms in and of itself, given nothing but food and air. To Pasteur it was proof that the air is filled with microbes which settle onto tiny dust grains and float about. To prove his point, Pasteur filled seventy-three flasks with broth, boiled them to kill any life and expel any air that might be in them, and sealed the flasks. He then took them to the village of Arbois, in the foothills of the Jura Mountains, and broke the seal on twenty of the flasks. The air rushed in and he carefully resealed them. In a short time, eight of the twenty showed signs of life.

Next, he climbed to the top of Mont Poupet, 850 meters high, and broke open and then resealed another twenty flasks. Only five of these responded with living growths. Finally, he traveled to the Alps, to the village of Chamonix, and climbed as high as he could on Mont Blanc, where he once again began to break the flasks. But now came an unexpected difficulty, which in the end proved propitious. He wasn't able to reseal the flasks with the torch he had brought, and so after breaking only thirteen of them he brought them back unsealed to the dirty, dusty inn in which he would spend the night. The next morning his assistant arrived with a better torch, and Pasteur sealed the thirteen flasks and once again he set off up the mountain. This time he broke and resealed the remaining twenty flasks in the clear, undisturbed mountain air.

All thirteen flasks which had lain open in the inn later showed bacterial growth. Of the final twenty which had sampled nothing but the pure mountain air, only one responded with a bacterial mold.

This showed that it wasn't air—or any mysterious force in

the air—that produced life inside the flasks, for all the flasks had been exposed to air. Instead, the molds were produced by microbes floating around in the air on dust particles. Close to sea level there were many such particles, and as one went higher and higher their frequency decreased, since gravity kept most of them down close to sea level. When the seals on the flasks were broken the ambient air rushed in, sucking in with it whatever dusts and associated bacteria were around.

Not everyone agreed. Dr. Felix Pouchet, Director of the Museum of Natural History in Rouen, believed that spontaneous generation of life was a common occurrence; all that was needed was oxygen for bacteria to suddenly materialize. He conducted the same series of experiments, and got quite different results. He exposed broth-filled flasks to the air on the plains of Sicily and at various heights on Mt. Etna and on ships at sea, and in all of them bacterial molds inevitably grew. So Pasteur's altitude experiment must have been somehow flawed, he reasoned. And surely there was less dust at sea than over land, yet his seaborne flasks were just as full of life as those opened over land. This proved that all air was "equally favorable to organic synthesis," as he put it, and therefore that it was a gas in the air—presumably oxygen—rather than heavy microbe-laden dust which caused the growth in the flasks. "With a cubic decimeter of air, taken wherever you like, I affirm that you can always produce legions of microzoa."

But what was wrong with Pasteur's results? Pouchet was sure that there was some experimental error involved. He didn't know what it could be, but he could prove his point by climbing even higher than Pasteur had ever done, exposing his flasks to the air up there, and getting the right result. So he did it. Together with two colleagues he climbed past the pass of Venasque, higher than Pasteur had ever gone, and still higher and higher. Finally they reached the great glacier of the Maladetta, three thousand meters above sea level and one thousand meters higher than Pasteur's highest, and there they waved their guides away, waited while the dust settled, and in the quiet clear mountain air they exposed and resealed their flasks. And sure enough, all the flasks soon showed signs of bacterial life.

The conflict between the two sets of experiments was so drastic, and the issue so important, that the entire scientific community of France soon took one side or the other. On April 7, 1864, Pasteur appeared before the intellectual nobility of Paris to give a public lecture at the Sorbonne. Anyone who was everyone was there: George Sand, Alexander Dumas *père*, Princess Mathilde, the scientists of the French Academy, and all the beautiful people, *le tout Paris*. Pasteur swept them all along on the tide of his passion.

Pouchet was wrong, he declared, because of carelessness. Contamination by microbes is all around us, and Pouchet had not taken the elementary precaution of sterilizing not only the contents of the flask but also the glass-cutters with which he had broken the seals. When he had cut off the necks of the flasks it didn't matter whether he was on sea or on land, at sea level or on a mountaintop, because microbes in the dust that was inevitably on the glass-cutters were immediately sucked into the flask along with the surrounding air, and that is why the contents of his flasks prospered.

To prove his point Pasteur had conducted a second series of experiments, which he now demonstrated. He held up a curiously shaped flask. It had the customary spherical base, reaching upward into a thin neck. But in this case Pasteur had stretched the neck out into a long S-shaped curve before sterilizing the contents. The flask—and many more like it—had never been sealed, he told the audience, and it had held the broth for all of four years; raising the flask up high, he asked them to be his witnesses: the fluid inside was clear, there was no sign of bacterial activity. "The liquid remains as limpid as distilled water," he claimed, and they looked and saw that he was right.

Why had it not been contaminated? Because the curving shape of the pulled-out neck allowed air to circulate but kept out the tiny but heavy dust particles with their bacterial load. "The difference is this," he explained. "[In a normal flask] the dusts suspended in air and their germs can fall into the neck of the flask and arrive into contact with the liquid, where they find appropriate food and develop; thence microscopic beings. In the second flask, on the contrary, it is impossible that dusts suspended in the air can enter the flask; they fall on its curved neck instead." Inside the spherical base of the flask where the nutritious broth awaited, no bacteria entered—and, of course, none were spontaneously generated by the air that did enter. "And therefore, gentlemen, I point to that liquid and say to you," he cried, his voice rising to a crescendo, that "never will the doctrine of spontaneous generation recover from the mortal blow of this simple experiment. No, there is no circumstance known today whereby one can affirm that microscopic beings have come into the world without germs, without parents resembling themselves. Those who claim it are the playthings of illusions, of badly done experiments, tainted with errors that they did not know how to recognize or that they did not know how to avoid."

Applause. Tremendous applause. A standing ovation. End of argument. Spontaneous generation was dead. One of the great moments of science.

But in truth, Pasteur was possibly wrong on one count and misleading on another. First, in his explanation of Pouchet's experiments he may have been in error. René Dubos, a well-known twentieth-century bacteriologist and science writer, has analyzed the experiments from a modern perspective and thinks that it may indeed have been oxygen that activated—if not created—Pouchet's bacteria. His glass-cutters are not likely to have been coated with sufficient bacteria to explain the results, but another difference between his and Pasteur's experiment may do so. Pasteur used an infusion of yeast in his flask, while Pouchet used a hay infusion. In the intervening years we have learned that some bacteria form spores which can survive quite high temperatures; the microbes in yeast are killed at a much lower temperature than those in hay. Pouchet's mistake may have been in choosing the wrong medium: although he heated his flasks as hot as Pasteur did, they were not sterilized; and when the seal was broken and fresh air came in they revived and blossomed.

That explanation is only a possibility, but there is another sense in which Pasteur's results were certainly misleading. He himself never claimed that spontaneous generation has *never* occurred; he claimed only that the experiments of Pouchet and his followers were wrong. He realized, though his audience subsequently forgot, that spontaneous generation *must* have occurred at some time in the past; the ultimate origin of all life anywhere in the universe must lie in such a process. Today we realize that this must be so, for we now know that the universe began in a Big Bang of such incredible proportions that no atoms or molecules could have survived, if any were there to begin with. The whole universe was nothing but pure energy out of which the particles of our material worlds soon emerged. At that beginning there was no possibility of life, and yet today life exists. No matter how we argue the point, there is no escaping the conclusion that somehow, at some point in time if not at many points, life emerged out of nothing but elementary chemical compounds and energy.

"It is easy to do experiments, but hard to do them flawlessly," Pasteur said, and he was right. He might have added, "It is easy to interpret experiments, but hard to interpret them correctly."

2.

Pasteur was a devout Catholic, and he saw in the debate with Pouchet much more than a scientific quarrel. If his adversaries were right, he argued, then "the idea of a God is unnecessary."

No wonder that he fought so hard to prove his point: he was not merely experimenting, he was rescuing God from irrelevancy.

He won the debate and saw spontaneous generation into its grave, where it lay dead and unmourned for nearly a hundred years until another excellent experimenter and his graduate student began an equally elegant series of experiments.

It was 1953 when Harold Urey, who had already won a Nobel Prize for his experimental proof showing the existence of the heavy isotope of hydrogen, and his student, Stanley Miller, published their first results. At that time the cosmologists were still wrestling with the question of the origin of the universe. The Big Bang had been proposed but was not yet universally accepted; its opponent was the Steady State theory, in which the universe was seen as having had no beginning, having always existed in much the same state as it is today, being infinite in both space and time. In such an infinite universe, life might have existed always. If the universe itself had never been created, the same might be true for life.

Miller and Urey attacked the problem of spontaneous generation of life by designing an experiment based on some theoretical ideas put forward by a Russian biologist, A. I. Oparin and by the British physiologist J. B. S. Haldane in the years 1924–1936. These ideas were in turn based on newly developed knowledge of the composition of the universe and the formation of the solar system (to which Urey had made important contributions). By then we knew, from analyses of the wavelengths of light coming to us from the stars, that hydrogen was the most abundant element in the universe. Helium, an element unimportant to life, is next most abundant, followed by oxygen, carbon, and nitrogen. These four elements form most of the material of all living things. From Friedrich Wöhler's work a hundred years previously we had known that these elements can be put together by simple chemical reactions to form the complex molecules called hydrocarbons, which are the building blocks of life.

The simplest of these hydrocarbons, methane, occurs naturally in the atmosphere of the giant planets. We don't find it (beyond trace amounts) in our own atmosphere because it reacts with oxygen to form carbon dioxide and water. But the oxygen in our atmosphere was formed relatively late in our history by photosynthesis carried on by plants, Oparin and Haldane reasoned, and so when the Earth formed there would be no oxygen to eat up the methane. So in the beginning much of the Earth's carbon might have been in the form of methane or even more complex hydrocarbons. The nitrogen could have been in the form of ammonia, since ammonia is seen today belching forth from volcanos. These two molecules, plus water, could have formed into progressively more complex hydrocarbons that eventually might have—purely by chance—happened on the possibility of reproducing themselves.

Would these molecules be alive?

Well, what *is* life? For most of us, life is like pornography: we can't define it, but we know it when we see it. A cat is alive, a rock is not. Grass is alive, fire is not. But what defines the difference?

Not mobility, for grass can't move. Not the ability to take nourishment from the environment, for fire can do that. Not the ability to reproduce, for fire can do that too.

Let's look at an evolutionary summary of life to help us, as illustrated below.

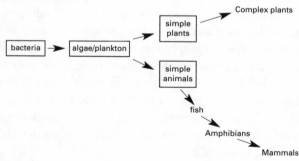

Creatures to the right—the complex organisms we see around us—are certainly alive. As we move to the left we find simpler and simpler creatures. We enter the realm of microbes. Are they alive? Bacteria certainly are. How about even simpler things, such as viruses, or viroids? These can be composed of nothing but a single molecule and, on their own, exhibit none of the signs of life. But when they enter a host cell they suddenly take over the host's machinery and use it to reproduce and then they go swarming out and into other cells, and they look like the hordes of Genghis Khan; at any rate, they look alive.

The point is that before we encountered microbes there was little difficulty in deciding whether something was alive or not. But the microbial world is a strange kind of universe; they do things differently there. Things can be *sort of* alive, and sort of not.

Let's look at the whole chain of evolution as if it were spread out on the page. On the beginning end of the chain we have the chemical elements: carbon, hydrogen, oxygen, nitrogen—all the different kinds of individual atoms. On the opposite end we have people and hippopotamuses. From this end, going backward in time, we find simpler and simpler organisms. From the other end, going forward, we find atoms forming naturally into molecules, and these forming into organic molecules, and perhaps, Oparin and Haldane postulated, there is no barrier or chasm in this sequence: perhaps the molecules just get more and more complex and eventually they become something that might be alive and finally they *are* alive.

Neither man ever did an experiment to test his thesis. It was Miller and Urey who finally attacked the problem experimentally. Urey reasoned from geochemical evidence that when the Earth formed, the volatile compounds which came bubbling out to form the atmosphere and oceans would be composed largely of hydrogen, methane, ammonia, and water. He visualized a turbulent atmosphere that would result in vicious thunderstorms. So Miller mixed these four compounds together in a flask and zapped it with electricity. When he analyzed the results he found that the flask now contained complex organic molecules, including amino acids (the stuff of which proteins are made).

Since that time further research has suggested that perhaps Urey wasn't quite right about the components of the Earth's initial atmosphere, but he wasn't far wrong either. Certainly carbon, hydrogen, nitrogen, and oxygen were present, but the precise chemical state of each of these is arguable. The carbon may have been methane or carbon monoxide or dioxide, the nitrogen may have been the elemental gas or ammonia, the hydrogen may have been the elemental gas or water, and oxygen may have been there as the oxides of carbon or as water, but these details don't really matter because extensions of the original Urey-Miller experiment have shown that when various forms of energy—lightning or ultraviolet radiation or heat from asteroid impacts or volcanos erupting—are passed through mixtures of any of these postulated compounds, amino acids are formed.

The next step is to form these simple compounds into complex polymers, long chains of hydrocarbons, and this step—although easily done in the laboratory—has not yet been demonstrated to occur under the conditions likely to have been prevalent on the early Earth, nor has life ever been seen to occur spontaneously today. But on the primitive Earth there was much more time to accomplish the former than any of our scientists have had, and the latter is not expected to occur because long before an organic molecule could grow into life today it would be gobbled up as food by the life forms already present.

This problem would not be encountered on the early lifeless Earth, of course, and so we envisage the forming oceans, lakes, and ponds as being increasingly pumped full of organic molecules until they formed a rich souplike mixture. The first life form "need not have been very efficient," as Norman

Horowitz has said. "Since it lived in a Garden of Eden with no enemies and no problems of food supply, it had only to reproduce itself fast enough to stay ahead of its own chemical decomposition. Furthermore, the chemical events that preceded the origin of life took place in a vast arena of space and time," an arena impossible to duplicate in the laboratory.

So we have not yet proven that, given the right conditions, life will form spontaneously. But since life did not exist in the beginning of the universe it *must* have been created at some point, and in all the work that has been done since the Miller-Urey experiment there has been nothing to show that a barrier exists between the living and nonliving worlds. Rather, we have continually approached this unexplored region from both right and left, coming ever closer to meeting in the middle: chemical compounds are synthesized to increasing levels of complexity, and living things are discovered with increasing levels of simplicity. All life on Earth seems to be the result of a chain of evolution beginning with simple atoms under the right conditions, and progressing to more and more complex organisms, including us.

3.

What, then, are the "right conditions"? There are just three of them. First, we need a source of energy. The second law of thermodynamics tells us that all natural systems tend toward disorder unless energy is pumped into them. Since life is a system of increasing order, it cannot exist without an outside source of energy. But this is no problem in our universe. Planets are expected to form around stars, and the radiation from the parent star is sufficient, as might also be volcanic energy.*

The second condition is time. The life-forming sequence of processes is an exceedingly complex one, and we need to mix the components together for a long time before we can hope to

*These points are discussed further in Chapters 4 and 5.

see life emerge. Again, this condition is easy to come by. Stellar and planetary lifetimes are on the order of billions of years, so no problem there. (Recent discoveries have pushed the original presence of life on Earth back to about 3.8 billion years ago, and we know that the Earth was bombarded by a rain of meteorites until about that time, leaving virtually no time lag at all before life appeared, according to some researchers. But that number of 3.8 billion for both the appearance of life and the end of the meteorite bombardment is not a sharp one; the bombardment ended gradually, and the date for the appearance of life has at least a 10 percent margin of error, leaving something like a couple of hundred million years for life to form—assuming that it couldn't form until the end of the bombardment, which is not at all a universally accepted assumption.)

The final condition is the presence of liquid water. We may be wrong here, since all we know about life comes from the study of life on this one planet, but from all we do know it seems that liquid water is essential. Aside from the observation that all living things are composed mostly of water, there are good theoretical arguments for it as a prime necessity.

First, in order to mix the chemical compounds together and give them a chance to form chains and rings of higher complexity, they have to be dissolved in something. They could conceivably form complex molecules just by bouncing together occasionally, but this would be a much slower process than if they were swirling around in the comparatively intimate contact of a solution, and while geological time is exceedingly lengthy, it is not infinite. If life is to form on a planet it must do so within the planet's lifetime of not more than a few billions of years. After about ten billion years a typical sun-like star ends its quiet, steady burning and rapidly evolves into a red giant; soon after, it declines and either fades away or explodes. Either way, the timespan of a life-bearing planet is limited; if life is to form, it must do so within this time frame.

Second, water is both the most abundant solvent in the universe and the best; that is, it will dissolve more things better than anything else. This property is linked to its structure: as the hydrogen atoms are bound to the oxygen, their electron cloud is pulled toward the attractive oxygen so they are left with partial positive charges. This enables them to attract negatively charged molecules in a weak but important "hydrogen bond." Not only does this help bring other compounds into solution, but it holds them with a weak, and thus a mutable, bond. This allows for things to change, and change is the essence of life.

So while life can be visualized without water, the life-forming process is so difficult that it just seems more likely to happen where you have a good supply of liquid water.

4.

Which brings us back to Mars.

Between 1971, when *Mariner 9* sent back to Earth pictures of ancient riverbeds on Mars, and 1975, when *Viking I* was launched from Florida toward the red planet, excitement had been mounting sky-high, so to speak, and the entire emphasis of the mission had changed. *Viking* was going to measure many things, to be sure; it would sample both the atmosphere and ground of Mars and measure a variety of chemical elements, it would send radio waves around the sun and test Einstein's general theory of relativity, it would carry out physical experiments to ascertain the solid-state characteristics of the planet. It would do all these things and more, but uppermost in everyone's mind was a complex series of experiments that would search for life on Mars.

During those intervening four years the best minds on Earth were arguing, testing, searching for the definitive experiment that could be shrunk to a manageable size and put on board a spacecraft, sent across 200 million miles of space and

operated by remote control, and which would send back the
ultimate answer: yes there is—or no there is not—life on Mars.
After receiving proposals from dozens of scientists, having
them reviewed and critiqued by dozens more, after meetings
and conferences and seminars which nearly reached the level
of no-holds-barred wrestling matches, NASA agreed on the
experiments to be included on Viking and sent to Mars to
bring back that elusive answer.

And what was that answer? As Leonard David, one of the
founders of the Mars Underground,* put it: "*Viking* went to
Mars and asked if it had life, and Mars answered by replying,
'Could you please rephrase the question?'"

The Viking spacecraft was designed to go into orbit around
Mars and then send a lander to the ground. A mechanical arm
with a scoop would deploy, dig up a little Martian soil, and
bring it into separate chambers in the lander where four different experiments would take place, three of which would look
for definitive signs of biological activity. Everyone would have
liked more than these four, but space was limited. Ten separate
groups were working on their experiments almost up to take-
off time although they knew there wouldn't be room for all of
them; finally the list was whittled down to four.

The nonbiological experiment was a gas chromatograph
mass spectrometer, designed and operated by Klaus Biemann
of the Massachusetts Institute of Technology, which would
look not for life but rather for its constituents, organic molecules. The general feeling was that there had to be organics
there since they are common in meteorites and comets and
even in outer space. They seem to be formed easily by natural
processes and there was no obvious reason why Mars should be
different. But they are the precursors to life; that is, if there are
no organic molecules there can be no life, and so the obvious
first step is not to assume they're there but to measure them
and find out for sure.

*A semiprofessional group dedicated to the human exploration of Mars.

The next three experiments would look directly for life processes, which meant making some assumptions about what those processes might be. The pyrolytic release experiment, designed by Norman Horowitz of Cal Tech, would detect the incorporation of CO_2 into organic compounds, looking for plantlike life. Radioactively labeled CO_2 (in which the carbon atom is carbon-14) was added to the atmosphere within his chamber, and the soil sample was illuminated by a xenon lamp to reproduce the intensity and wavelengths of sunlight on Mars. After allowing a few days for photosynthesis—in which the putative organisms would incorporate the radioactive CO_2 into their bodies—Horowitz would tell *Viking* to sweep out the gases in the chamber (to get rid of any radioactive CO_2 left over), and then to heat the soil. If it contained plantlike microbes that had absorbed the radioactive CO_2 and fixed the carbon atoms into their own structure, these atoms would be vaporized by the heat and would then be swept into a Geiger counter. If there were no microbes, all the radioactivity would have been swept out on the first sweep, before the heating. If any showed up in the second sweep it would mean that some of the atmospheric CO_2 had been trapped within microbial bodies and not released until those bodies were vaporized. If Mars were a dead planet, all the radioactivity would be seen in the first sweep; Geiger counts in the second sweep would indicate life on the planet.

The third experiment was similar, but would search for animal-like microbes that respirated instead of photosynthesizing. Designed by Gilbert Levin of Biospherics, Inc., and called the labeled release experiment, it would also use radioactive carbon-14, but this time to detect the decomposition of organics. In this case a nutritious broth of organic foodlike chemicals (in which some of the carbon would be radioactive) would be dripped onto the Martian soil that had been brought into the lander and deposited in Levin's chamber. The idea was that if there were animal-like microbes in the soil they would eat the food and thereby incorporate the radioactivity into their bodies. There the organic molecules would be metabolized and finally breathed out as CO_2. The radioactive carbon would have been injected into the chamber only as part of the liquid organic food. If any of it showed up in the atmosphere in the chamber, it could only be because it had been changed through a metabolic process into CO_2, which would be a positive signal for life on Mars.

The fourth experiment had been designed by Vance Oyama of the Ames Research Center in California. It was similar to Levin's in that it looked for animals but dissimilar to both Levin's and Horowitz's in that it did not use radioactive carbon. Referred to as the gas exchange experiment, it would measure the uptake and release of various gases, including oxygen as well as CO_2. It therefore could investigate a broader range of reactions than Levin, but with less sensitivity. The idea was that perhaps living creatures on Mars would be so different from those on Earth that they would not metabolize carbon in the same way. Still, they had to interact with their environment in *some* way; they had to take nourishment and release waste products if they were to be considered alive. So Oyama would look at a wider range of possibilities than the other two. The disadvantage of his design was that he wouldn't be able to see individual reactions as the other two could: when a Geiger counter detects a radioactive signal, it is seeing the decay of a single atom. No other technique has this sensitivity.

This group of experiments was the best compromise that the best minds in the country could come up with. It wasn't perfect. Horowitz, for example, claimed that both Oyama's and Levin's experiments were flawed because they dripped liquid food onto the Martian soil; microbes on Mars, he argued, if they existed at all did so without much water in their natural lives and could well be drowned by the food *Viking* was feeding them. These two experiments were "terrestrial in orientation," Horowitz argued, "so much so that neither of them could operate under Martian conditions . . . [They] seemed to be designed for Lowellian or pre-*Mariner-4* Mars rather than for the real Mars, and serious questions were raised about the wisdom of including them in the Viking payload."

Others agreed and went even further, protesting that without knowing anything about the chemical possibilities on Mars we could have no possible idea of what form life could take there, and all the experiments were too narrowly focused: the ground could be teeming with microbes that didn't react in the way the experiments assumed they would. They might not photosynthesize carbon, might not like the kind of food *Viking* carried, might be totally different creatures than the kind we were looking for.

All this was true, but what else could be done to find them? The only true test of life, everyone agreed, was that life must reproduce and also mutate; that is, it evolves. But how could you send an experiment that would take the long time necessary to see evolutionary change on Mars? That was impossible. The only really valid alternative suggestion was that it was too soon to go to Mars to look for life; we were skipping over the basic questions. What should be done first was to send a fleet of spacecraft there to search out the physical and chemical conditions in detail. Then, when we knew more about the Martian environment we could better visualize what life there might look like, what chemical reactions it might undergo, and where it might best be found—only then should we go looking for it.

But that suggestion ignored the political pressures NASA was under. It is not only a scientific organization but a political one, and it would ignore Earth's social environment only at risk of its own existence. It could no more make its decisions based on science alone than Bambi could wander free and easy across a meadow prowled by hungry wolves. This had been true at the time of its greatest triumph, the race to the moon; it was true at the time *Viking* was planned; and it is just as true today.

So no one tried seriously to get funding from Congress for the proper series of experiments. Instead it was going to be quick and dirty, a triumph of American know-how—slip it in and send the data back and announce the answer: life on Mars.

And oh yes, there was one more experiment on board. Just before the final package was approved, Carl Sagan, of Cornell University, spoke up with another argument. We were sending a sophisticated group of experiments that would sample the soil, bring it aboard, and search in it for microbes. Wouldn't we look silly to future generations, he suggested, if all the while a group of green-skinned, three-armed, two-headed Martians seven feet tall was standing next to the spacecraft, scratching their heads and wondering what the hell was going on? Sagan suggested putting a camera on board, something to just look around and see if there was anything alive up there.

At first there was laughter and then there were arguments. What would a camera show them? Did he expect to see Martian elephants? "Probably the pictures will be unspectacular," he admitted. "Rocks, lava flows, sand dunes. [But] an occasional scraggly plant would not be unexpected. And there are other possibilities—fossils, footprints, minarets . . ."

The argument was vigorous. Every ounce on *Viking* was important; putting something on meant taking something else off. And what was the point? The environment up there was so hostile to life that the best they could expect was some sort of simple microbe. The odds against any macro life form existing were immense. Sure, Sagan agreed, which means that the odds aren't infinite. He was expert at turning an opponent's argument around and rendering him helpless. If the odds aren't infinite, then it's possible.

Finally, supported by nonbiological points of view which agreed that it would be wonderful to be able to look around and see the surface *Viking* was sitting on, he won the argument and the camera was included (at the expense of someone else's experiment, which is now lost to history). The camera was the least sensitive of the biology experiments, since it could detect only large and admittedly unlikely forms of life. But it was also the only one that did not have to make assumptions about whether or not Martian life respired or photosynthesized or used unusual chemical processes. In the end, everyone agreed that it was a good idea.

5.

Viking 1 lifted off from Cape Canaveral in Florida on August 20, 1975, after a two-week delay on the launching pad to correct a series of minor glitches. It was followed three weeks later by *Viking 2*, a duplicate machine which would go to a different spot on Mars and repeat the same experiments. Nine months later, on June 9, 1976, *Viking 1* reached Mars and went into orbit (*Viking 2* got there on August 7). Immediately the political pressure began to build. The Fourth of July, 1976, would be the two-hundredth anniversary of the birth of the nation, and every politician from President Gerald Ford on down to the local school board members wanted to celebrate with a *Viking* landing on Martian soil.

But the lander team was patient. The landing was probably the trickiest part of the whole mission. They couldn't "fly" down to the surface, steering the craft by remote control from a television screen because Mars was nearly two hundred million miles away. It would take a signal traveling at the speed of light about eighteen minutes to reach Earth. That meant if they tried to fly the craft down and there was a large rock in the way, it would take eighteen minutes for the view of the rock to reach them and another eighteen minutes for the message "turn left" to reach the spacecraft, and by that time it would have crashed into the rock. So they had to turn the landing over to an onboard computer, which meant they had to pick the flattest, most unobstructed landing place they could find from orbit and hope there would be no obstacles they hadn't seen.

The biology experimenters wanted to land as close as they could to where there was evidence of water flows, but the evidence consisted of eroded streambeds and crumpled rock and these were exactly the kinds of places the landing team didn't dare try. For a full month they analyzed data from cameras and infrared sensors aboard the orbiting spacecraft together with radar images from Earth, sorting through the terrain below as *Viking* floated around and around Mars, and the two-hundredth Fourth of July came and went. Finally they picked their spot, a reasonably flat-looking space on the northern plain of Chryse, and at 1:05 a.m., July 20, 1976, they took the irrevocable step and told *Viking* to send the lander down to Mars.

They also sent out the word to call everyone on the *Viking* team and tell them what was happening. Suddenly a dozen small California roads began to fill, as if a thousand tributaries were mysteriously, in the middle of the night, rising and flowing into the main stream. The main stream was Oak Creek Drive, which runs through the foothills of the San Gabriel Mountains to the California Institute of Technology's Jet Propulsion Laboratory (JPL), headquarters of the *Viking* mission. Toward it flowed scientists and engineers, reporters and science fiction writers, the nerve blood and the flotsam and jetsam of the first project to reach across millions of miles in the most remarkable exploration of all time.

Aboard *Viking*, in orbit 200 million miles from Earth and 2400 miles from Mars, the countdown had started, and as the stream of cars began to reach the JPL buildings the lander fired the explosive bolts that held it to the orbiter, compressed springs pushed it away, and its computer came to life as it fell out of orbit. Quickly the thin Martian atmosphere began to build up as the lander fell through it toward the surface, and the atmospheric shield began to glow from the heat of friction.

As it fell its onboard radar measured the quickly shortening distance to the ground, and at an altitude of ten miles the computer pulled its ripcord and a parachute blossomed out to slow the descent. Two and a half miles from the surface the lander stuck out its three legs for the landing, and fired its retrorockets for the final gentle touchdown.

In the auditorium at JPL the crowds too began to build up, and the atmosphere in there began to glow from the excitement and staccato babble of voices calling to each other. The people could not know what was happening up there in the thin atmosphere of the Martian sky; they only knew that the die had been cast, that all their work of years and years was now plunging down from orbit to land—or to crash—on an alien world. The television monitor screens at JPL had been preprogrammed for this event. They couldn't show the view from the lander as it fell because of the eighteen minute delay. Instead they showed a chart, a plot of altitude on the vertical axis against time on the horizontal axis, and on the chart were three curves. The middle curve showed the proper sequence of events that would be followed if the lander's rocket engines fired precisely the right amount at precisely the right moment to slow it down just enough to touch gently on the surface of Mars. The upper curve was the limit of safety if the engines fired too soon; if the lander's path was higher than that it would run out of rocket fuel while still high above the planet, and it would then fall out of control and crash. The lower curve was the opposite limit: if it deviated below it, it would be traveling too fast when it contacted the ground, and then too it would be broken into pieces.

Those curves were not the whole story. If Lander came down on an uneven surface it would be tilted, and if the tilt was too great the antennas would not deploy properly and communication would be lost. Still in everyone's mind at JPL was the story of Russia's *Mars 6*, which three years earlier had landed on Mars but had stopped signaling just seconds before touchdown; no one ever knew what had happened to it. If Lander came down on a boulder it would flip on its side and be silent forever; perhaps that was what had happened to *Mars 6.* The cameras on board Orbiter had picked the best possible spot for a landing, but they were not sensitive enough to see boulders a few feet in diameter, and one such obstacle would be enough to destroy the mission.

So all in all, things were pretty tense in the San Gabriel foothills as the invisible Lander dropped through the eighteen-minute delay toward the planet below. The computer screens at JPL told them second by second what was supposed to be happening: now the legs should be extended, now the parachute should be jettisoned, now the retrorockets should be firing, now the Lander should be touching down gently . . .

And then for eighteen minutes, silence. Silence from the spacecraft, silence in the auditorium, silence in the hundreds of small cubicles throughout the JPL complex . . .

Waiting for Lander to communicate. Waiting for the signal, dreading the silence that might simply continue on into eternity . . .

"Touchdown! We have touchdown!"

At 5:12 a.m. the signal came in loud and clear from Lander: it was on the ground of Mars, and all systems were Go. Cheers and shouts erupted, the applause was deafening, everyone was on their feet laughing and slapping each other's backs. *Viking* was on Mars.

At 5:54 a.m., while the champagne was still flowing in the JPL auditorium, the first life-seeking experiment to deploy on the surface of an alien planet went into operation. The blank television screens scattered around the room began to flicker into life as the *Viking* camera sent back its first picture of Mars, bit by bit. The camera was mounted on top of Lander in a projecting tube, and it looked out through a narrow slit that allowed only a thin line to be viewed. As the slit rotated the full picture was built up, line by line, and the transmissions came through in this manner until a whole picture was built up.

The first picture was black and white, because color took longer to transmit and everyone was too impatient to wait. It showed a downward view from the camera, looking at the landing pad resting on Martian soil. The detail was incredible, enough to take your breath away as you sat looking to see the results of years of planning, guessing, working, dreaming, hoping and praying.

You could see the rivets on the landing pad, which was a great relief because it meant the cameras were operating with the clarity they were designed for, so that if a tiny lizard or flower were there we'd be able to see and identify it. But in that soil around the pad there was no lizard, no flower.

For the next photograph the camera looked up and showed a panoramic view of a sandy, rocky, barren desert. "You can almost imagine camels coming up over the dunes," Thomas Mutch, leader of the imaging team, said as the picture took shape. As it built up, the room fell momentarily hushed and quiet; everyone was secretly waiting to see in the next vertical strip those camels, the first sign of life. Or at least the outstretched limb of a bug, the branch of a bush, maybe the spires of a city . . .

Line by line the picture took shape, showing the sand, the rocks, the desert, and nothing else. Stretching out to the Martian horizon, nothing but sand and rocks. Reflected in the television monitors, the auditorium at JPL itself seemed cold and quiet and empty. Carl Sagan broke the silence, saying that it didn't mean any more than if an alien rocket had come down in our Sahara Desert and looked around with its cameras: it too wouldn't see anything alive, while just over the horizon would be Cairo. Gerald Soffen, chairman of the *Viking* science steering group, said that if you showed these first pictures to a biologist and told him it was a desert on Earth, he would tell you that there might well be hundreds of different kinds of living organisms hiding in that soil. Harold Massursky, a senior astrogeologist at the United States Geologic Survey and head of the site certification team that had the responsibility of finally deciding where to land the lander, pointed out happily that as it descended its mass spectrometer had worked perfectly and had found more than two percent nitrogen in the atmosphere, far more than anticipated, greatly improving the possibilities of life on Mars.

Still, mingled with the intoxicating exhilaration of flying the spacecraft across millions of miles of space and landing it safe and intact on the ground of Mars, there was a lingering miasma of disappointment. No one wanted to break the spell at that magic moment, but later Gerald Soffen would look back and reflect rather sadly that "the orbiter camera could have seen cities or the lights of civilization. The infrared mapper could have found an unusual heat source. The water-vapor sensor could have detected watering holes . . . The entry mass spectrometer could have identified gases that were wildly outside the limits of chemical equilibrium. The seismometers could have detected a nearby elephant . . ."

Could have, but didn't.

In the days that followed, the full assortment of packaged experiments began. No time was wasted, for in addition to the impatience of the men who had been working years for this moment no one knew how long the lander would continue to operate and send its data back to Earth. It had been designed to work for ninety days; in the event it lasted for two years. But no one knew what the future held, and quickly they got to work.

The order of the experiments had been carefully worked out before *Viking* ever lifted off from Florida; it had been argued over, bickered over, appealed to higher authority, and finally decided and written irrevocably into stone. Certain things had to be done before other things were attempted; certain things were judged to be more important than others; certain things needed more data transmittal time, had physical relationships with other things, had more or less chance of success. It was all a complex jigsaw puzzle that had to be put into place perfectly, with everything done in the proper order no matter how one's nerve's jingled and jangled as each of the experimenters waited their turn.

And it was evening and it was morning, the first day. And the second day, and the third day . . .

6.

On the eighth Martian day the lander's sampler arm finally reached out and scooped up a sample of Martian soil, lifted it up and dropped it into the entry port. A carousel rotated and dispensed some of it into each of the three biology experiments, while the arm reached out again to get more soil for the mass spectrometer. This time, however, the arm failed; the mass spectrometer would have to wait while the biology packages began their routine.

In Norman Horowitz's pyrolitic release chamber the radioactive CO_2 was pumped in above the Martian soil. The faux Martian sunshine—his xenon lamp—was turned on, and Dr. Horowitz sat and waited for the microbial Martian plants to look upward and begin to photosynthesize. (He would also run a variation experiment without the artificial sunlight, for no one really knew what conditions the Martian microbes might like.)

In Gilbert Levin's labeled release chamber the radioactive chicken soup began dripping onto the Martian soil, and Dr. Levin sat and waited for the little Martian animals to sniff around it and begin eating.

In Vince Oyama's gas exchange experiment the nonradioactive chicken soup was spritzed into the chamber, to give the little beasties just the merest hint of moisture so as not to drown them right away, and Dr. Oyama sat and waited . . .

But not for long. Just two days later Oyama's instruments

recorded both oxygen and CO_2 peaks building up in the at-mosphere of his chamber. This was a positive signal, but it was confusing. The CO_2 couldn't be from plants because there was no sunlight, natural or artificial, in his chamber. And animals don't release oxygen, they suck it up. And it happened too soon for any expected biological response, and there was much too much gas given off for any reasonable abundance of living creatures to produce.

The next day, Levin's labeled release experiment also began to give a positive result: there was a surge of radioactive gas which then leveled off. This indicated CO_2 release from the soil, which was exactly what was supposed to happen if there were microbial animal-like organisms living there.

Five days later, there was also a positive result from Horo-witz's pyrolitic release package, indicating photosynthesis was taking place in his chamber. So all three of the life-seeking ex-periments had found what they expected to find—or nearly so—if there was life on Mars.

But there was not wild exhilaration in the laboratories back on Earth. There were frowns mixed with frozen smiles. The ex-periments were *too* positive and had come too soon: they were much quicker and greater than anyone had visualized. And worst of all, by this time Dr. Biemann's mass spectrometer had received and analyzed the Martian soil, and there weren't any organic molecules there. The ground was more sterile than any place on Earth.

Now the one thing everyone had been sure about was that there would be organic molecules on Mars. They are nearly ubiquitous in the universe. There was absolutely no reason to think they might not exist on Mars. The mass spectrometer had been included merely to measure exactly what organic molecules were there, and in what proportion.

So Mother Nature had done it to us again, reminding us that the only thing predictable about her was her unpredict-ability. Before the mission had set off, the group of scientists with onboard experiments had met together with those responsible for the planning of the flight, and all had agreed on the criteria for each of the experiments deciding that there was or was not life on Mars. Now they had reached the planet successfully and, in line with their hopes but beyond all reasonable expectations, all the experiments had worked successfully, and three out of the four had given positive results. (No one had really expected the camera to find any sign of macro life.) But they had never considered the possibility that the experiments might give results that were *too* positive, or that the mass spectrometer results would be so totally negative, and now they were hopelessly confused.

On July 31 the *Viking* team held a press conference to tell the world what was going on. But it was difficult because they didn't *know* what was going on. As Dr. Harold Klein, the leader of the biology team, summarized it, "If it is a biological response, it is stronger than anything we have obtained from terrestrial soil. It would mean that biology on Mars is more highly developed, more intense, than life on Earth."

Everyone stopped and looked around at each other. If there are no organic molecules on a planet, there can't possibly be life there, let alone life "more highly developed, more intense, than life on Earth." Oyama looked at his results again, and concluded that Mars has an "active surface material that is *mimicking* life." Dr. Horowitz was beginning to hedge his bet: "The data is *conceivably* of biological origin." Dr. Levin and his colleague, Dr. Patricia Straat, stuck to their guns: "The response that we are getting is consistent with the kinds of response we are used to seeing in terrestrial soils (which is due to the living things in the soil)."

And so the days proceeded into months, and the experiments were repeated on both *Vikings*. But as more data poured into the computers back on Earth, they only made things more confusing (which is not what data are supposed to do). The *Viking 2* instruments showed much the same results as had

Viking 1: no organics, but lots of positive results from the three biology packages. By stretching your imagination to the breaking point you might possibly have maintained that though there were few organic chemicals on Mars, *Viking 1* had dug into the equivalent of a microbial anthill, and that's why there was such a strong positive response; or alternatively that Mars was loaded with organics but *Viking 1*'s mass spectrometer was faulty or had picked up a strangely sterile piece of soil. But the continuing stream of data did not allow this. The mass spectrometers on both *Vikings* had picked up traces of the rockets' exhaust gases (which had purposely included molecules foreign to biological signals), so they were both working properly and there were no organics at either landing site. And the three biological experiments had returned basically the same positive information from *Viking 2*.

Therefore the processes they were looking at were planetwide: There were no organic chemicals on the surface of Mars, but something up there was reacting with the food *Viking* was distributing. Finally—slowly and reluctantly but inevitably and finally (and with a continuing minority dissenting opinion expressed by Levin and Straat)—the consensus of the biology team turned to Oyama's opinion: the surface of Mars is coated with an active chemical material that mimics life.

What could it be? A series of chemical experiments provided the answer. There is no ozone layer in the Martian atmosphere as there is on Earth, to protect the surface from the sun's deadly ultraviolet rays. This radiation is capable of splitting and destroying any organic molecules exposed to it. It is further capable of splitting water molecules into H (hydrogen) and OH (hydroxyl), providing one explanation for the lack of water on Mars and also providing another answer to the lack of organics, for OH is a powerful oxidizing agent. Here on Earth we think of oxygen as necessary for our life, but it and all oxidizing agents are powerful chemicals that can destroy organic molecules. Living organisms have learned to protect themselves from oxidation, to govern the process and use it to extract energy, but to nonliving organic chemicals it is a destructive poison.

Not only do the radiation-plus-oxidation processes account for the paucity of organic molecules on Mars, but they also provide an explanation for the seemingly positive biological results. A strongly oxidizing surface layer on Mars would react quickly—more quickly than life could—with the organic chemicals in the "food" *Viking* provided for the putative Martian microbes. The food would be oxidized immediately on contact, and the result would be bursts of oxygen and carbon dioxide: exactly the results the *Viking* experiments showed.

Dr. Horowitz had argued from the beginning that the experiments were too closely based on analogies with terrestrial life. He was right—no one had expected a Martian surface so chemically exotic—but he fell into the same trap himself. When his first experiments indicated photosynthetic life he rearranged his protocol and repeated the experiment, but with the artificial Martian sunlight (his xenon lamp) turned off so that the chamber and its soil were in total darkness. He got essentially the same result as before, and so concluded that it could not be a biological signal since "soil samples from the surface of the Earth typically fix much more carbon in the light, owing to the presence of photosynthetic organisms." Which is the sine qua non of terrestrial explanations. Finally, he concluded that "*Viking* found no life on Mars, and, just as important, it found why there can be no life. Mars lacks . . . any liquid water whatsoever."

And yet . . .

7.

Almost exactly twenty years after *Viking* landed on Mars, on August 16, 1996, a report of new experiments was published in *Science*, the official journal of the American Association for the

Advancement of Science. Its concluding sentence was: "Although there are alternative explanations for each of these phenomena taken individually, when they are considered collectively, particularly in view of their spatial association, *we conclude that they are evidence for primitive life on early Mars*" [emphasis added].

What had happened to turn the whole *Viking* argument topsy-turvy? The answer was that *Viking* had carried a small number of experiments of limited resourcefulness a couple of hundred million miles. The experiments had been planned on Earth and could not be seriously modified when conditions on Mars turned out to be different from those we had expected; we were locked into an experimental protocol that turned out to be unsuitable for that planet. "*Viking* went to Mars and asked if it had life, and Mars answered by replying, 'Could you please rephrase the question?'"

What we really needed was Mars itself here on Earth; we needed pieces of that planet which we could take into our laboratories and investigate with all the sophisticated technology at our disposal, changing and rearranging the experiments as each one brought forth new information that would lead to new planning strategies. But Mars was too far away for us to go there with a rocket large enough to carry fuel for a return trip; we *couldn't* bring back samples as we had from the moon.

And then, suddenly, we had them.

4 | The First Martians

*If this discovery is confirmed it will surely be one
of the most stunning insights into our universe
that science has ever uncovered.*
PRESIDENT CLINTON, August 1996

1.

It began nearly twenty years ago, when two graduate students at Harvard noticed something unusual about three meteorites.

We had come a long way by then since Thomas Jefferson was more ready to believe that a couple of Yankee professors would lie than that stones could fall out of the sky. Research on meteorites had progressed rapidly, especially since the 1950s, pushed by a small but enthusiastic group of researchers. There was a strong scientific consensus that most meteorites were asteroids perturbed out of their orbit beyond Mars and sent into Earth-crossing paths, so that eventually they were swept up by Earth's gravity and fell to the ground. Using newly developed isotopic techniques as well as more established methods of chemistry and mineralogy, we had been able to show that they were not the result of a shattered planet, as had once been thought, but were the primordial remains of the material out of which the solar system had been built and which had never formed themselves into a planet (presumably because of the disturbing effect of Jupiter's gravity).

There are three major classes of meteorite. Iron meteorites are pure metal, mostly iron and nickel, while the stone meteorites are broken into two categories: The chondrites have small spherical bodies, chondrules, which formed by some as yet unknown melting process in the solar nebula before the planets were created. The achondrites are similar to volcanic rocks on Earth.

Each of these major classes can be broken down into sub-categories. One of the most interesting is that of the carbonaceous chondrites, discovered when the first one fell at 5:30 p.m., March 15, 1806, near the village of Alais, in southern France. It was a stone meteorite, certainly, but was strangely friable, and—when subjected to a hot flame—gave off copious quantities of steam and left a tarry residue. In 1834 Jöns Jakob Berzelius, a Swedish chemist, analyzed it and found that much of the carbon was in organic form; that is, in the form of the hydrocarbons which form the basis of all life. But his student, Friedrich Wöhler, had recently shown that organic compounds can be produced by inorganic chemical reactions, and so he did not conclude that they were the residue of living creatures.

A hundred years later, someone did. He did even more: he found the creatures still living in the meteorite.

Charles B. Lipman was a Russian immigrant who as a child was interested not in space or astronomy but in biology, which he studied diligently enough to earn a Ph.D. and eventually to become Professor of Plant Physiology and Dean of the Graduate School at the University of California. He became interested in ancient life forms, and published a startling series of papers in which he claimed to have cultured and grown bacteria that had resisted death for millions of years. These hardy souls (if souls they have) had evidently encrusted themselves in spores and resided in suspended animation within grains of coal. When Lipman crushed the coal grains and liberated them, warmed them and fed them in a pleasant culture dish, they revived, procreated, and grew into a flourishing colony. Since the coal was known to be millions of years old, so too were the bacteria that had been inside it.

Or so Lipman reasoned. He next turned his attention to even older materials. The first results on radioactive dating of meteorites had recently been published, showing them to be billions of years old, and so he tackled them in the same way. He had been trained in bacterial research, and appreciated that earthly bacteria are tricky little things which can easily slip in, unwanted, to rack one's lungs or disturb one's experiments, and so he carefully sterilized the surface of each meteorite before crushing it, then soaked the residue in water to coax the microbes out and fed them a nutritious broth. In 1932 he reported the results of his experiments to the American Museum of Natural History. "From somewhere in space a few surviving bacteria" within the meteorites—all kinds of ordinary chondrites, not just carbonaceous ones—have come falling to Earth. These were similar to those we see on Earth: there were elongated bacilli that looked like stretched balloons and smaller, spherical cocci.

Lipman turned his attention to the interesting observation that life which has originated in space is so similar to earthly life. Others turned their attention to the same observation, but with a different emphasis. Michael Farrell of Yale was irritated by Lipman's claim; he argued that in the previous years no one had been able to duplicate Lipman's work on coal bacteria unless the coal they used had cracks in it. The implication was clear: the cracks provided a pathway for terrestrial bacteria to slip in, and all the precautions Lipman had taken to sterilize the surfaces of his coal or meteorite samples, then to crush them and incubate the interiors under sterile conditions, meant nothing. Bacteria had infiltrated the interior, where they could hide from the exterior sterilization, and the sterile conditions of the lab signified a barn door closed too late. A few years later, in 1935, a clever series of experiments by another worker effectively proved Farrell's assertions, and the case for meteoritic bacteria was closed.

But not for long. Sometimes it seems that each generation finds its own extraterrestrial life. In 1961 a Hungarian professor of chemistry at Fordham University, Bartholomew Nagy, and a few coworkers shocked the world anew with two reports. First, on March 16, he presented a paper to the New York Academy of Sciences. Collaborating with a research petroleum chemist from the Esso Research laboratories, he had analyzed hydrocarbons from the Orgueil carbonaceous chondrite. They found molecules that seemed quite like those found in living organisms; one in particular looked a lot like cholesterol. These molecules couldn't be earthly contamination, they argued, for their abundance in the meteorite was greater than in terrestrial soil; contamination always flows from greater to lesser concentrations, and there was no way for the meteorite to suck up concentrations higher than in the materials that supplied the contamination. In addition, they had been aware of the Lipman controversy, and had cleaned their equipment with acid, baked it out in vacuum, and had used all the techniques that Esso had learned were necessary to avoid contaminating samples in the laboratory. "We believe that wherever this meteorite originated, something lived," they stated.

At the resulting press conference, according to *Life* magazine, the conversation went like this:

"You say you have found traces of life. What kind of life?"

"Well, as I said, the chemical compounds in the meteorite are similar to those found in butter and living things in general. But we don't know specifically what kind of plant or animal life. We know only that there was life there."

"Okay, sir, but are there any of these chemicals in a real an-imal? I mean, for example, are these chemicals in me?"

"Yes, they are."

Which impressed the *Life* reporter, but which means noth-ing. The most abundant chemical compound in "living things in general," including the *Life* reporter, is H_2O, but finding water in a meteorite doesn't mean that life existed there. The important question is, could the compounds have been made by nonliving processes? "We calculate that the odds against these compounds being made at random by non-living proc-esses are in the magnitude of a billion to one," claimed Dr. Warren G. Meinschein, Nagy's collaborator from Esso.

Which meant that, to the Nagy group at least, life had at one time existed in this meteorite. To the rest of the scientific community the answer was not quite so clear—the chemical compounds they identified *could* have been made by some nonliving process—but at the least the data of Nagy and coworkers were intriguing. Or perhaps, as Dr. Farrell might have felt, the data and the presentation and the press confer-ence were irritating.

One person who was irritated was Prof. Edward Anders of the Enrico Fermi Institute for Nuclear Studies at the Univer-sity of Chicago. Anders was one of the world's leading mete-oriticists, and one tough cookie when he was irritated. I (D. E. F.) once saw him at a meeting of the American Geophysical Union bring a graduate student to tears over the question of whether certain minerals found in meteorites were created by internal pressures or by shock. To paraphrase Winston Church-ill's description of Field Marshall Montgomery, he is in con-flict indefatigable, in defeat indomitable, and in victory insuf-ferable. He now took arms against what he saw as a premature claim, and began his own investigations of "life" in carbona-ceous chondrites.

For the moment, however, the impetus lay with the pro-lifers. Another voice was soon to be heard, with an even more spectacular claim, although the scientist involved was reluctant to make it public. When a reporter, following up on the Nagy story, contacted the meteorite curator at the Smithsonian from whom the Nagy group had received Orgueil, he was told that a piece of another carbonaceous chondrite—Murray—had been given for similar investigations to a Dr. Fred Sisler of the United States Geological Survey. When Dr. Sisler was contacted he said he wasn't ready to publish his data yet, but yes, he was investigating living organisms in meteorites.

"You mean, organisms that *previously* were alive, don't you?"

"Well, no. I mean living organisms. Organisms in the meteorite that are currently alive."

Whoa, thought the reporter. *What have we here?*

What Dr. Sisler had was a bunch of little wiggling things that he had cultivated in the laboratory, things that had come out of the Murray meteorite. Working in conjunction with Walter Newton, who had a germ-free sterilized laboratory in the National Institute of Health, he had crushed a chunk of Murray and put it in a sterile but nutritious broth. After several months of careful incubation in one of the world's most sophisticated germ-free environments the solution turned cloudy, indicating the growth of microscopic somethings in there. Looking at the solution under a microscope, Sisler found it was crowded with wiggling bacteria.

Or at least with something which resembled bacteria. They looked like elongated balloons, which bacilli do. But at 200 nanometers (billionths of a meter) they were smaller than terrestrial bacteria, which are generally in the range 500 to 1000 nanometers, and he was unable to classify them as any known terrestrial microbe. So they were sort of like earthly bacteria, but different, which is just what you might expect for extraterrestrial life: if they were exactly like Earth creatures you would have to suspect contamination, and if they were too different you wouldn't be likely to recognize them as alive.

In fact, Dr. Sisler cautioned, he wasn't sure that they were alive. Something that small could appear to be wiggling simply because of atoms or molecules bouncing against it. And although he was sure their laboratory was sterile, there was always the possibility that the meteorite, which had fallen in 1950, had been infested by a previously unknown earthly bacterium before it reached his lab. (Sisler never published, which obviously means he discovered something wrong with his experiments. But we don't know what; that's all there is, there ain't no more.)

In November of 1961 the Nagy group was back in the news. They published in *Nature* the results of a new study which shook the scientific community to the core. Titled rather innocuously "A microbiological examination of some carbonaceous chondrites," it got right down to business with the opening sentence: "Microscopic-sized particles, resembling fossil algae, were found to be present . . ." What they found under the microscope, they announced, were five different types of "organized elements" that resembled terrestrial life forms. They found small circular forms, some covered with spines, others with shapes like shields, while others were cylindrical and one type was hexagonal with whiskery protuberances and "vacuole-like bodies in the interior."

And away we went! The New York Academy of Sciences hosted a special meeting on the East Coast, North American Aviation hosted another one on the West Coast, *Nature* and *Science*, the *London Times* and the *New York Times*, all reported discussions and theories and opinions, and the tabloids and television news programs carried rumors and unsupported allegations, and the media celebrated Christmas all year long.

Slowly, however—and out of sight of the media and the public—the antilifers were gathering strength and evidence.

Ed Anders let loose the first salvo at the 1962 spring meeting of the American Geophysical Union in Washington, D. C.. The carbonaceous chondrites are friable, he pointed out; you can crush them in your fingers without effort. This means they are not solid rock but are porous, with numerous tiny passageways for specks of dust and bacteria to penetrate. Daily changes in barometric pressure would create waves of air movement which would push ambient airborne particles into and out of the meteorite, flushing it with whatever is floating around in the air. And between the time of its fall and its disbursement to the Nagy group, Orgueil had been stored in the American Museum of Natural History in New York City. In the 1950s you could take a walk in New York and return to your hotel with a thin film of dust on your skin, dust that was literally crawling with bacteria and was loaded with all sorts of organic compounds. As Anders pointed out, it didn't matter how careful you were in the laboratory if the barn door had already been left open before you ever got your meteorite to study. Automobile exhausts alone would produce the same mix of hydrocarbons found by Nagy et al.

Not so, they countered: the meteorite had more than a hundred times more organic molecules than could be produced by contamination. But how did they make such a calculation? The details were not convincing, because until this time no one had spent much time analyzing contamination processes at such low levels.

Even if the results were not due to contamination, Anders plowed on, how did anyone know they represented life processes? Scientists had before then found organics in many places devoid of life, in particular in the laboratory (see our previous discussion of the Urey-Miller experiment in Chapter 3).

Well, okay, but how about the "organized elements"? These were more difficult to dismiss as contamination since, unlike the organic chemical compounds, they didn't seem to give a precise match to anything terrestrial. And thus the debate continued in journals and meetings.

Nagy's style of presentation wasn't up to Anders's. I (D. E. F.) remember him coming to Cornell at this time to present a seminar on these tremendously exciting discoveries. The auditorium was filled as he was introduced and greeted with expectant applause. His first words, spoken in a soft voice with a heavy Hungarian accent, were "Could we have the lights out, please?" He then proceeded to show several hundred slides in the next hour, accompanying each with a description in the same dull accented monotone, without ever turning the lights back on. I have never, before or since, slept so soundly.

Yet it was not his speaking style that doomed him, but the indefatigable research of Anders and others. One of the claims that the organized elements were fossils of living creatures lay in the fact that they accepted staining with a particular chemical known as Gridley's compound in much the same manner as do terrestrial fungi. In 1963, with Frank Fitch, a professor of pathology at the University of Chicago, Anders published a study showing how the Type 5 organized element—the one that most people had accepted as most likely to be an alien fossil—could be reproduced by Gridley staining of nothing more exotic than ragweed pollen. The clue had been that the Gridley agent had long been known to have "a violent and disruptive effect" on cells. Anders and Fitch tried it on some organisms likely to be contaminants of the meteorite, and rang the bell with ragweed pollen. They found that the staining caused both swelling and shrinking, inducing collapse and rupture and resulting in an "organized element" which appeared quite different from the normal grain of ragweed pollen and was indistinguishable from the Type 5 element.

In 1964 Anders presented to a scientific meeting a picture that brought the house down in laughter. It showed a sample of the Orgueil meteorite with a plant growing out of it. Clearly this life form was not indigenous to the meteorite; in fact, Anders suggested, it was the remains of a hoax intended to embarrass Louis Pasteur who, at the time Orgueil fell, was presenting his arguments against spontaneous generation of life. Now Anders used it to demonstrate how easily terrestrial organisms could penetrate the meteorite. In summary, his arguments were similar to the description given by George Bernard Shaw of a contemporary musical piece: "There is much that is original and interesting in this work. Unfortunately, what is interesting is not original, and what is original is not interesting." Substitute the word *indigenous* for *original* and you have the anti-life-in-meteorites argument.

It was a battle won by attrition. As T. S. Eliot wrote:

> *This is the way the world ends*
> *Not with a bang but a whimper.*

There was no one overpowering moment, no bang, at which everyone agreed that the evidence for life in the carbonaceous chondrites was false, because it is rather difficult to definitively prove a negative. But on Sunday, September 28, 1969, a new carbonaceous chondrite fell on the town of Murchison in Australia. It was gathered quickly and delivered safely to museum curators, so that it had little or no chance to pick up terrestrial contamination before being put in a sterile environment. When investigated, there was found to be no hint of biogenic molecules or organized elements in it: "None of the five common [compounds] present in biological organisms could be found." As the absence of evidence piled up, the arguments of Anders and his supporters and coworkers took their effect, and as the years slipped by there was no competing evidence to disprove them. In a review written in 1973 for *Science* Anders began by saying simply, "The intense controversy . . . has subsided. Most authors now agree that this material represents primitive prebiotic matter, not vestiges of extraterrestrial life." No one argued with him. In a review the previous year for *Scientific American*, titled "Organic Matter in Meteorites," the authors hadn't even bothered to mention the work of Nagy and his group. In 1975 Nagy himself wrote a comprehensive book summing up everything that was known about carbonaceous chondrites, concluding that it was unlikely that he had in fact discovered alien life forms in them.

But, as we said, it seems that every generation must find its own extraterrestrial life, and as we go to press in 1998 the suspicion is strong that this time we've truly found it.

2.

This new story begins in 1980 with the sentence with which we began this chapter. The two graduate students, Harry McSween, Jr. and Ed Stolper, were studying the other group of stone meteorites, those without chondrules. Like terrestrial volcanic rocks, these achondrites had been melted at some time since their formation and thus no longer held primeval information dating back to the solar nebula, and so most meteoriticists thought them much less interesting than the chondrites. But McSween and Stolper noticed something unusual about three of them.

By this time we had been able to define three distinct types of ages for meteorites, based on different processes affecting particular isotopes. The "solidification (or formation) age" is the time at which they became solid bodies, and is very well established at 4.6 (plus or minus about 0.1) billion years for most meteorites. Since the Earth and moon share this common age, we feel confident in defining it as the time of formation of the solar system as a whole. The "cosmic ray age" or "exposure age" is the time at which the meteorites became exposed to cosmic rays, and is much less than the solidification age: typically a few tens of millions of years or less for stones. Since cosmic rays can penetrate to a depth of about a meter in solid rock, this is interpreted as indicating that the meteorites were originally parts of larger bodies (presumably asteroids or something similar) and that the collisions which broke them into meter-sized fragments were responsible for knocking them out of their asteroidal orbits into Earth-crossing ones. Finally, since the Earth's atmosphere shields them from cosmic rays once they land here, we learned to obtain their "terrestrial age" by measuring the decrease in cosmic-ray-induced radioactivities. These ages are typically hundreds to thousands of years for "finds" (those meteorites that were not seen to fall but were dug up, typically by farmers tilling the soil).

What the two grad students noticed was that nearly half of the mass of each of these three meteorites—*Shergotty*, which fell in 1865 in India; *Zagami*, which fell in 1962 in Nigeria; and a third which was found lying on the ice in Antarctica in 1977—is a kind of glass formed by a shock. Furthermore, they have unusually short solidification ages, less than about 1.1 Giga-years (Gyr or billion years).

If they were terrestrial rocks, such short ages would be no problem. Every year, somewhere on Earth an igneous rock erupts from a volcano and solidifies, so that although the age of the Earth is 4.6 billion years, any rock you might pick up could have an age ranging from zero up to that limit; an Earth-rock 1.1 Gyr old would be no surprise.*

But these Shergottites (as they came to be called) are meteorites; they come from somewhere else. McSween and Stolper wondered about that. Where could they have come from? They couldn't be asteroids because bodies as small as asteroids couldn't generate enough internal heat for volcanic activity as recently as a billion years ago. Nor could they have come from the moon; volcanic activity there ceased nearly four billion years ago because a body that small can't retain its internal heat any longer than that.

So where else is there a planet large enough to have retained enough internal heat to melt stone as recently as 1.1 billion

---

*An Earth-rock older than about four billion years would be a great surprise, however, since the Earth's surface during the first half-billion years was being continually reworked by volcanism and meteorite impacts and erosion, and so no rocks have been found older than that.

years ago? The large Jovian planets, certainly; but their gravity is so intense you couldn't possibly eject a rock from them. Mercury, like the moon, is too small. The remaining possibilities are Venus and Mars. For dynamic reasons Venus doesn't look likely (its gravity and thick atmosphere make it difficult to conceive of an asteroidal collision knocking large chunks off it and out into space). In 1980 McSween and Stolper wrote in *Scientific American*: "Although most meteorites are believed to have come from comparatively small bodies, these three meteorites exhibit properties that indicate their source was a large planet, conceivably Mars."

The dynamic reasons are tough, though, even for Mars. Like many meteoriticists, I (D. E. F.) didn't take their suggestion seriously. (In private conversation, they later said that although they had conceived their idea earlier they didn't publish it until 1980 because by then they had faculty positions; they had been afraid that coming out publicly with such a far-out idea earlier might be detrimental to their careers). For a few years the idea hung around in purgatory; no one could prove it true, and no one was bothering to prove it false. Not many people were interested.

Until 1983, when Don Bogard and Pratt Johnson of the NASA Johnson Space Center in Houston published their analysis of Elephant Moraine 79001.

3.

Elephant Moraine 79001? What kind of a name is that? Until the 1970s all meteorites were named after the town or area in which they fell—thus Murchison from a town in Australia, Alais from a French village, and so on. The likelihood of more than one meteorite falling on any one province is so small that our system of nomenclature didn't have to worry about it. But that all changed in 1969 when Renji Naruse, a Japanese glaciologist on a field trip to Antarctica, found a meteorite lying black and starkly visible on the ice in the Yamato Mountains. During the rest of his sojourn there he found eight more. Sometime later he fell into conversation with an ardent meteoriticist, Bill Cassidy of the University of Pittsburgh, "and after about a half hour," Cassidy later recounted, "a light came on over my head and I realized there must be a concentration mechanism at work there."

What Cassidy envisaged was a mechanism in which ice flowed across vast areas and piled up somewhere, depositing in one convenient place all the meteorites that had fallen on the original large area. The ice would slowly evaporate as it was worn away by the dry winds, and the meteorites would be left behind. It would be as if a large broom swept over the Antarctic continent and brushed the cosmic debris into a corner, waiting for a dustmaid to pick them up and take them away.

Excitedly, he wrote up a research proposal, sent it off to the National Science Foundation, and began preparations for a trip to Antarctica. But the NSF said no. The idea didn't sound very interesting to them. The meteorites Naruse found could be chalked up to beginner's luck, they said. It wasn't worth the hundreds of thousands of dollars it would cost to go looking for more, which probably weren't there anyway.

While the NSF was saying no, Japan was saying yes. Takesi Nagato got Japanese funding and went down to Antarctica, where he found 663 meteorites. As soon as Cassidy heard the news he called NSF again, and this time they said they might reconsider.

And so they did. In 1976 Bill Cassidy led the first U. S. expedition in search of Antarctic meteorites. By 1996 the United States and Japan had collected nearly ten thousand meteorites each. So many were found that the nomenclature had to be revised. And so Elephant Moraine 79001 is the first meteorite catalogued in 1979 from the region known as Elephant Moraine.

By this time meteoriticists had identified several other me

teorites as similar to the Shergottites and likely to share their place of origin. These were the Nakhlites (named after Nakhla, which fell in Egypt in 1911) and the Chassignites (named after Chassigny, which fell in France in 1815), and several of the Antarctic meteorites (with names like Elephant Moraine 79001 or Allan Hills 84001). Together they comprised eleven different specimens, and are collectively referred to as the SNC Shergottite-Nakhlite-Chassignite or SNC, (pronounced "snick") meteorites. When Bogard and Johnson measured the rare gases in Elephant Moraine 790001, it became the first meteorite to be positively identified as a chunk of the planet Mars.

Bogard and Johnson weren't all that certain at first. The title of their publication in *Science* was "Martian Gases in an Antarctic Meteorite?" But the data they presented soon made the question mark unnecessary. They measured the abundances of the rare gases helium, neon, argon, krypton, and xenon, and compared the elemental ratios and the isotopic abundances within each element to those measured in the Martian atmosphere by *Viking*, and found an excellent match. Furthermore, when they compared their data to any other possible source, they showed a definite mismatch.

The idea is that an asteroid must have hit Mars and knocked 79001 loose, and when this happened the shock of the impact melted the Martian rock and, before it had a chance to solidify, it flew through the planet's atmosphere and absorbed some of the rare gases present in the air. When later it fell on Earth those gases were tightly held within the now solid rock, and released only when Bogard and Johnson melted it again in their laboratory.

Their report sparked further research into the other SNCs, and all the incoming data supported that conclusion. Today there is no doubt that the SNCs are remnants of the planet Mars, and most researchers today refer to them simply as Martian meteorites.

4.

All of which is very interesting, but the big news came out of the blue at 1 p.m. on August 7, 1996, at a special press conference held at NASA to announce the results of a paper that would be published in the next issue of *Science*. In the previous few years there had been no hint of exobiological activity, no new NASA program aimed at Mars. On the contrary, since *Viking's* negative results twenty years ago it was considered rather *de trop* for scientists to talk of searching for life on that or any other planet. As recently as 1986 Norman Horowitz had proclaimed: "Viking found no life on Mars, and, just as important, it found why there can be no life [there]."

But in the first few days of August rumors had begun to seep out of Washington about life on another planet. The source of the rumors was not all that reputable: it turned out to be a young hooker who had hooked Dick Morris, President Clinton's political adviser. She was claiming that Morris was in the habit of whispering sweet nuggets of White House secrets into her ears, to impress her with his importance; one of the nuggets was that NASA had found life on Mars but was keeping it a secret.

Why would NASA do that? If they had made such a fantastic discovery, why wouldn't they milk it for all the publicity they could get? Well, they would—but not quite yet. The way you do things in science, if you want to do them right, is if you've got something to say you write a paper and submit it for publication before you start blowing your horn about it. When you do, the editors send it out for review to the best people they can find, and these people pick it to pieces as best they can. Then the editors send it back to you and you rewrite it or redo parts of the experiment if necessary or dump the whole thing if the reviewers find something basically wrong and unfixable. So, by the time it gets published—if it ever does—the work has been vetted and inspected and checked out, and it's pretty likely to have some validity.

This is what the people at the Johnson Space Center (JSC) in Houston had done. They wrote their paper and sent it to *Science*, one of our best journals, and it had been reviewed by five scientists, rewritten, and then reviewed by four more, and finally it had been accepted for publication. It was due out on August 16, and until that time nobody at the JSC was saying a word because that's how the game is played.

But now the hooker was not playing by the same rules and the rumors were slipping out, so the top people at NASA (who run the JSC) decided they had better make a formal statement before the rumors got too ridiculous.

The media people were greeted at NASA headquarters in Washington by a panel of six scientists who gave them the news that burst out of the briefing and in more detail a week later from the pages of *Science*, and from there it spread across the newspapers and television screens of the world: "Search for Past Life on Mars" was the title of the paper, and scientific hysteria was the immediate result.

A group led by David McKay and Everett Gibson, veteran lunar and meteorite researchers at the JSC, and including workers from Lockheed-Martin, McGill University, and Stanford, had analyzed another SNC Antarctic meteorite, Allan Hills 84001 (ALH84001), and found a variety of curious experimental results. Any one of the results could be explained in purely inorganic chemical or mineralogical terms without the intervention of biological processes, they said, but putting them all together left any one of these explanations unworkable. As the team reported in the final sentence of their paper, "We conclude that they are evidence for primitive life on early Mars."

About noon on December 27, 1984 a young woman named Roberta Score was hopping onto her snowmobile in the Allan

Hills ice field of Antarctica among fifteen-foot-high wind-carved ice waves when she spotted what looked like a bright green potato lying on the ice. It was a meteorite, one of several hundred picked up during that search season, but because of its unusual color it was the first to be curated back at Houston, and so became ALH84001. It turned out, back there in Houston, to be not so unusual after all: on the Antarctic ice fields everyone was wearing dark glasses, and either the glasses or the bright glare reflecting off the blue ice made this particular meteorite appear green. But in normal light it turned out to be the usual meteoritic dull gray, and on inspection was diagnosed as a diogenite, one of the more common types of achondrite, and filed away for the usual future research.

Six years later David Mittlefehldt, working in the geochemistry lab at Houston, found that it was more oxidized than any known diogenite. He realized that something was wrong, but he didn't know what and it didn't seem important enough to worry about. Another three years passed until, in the spring of 1993, while working with what he thought was another meteorite that had unusual properties, he realized that his specimen was actually a mislabeled piece of ALH84001. Putting together the data from three years previously with what he was measuring today implied that this meteorite, like the SNCs, was from Mars. "Within a split second I knew I had a new martian meteorite, and a unique one at that. . . . The rest, as they say, is history."

At the time, it didn't seem particularly like history in the making. We now had twelve Martian meteorites instead of eleven, that was all. But then, this wasn't just another Martian meteorite; it had some particular characteristics that made it different, which was why he walked across the hall to where a young postdoc was working, and said, "Hey Chris, come take a look at this."

One of the things that made it different, Mittlefehldt had

found, was that 84001 had a lot of carbonate in the form of tiny, unusual, distinct globules. Chris Romanek had just obtained his Ph.D. at Texas A & M and had come to the Johnson Space Center on a National Research Council fellowship to work on laser instrumentation development in Everett Gibson's lab. His doctoral work had been done on terrestrial carbonates (a common and important mineral of the form $CO_3^=$), and he was interested in developing new experimental techniques to delve into their history and evolution. So Mittlelfehldt thought he'd be interested in the extensive and unusual carbonates in 84001.

He was indeed, and when he and Mittlefehldt discussed them with Gibson they decided to put together a consortium of scientists to work on different aspects of this unusual meteorite. Romanek and Gibson measured the carbon isotopic abundances and found they were enriched in the rare isotope carbon-13 relative to Earth, but were identical to the abundances found in other Martian meteorites, showing that they had been formed on Mars rather than later while the meteorite was lying on the Antarctic ice. The oxygen isotopic abundances indicated that the carbonates were formed by precipitation from a fluid (presumably water) at low temperatures, in contrast to Mittlefehldt's original conclusion that they had formed at very high temperatures (up to about 700° Centigrade). If Mittlefehldt had been right, it would have meant that the globules had formed deep within the Martian mantle and would be clues to what was going on inside the planet. But if this newer work was right, the globules would have formed on the surface of the planet at a time when the crustal rocks were in contact with abundant water.

Today Mars has no surface water. But some preliminary experiments on 84001 had already indicated that its age was much greater than the other SNCs: it had probably solidified about four billion years ago. At that time there might well have been lakes and rivers on the planet. As Romanek and his coworkers put it, "If the minerals . . . formed from fluids in equilibrium with the martian atmosphere, they should prove instrumental in describing both the operation of near-surface processes on Mars and the evolution of the martian atmosphere."

Dr. David McKay, a geologist at JSC, was interested in lunar and meteoritic regoliths, or fractured surface rock. He thought that an igneous rock from the crust of Mars with fractures in it could have new and important evidence of regolith activity on that planet, and so he began investigations with that in mind. Meanwhile, Chris Romanek began thinking about a meeting of the Geological Society of America he had attended a couple of years earlier in Cincinnati, where Dr. Robert Folk of the University of Texas had pointed out that carbonate minerals "have a great affinity for organic compounds," and that "bacterial bodies are readily preserved in full three dimensions in carbonate deposits." Folk found these creatures by careful acid etching of the carbonates, dissolving away the surface structure and revealing the complex bacterial fossils within. To see details this small he used a scanning electron microscope (SEM), which uses a beam of electrons instead of light to get thousands of times better magnification.*

---

*Normally we see an object when light interacts with it and is scattered back to our eyes. Due to the complex wave-particle nature of light it's impossible to completely understand this interaction without mathematics, but think of it this way: Light has a characteristic wavelength. If you were attempting to find a small bump in a smooth wall by throwing balls at the wall and seeing how they bounce off, a basketball wouldn't work; it wouldn't be affected by the small bump (it wouldn't "see" it) and would just bounce straight back. But a Ping-Pong ball would bounce off at a different angle if it hit the bump, thus pinpointing its location. Quantum theory tells us that electrons at high velocities can also behave as waves, with a wavelength much smaller than that of light. So they constitute a "smaller ball" and can show us much smaller objects. (In the SEM the scattered electrons paint a picture on a screen much in the manner of television, and thus we see the magnified object.)

By 1994 the best SEMs could give magnifications of better than 100,000 times.

Romanek thought, "Why not try that with the ALH carbonates?" He didn't really expect that carbonates on Mars would "have a great affinity for organic compounds," let alone that "bacterial bodies would be readily preserved" in 84001, because the *Viking* results twenty years earlier had shown that there *weren't* any organics on Mars and therefore there was no life there, bacterial or otherwise. But the idea was so entrancing.

He had about a year's experience working with the SEM, which wasn't enough to qualify him as an expert, but he knew how to work up a sample, mount it, and view it. So he did. And when he did, he couldn't believe what he saw. The photographs were stunning. He had never really believed he would see anything like that. He had done it almost on a whim, just so that later he could sit alone in the twilight and look out across the Texas plains and fantasize quietly about how wonderful it would have been if he had found fossil bacteria in a rock from Mars. But now he had found them . . . or thought that just *maybe* . . .

He took the photographs to Everett Gibson, half expecting to be laughed out of the lab, but Gibson pounced on them enthusiastically. Looking at them, you certainly couldn't say that what you were seeing were bacteria. But you also couldn't say they *weren't*. There were strange wiggly shapes there, and if you squinted at them just right they sort of looked like maybe they might be something that had once been alive. What they were looking at could best be described, Gibson says, as "similar to terrestrial carbonate-bearing rocks where biology is operating." Romanek and Gibson showed the pictures to McKay, who supervises the geology group's SEM and who knew more about the experimental convolutions than either of them. He too was fascinated, and they decided to pursue this further. In the meantime, they needed some information that the SEM couldn't provide.

So they walked down the hall to talk to Kathie Thomas-Keprta.

Thomas-Keprta, an employee of Lockheed-Martin at the JSC (as is Mittlefehldt), is in charge of one of the electron beam laboratories, and her specialty over the past few years has been looking at carbon in extraterrestrial samples. Carbon, aside from being the basic chemical in living organisms, is one of the most common elements in the universe and is involved in all sorts of inorganic processes and structures. She had studied it in a variety of cosmic samples that had no relation to biology, and was accustomed to seeing it without thinking of life. When McKay and Gibson told her what they thought they were seeing in 84001, she was not particularly impressed. "They had a theory that somehow this Martian meteorite contained life on Mars. I thought the entire group needed some help. Yes, I did. I was convinced that I could convince them that they were wrong."

Her experimental technique is transmission electron microscopy. The difference between this and the SEM is in penetrating power. The SEM looks at the surface, while the transmission electron microscope (TEM) sends its electron beam right through a thin sample to see what's inside. Allied with this capability is the principle of electron diffraction, in which electrons are scattered through different angles depending on their wavelength and the structure of the diffracting crystal. By measuring these diffracted electrons she was not only able to see individual grains in the meteorite but also to determine how each crystal was diffracting the electrons—that is, what minerals the grains were composed of.

When she focused on the dark rims of the carbonate globules, they showed an "Oreo cookie" appearance: the rims consist of white carbonates sandwiched between dark crystals of iron oxides and sulfides. She was able to identify these as iron oxide [magnetite (Fe_3O_4)] and iron sulfide [pyrrhotite (FeS)], and suddenly her ideas about Mars changed direction. These incredibly small crystals aren't normally found together. Oxides form under oxidizing conditions, but sulfides form under reducing conditions. Oxidizing and reducing conditions are opposites: if you have enough free oxygen (or imitators) for an oxidizing environment, it can't also be reducing. So how could these minerals be formed in intimate contact with each other?

By bacteria, is the suggestion. Living creatures, such as bacteria and humans, can manipulate their environment. Some terrestrial bacteria are known to form magnetite in order to help them find direction in the Earth's magnetic field, and some bacteria form iron sulfides as waste products. So in an environment with bacteria present, both minerals can coexist; without bacteria, it's hard to see how they could.

Luckily there was a specialist in magnetic bacteria in residence at JSC at just this time. Hojatollah Vali, a professor at McGill University, had come to the JSC for six months on an National Research Council fellowship to do research with David McKay, and he was as impressed as Thomas-Keprta was with the mineral identification. He had seen both oxide and sulfide minerals coexisting in nonbiological systems, he told them, but only when they were formed at very high temperatures. He also pointed out that their formation by inorganic means takes place in conditions in which carbonates do *not* form. So the mere fact of seeing them together with carbonates is unusual, and if the analysis by Romanek and Gibson was correct—that the 84001 carbonates formed at low temperature—he didn't see how they could possibly have formed together without the help of bacteria.

During this time they had sent samples of the meteorite out for a different kind of analysis. The Johnson Space Center had an ongoing collaboration with Richard Zare, a chemist at Stanford University, in which Zare's group was looking at carbon compounds (of nonbiological origin) in meteorite and cosmic dust samples provided by the JSC. They now thought it was time to bring in Zare's group to take a look at this particular specimen. Romanek separated out some globules, and Thomas-Keprta sent them on for analysis at Stanford. This time, the samples were sent in disguise. Careful to the nth degree, they didn't want to take even the barest chance of influencing the Stanford results, and so Zare wasn't told what he was expected to find or even what kind of samples he was looking at. Romanek labeled the specimens of 84001 with Walt Disney names; as far as the Stanford group knew, they were looking at Mickey, Minnie, and Goofy.

What they were looking for was evidence of hydrocarbons. What they had to look *with* was a bright graduate student from Britain, Simon Clemett, and a brilliant new piece of experimental apparatus, a double laser beam mass spectrometer, which allows a precise identification of whatever molecules are present in the samples. The results for Minnie, Mickey, and Goofy couldn't have been clearer: they showed that among the materials vaporized were organic molecules of a particular class, called polycyclic aromatic hydrocarbons (PAHs): hydrocarbons because they are made primarily of carbon and hydrogen atoms, aromatic because they stink, and polycyclic because their structure consists of a series of rings of six carbon atoms linked together in a hexagonal structure.

The importance of their identification in 84001 is twofold. First, their discovery is the first documented occurrence of organic molecules on Mars. *Viking* had shown that there were no organic molecules up there—or so we thought. Now we realize that *Viking* showed only that there are no organics on *the surface* of Mars, and this is reasonable since the intense ultraviolet radiation and oxidizing chemicals there would immediately chew them up. But the 84001 data show that organics do exist in the region from which the meteorite was excavated when it was knocked loose from the planet and sent on its long journey toward Earth.

The second line of importance is that PAHs are often—though not necessarily—linked to bacterial decay products. They are found, for example, in automobile emissions (and most people feel that fossil fuels are just that: fuels formed out of fossil life) and in the black crust you get when you barbecue a steak. On the other hand, PAHs can also be formed nonbiologically: they are seen in carbonaceous chondrites, which were never part of a planet on which life might have formed, and in the ubiquitous cosmic dust which is a remnant of comets and/or the leftover material from the solar nebula, and which is seen by no one as a reservoir of life past or present.

On Earth there is an enormous variety of PAHs, formed by both living and nonliving processes. In 84001 the variety is much less, and this is just what you would expect on a planet with a limited and simple variety of living organisms. But are the particular PAHs in 84001 positively known to be the decay products of bacterial action? No, they are not. This is disappointing, but not particularly surprising. The Martian bacteria, if such they are, are not earthly life forms; the PAHs they produce should be similar but not identical to those produced by bacteria known to us. The problem is in our ignorance, not in our measurements: we simply do not know very much about bacterial PAHs, since prior to 84001 there wasn't a terrific amount of interest in them.

While the Zare lab was carrying out these analyses, the folks back at the Johnson Space Center continued their work with the scanning electron microscope. The geology division's SEM was fifteen years old, and could magnify only to 20,000 times, but the materials science division at the JSC had just taken possession of a brand-new, state-of-the-art machine dedicated to studying the *Challenger* samples, to make sure that such an accident wouldn't happen again. McKay and Gibson got permission to use it, and with the better resolution at magnifications up to 150,000 they were able to look at objects down to 5nm (nanometers: one billionth of a meter) in size. In late 1995 they saw an object that almost literally knocked their eyes out.

Shown in the photosection, it is now famous—if perhaps over-hyped—as The First Martian To Be Seen On Earth. It's a 380 nm segmented structure that looks incredibly like the fossilized bacteria seen in terrestrial samples. When Gibson saw it, he—well, let him describe it: "One evening we were looking at phases and we came across a segmented structure. It was uniform and appeared to have a head and a tail and it was fat in the middle, and right then it began to dawn on me: could we be seeing evidence of biogenic activity in a microfossil life?"

It was time to take the plunge. They gathered their coworkers and all their data—the chemistry and the isotopes and the mineralogy and the electron microscope pictures—and decided to publish. Together they sat down and wrote the paper whose concluding sentence set off the biggest scientific uproar in many, many years: "We conclude that [these data] are evidence for primitive life on early Mars."

And the rest, as they say . . .

But where do we stand today? After all the press conferences and the media hype, after the data have been published and presented at meetings and argued over and criticized and praised, what can we now say? Is there—or was there ever—life on Mars?

5.

When I (D. E. F.) was a young assistant professor, we used to have a lovely conference every spring in New Hampshire. It was called the Gordon Conference on Nuclear and Cosmochemistry, and it was idyllic. Held at one of the picturesque New England junior colleges amid green trees and cool lakes, it featured informal discussions of current problems and a lobster buffet on the final night.

And then came Progress. In the 1960s we went to the moon and brought back chunks of it, and this was so spectacular that NASA felt it had to have its own conference. So they called it the Lunar and Planetary Science Conference (LPSC), and it absorbed the cosmochemistry half of the Gordon Conference, and instead of going to New England to hike along the lonely roads and swim in the clear lakes and argue meditatively, we now journey to NASA's massive complex southeast of Houston. We drive along NASA Road One—the ugliest street in America, crammed with endless construction and traffic and always-red traffic lights and billboards and malls and junk-food joints and dirt and dust—and we settle down to a formal, well-organized conference at which chili substitutes for lobster and everyone gets eight minutes to present their newest data.

At the 1997 LPSC the featured presentation was a series of papers bringing us the latest information on 84001 and its hints of life on Mars. So what has been proven?

1. ALH 84001 is a meteorite that was knocked off the surface of Mars, sailed through space for millions of years and fell to Earth several thousand years ago. No one seriously disputes this.

2. The rock solidified on Mars early in solar system history. Sometime later, liquid water flowed through it and deposited small globules of carbonate minerals. Or, on the other hand, liquid water was never present and the carbonates condensed from a very-high-temperature phase. This is a particularly bitter argument, with no quarter being given. Harry McSween, who with Ed Stolper was the first to argue convincingly that the SNCs came from Mars—and who might have been expected to enthusiastically support the life hypothesis since it would make his original argument that much more a part of scientific history—instead takes the opposite view. McSween argues that his interpretation of the oxygen isotopic abundances, using chemical equilibrium theory, predicts a carbonate origin at very high temperatures, greater than about 700° Centigrade, too hot to support life. Allan Treiman, of the Lunar and Planetary Institute in Houston, argues that the carbonate minerals were *not* in equilibrium when they formed and therefore McSween's argument is irrelevant.

McSween has another argument. With two other investigators he has found magnetite "whiskers" among the supposedly biogenic magnetite grains. Such whiskers—elongated, thin crystals looking under the microscope like their eponym—are known to form directly from the vapor phase; that is, at extremely high temperatures. No one disputes this, but Kathie Thomas-Keprta reports that the experimental observation may be wrong: her group has never seen whiskers in any of their samples.

A group from Cal Tech and McGill University measured the manner in which magnetic grains lined up in the carbonates and concluded that Mars must at one time have had a magnetic field, and that a low-temperature origin was necessary to allow the grains to line up like that. From the audience, Harry McSween denies this low-temperature interpretation, arguing instead that the shock which knocked the rock off the planet could have induced metamorphism, which could account for their observations.

In the next paper, McSween and coworkers from Tennesee (the University and Oak Ridge) present data on sulfur isotopes which rule out the possibility of life in 84001: living creatures on Earth fractionate the sulfur isotopes in a distinct manner, and this is simply not seen in the Martian meteorite. Case closed. But questions from the audience stress the experimental difficulties in their work. The ion beam used to analyze the sulfide minerals is nearly as wide as the grains analyzed, which are surrounded by carbonate which has oxygen-16. If the beam slips just a bit during analysis, two oxygen atoms in the carbonate would give a signal at mass 32, interfering with the signal from sulfur-32 and messing up the results. Also, Martian bacteria could fractionate sulfur in a different way since bacteria are known to adapt strongly to local environmental conditions, which are of course much different on Mars.

Another multiuniversity group, including Everett Gibson and Chris Romanek, then presents independent oxygen isotopic evidence showing that "the only explanation consistent with the observations is that the carbonates precipitated in a disequilibrium system . . . below 100°C." At the end of these talks the consensus seems to be solidifying that a low-temperature origin is probably right, but there's still room for doubt.

3. There is no doubt that the general scenario, in which 84001 solidified into rock a long time ago and sometime after that the carbonates were formed, is correct. But there are arguments about exactly when all this happened. The first estimates were that the rock solidified about 4.5 billion years ago and that the carbonates were formed at about 3.6 Gyr—roughly the same time as the first life appeared on Earth. But some later work then indicated more recent carbonate formation, at roughly 1.6 Gyr. In response to questions at the conference, Grenville Turner of Manchester, England, replies that he is "happy with a 1.3–1.9 Gyr age for carbonate formation," and that the age of 3.6 Gyr is only an upper limit. This is not critical to the question of life, although most guesses about the presence of water on the surface of Mars are that the further back in time we go the more likely we would be to find water there, and so the older the better for the prolife forces. (It is necessary that the time of carbonate formation be older than the time at which 84001 was knocked off Mars, in order for the carbonates to be Martian and not terrestrial in origin, and there is no doubt about this.)

4. The other ages are not in dispute. The meteorite was exposed to cosmic rays in space (meaning its diameter was by then reduced to about a meter or less) for about 16 million years. This number is not precise, but is average for a stone meteorite and is totally unimportant except to prove that 84001 *is* a meteorite and not a queer Earth rock. There is less than a quarter the expected amount of cosmic-ray-produced carbon-14 in it, so there must have been enough time on Earth before its discovery for the carbon-14 to run down, meaning that 84001 landed on Earth about 13,000 years ago. Again this is not an exact age, but it can't be off by much. The only im-

portance attached to it is that it gives a long time for terrestrial contamination to slip inside.

5. The big questions and arguments naturally revolve around the question of fossil life. If the carbonates formed at high temperature (much greater than 150° Centigrade or so) all bets are off; we have no evidence or even theoretical speculation that life can exist at such high temperatures. (You can always get into the game of saying that "The universe is queerer than we suppose," and go on from there to say that *anything* is possible. Which is true, but then you're out of science and into science fiction.) If the carbonates formed at low temperature—as most people now agree—they probably formed by precipitation from water, and the arguments can begin.

There are three sets of observations in contention, none of which are definitive for life. The question is whether or not the entire set of observations can best be explained by living organisms, or indeed whether a higher level of proof than "can best be explained by" is needed. The first observation consists of the SEM photographs that show within and around the edges of the carbonate globules clusters of tiny shapes that resemble bacteria on Earth. And yes, they do remarkably resemble terrestrial organisms. McKay says that one day his bright thirteen-year old daughter came into his office and he showed her the pictures without telling her what they were and asked her, "What's this look like, Jill?" She looked at them and immediately said, "That's bacteria." And when Ev Gibson took the photos home, his biologist wife Morgan glanced at them and asked, "What are you doing with bacteria?". But at the 1997 meeting of the Geological Society of America they were referred to as *dubiofossils*. A *fossil* is the remains of something that was once alive; a *pseudofossil* is something that looks like a fossil, but is not; the etymology of the term *dubiofossil* is obvious. (In the hierarchy of geological humor, this ranked pretty high at the meeting.)

Looking at them you find it hard to believe that they could

be chance aggregations of minerals, having nothing to do with organized life forms. But the renowned Texan microbiologist, Bob Folk (whose paper sparked Chris Romanek's first SEM investigations into 84001) has remarked, in connection with similar photographs (which he took himself) of terrestrial nannobacteria: "Shape is almost useless as an identifiable characteristic. . . . To be properly identified, the bacteria must be made to come back to active life, reproduce, and perform a characterizing metabolism." This has not been done with the Martian shapes, and it *cannot* be done since they are fossils only, not living bacteria. Shapes is all we got.

So what else can be done to convince us that they are indeed relics of living Martians? One of the first antishape arguments was that they are an artifact of the investigation process, for in order to view a specimen in the SEM it must first be coated with a conducting material. The argument was that the coating process could result in tiny lumps which by coincidence might line up together in a way that mimics the appearance of bacteria. To answer this, the NASA group called on Andrew Steele and his coworkers from England. At the conference they presented the results of their study done with atomic force imaging, a new technique which allows them to view the sample without any coating process. The result was that they saw the same shapes as the SEM had, removing all possibility that the coating process was responsible.

However, Steele also noted that the putative bacteria seemed to arrange themselves in parallel lines, and that disturbs him. "Bacteria are the most fascinating things," he says in private conversation. "They come running along and when they see another one they circle around each other and then shoot away. They are definitely communicating with each other. There are some who, if they are attacked by a virus, blow themselves up so the virus can't get inside to reproduce and attack their brothers. They do all sorts of fascinating things. But what they do *not* do is line up like little wooden soldiers."

"How about size?" we asked him. Size is another problem with the bacterial hypothesis. Normal bacteria are about 500 to 3000 nanometers long, while the known terrestrial nannobacteria range down to about 100 nm. The 84001 bacteria are mostly smaller than this; their *maximum* size occasionally reaches 100nm or a bit more. This makes them different from terrestrial organisms, but after all they're from Mars so you'd expect them to be different.

The real problem is not the length, but the volume (or diameter). There is an old poem about dogs which have little fleas which bite 'em

> *And these fleas have smaller fleas*
> *Et cetera ad infinitum.*

But not so. There is a lower limit to the possible size of living organisms, which is due to the inescapable fact that they have to be made out of molecules which are made of atoms which themselves have a finite size. And life is made up not of any old molecule, but of lengthy and complex organic molecules, involving thousands of atoms in complicated interlinked shapes, and this just has to take up a certain amount of space. You can't make the complex molecules on which life depends without enough space to pack them in. Steele and his colleagues agreed that the size of the 84001 shapes are okay down to about 50nm, because bacteria are clever creatures and they have all sorts of strategies to minimize their requirements and pack information into small volumes. But less than 30nm is a real problem because of molecular size. And if the bacteria-like shapes of less than 30nm are not real bacteria, then there's no reason to suppose that the larger ones—up to 100 nm—are real either.

On the other hand—Harry Truman, when he was president and dealing with scientists about the possibilities of developing the atomic bomb, used to pray for God to send him a one-armed scientist—on the other hand, then, Robert Folk has pointed out that "when bacteria become stressed, they condense down from their normal size and shape," and that "encasement in carbonate crystals may [be the cause of] that stress." So the bacterial shapes seen in the carbonates may represent organisms that were bigger when alive and thriving. And on the *other* hand—Sorry, Harry—these shapes may be only parts of bacteria: appendages or filaments that are the only things left from the living creatures that once inhabited the waters of Mars and were trapped in these carbonates, blasted off into space, and finally landed under our microscopes.

The second line of evidence is mineralogical in character: the "Oreo cookie" rims of the carbonate globules contain magnetite and pyrrhotite in intimate contact; while each of these can be formed without the presence of life, their linked presence can best be explained by bacterial action. The argument against this is twofold. First, Mars doesn't have a magnetic field so why should bacteria form magnetite up there? But the data presented at the Houston conference, referred to above, indicate that the magnetite crystals are aligned in such a way as to indicate that Mars *did* have a magnetic field at the time they formed.* As we have seen, Harry McSween and some others don't find this evidence particularly compelling. Another argument is that while bacteria do produce these crystals on Earth, they are also produced inorganically; and while in such circumstances they are usually not found together, who knows what might happen on Mars where the chemical conditions are so very different from Earth's?

The final argument for life on Mars rests on the chemistry of polycyclic aromatic hydrocarbons (PAHs), and the conclusions rest on differing interpretations of the data. There was nothing new on them offered at the Houston conference, but

*A few months later, in September, 1997, Mars Global Surveyor found a "surprisingly strong" magnetic field on Mars.

just a few weeks earlier John Kerridge of UCLA offered his opinions at the Seattle meeting of the American Association for the Advancement of Science. There is no question that PAHs can be produced by living organisms, but they can also be produced by processes having nothing to do with life. Kerridge pointed out that on Earth if a petroleum exploration company searching for oil found PAHs in a particular region, that would not be enough to convince them there was petroleum there; the presence of the PAHs would not raise shouts of fossil life and induce the geologists to start drilling.

But the pro-lifers argue that the particular PAHs found in 84001 *are* indicative of life. Kerridge answers that this just isn't so. If you look at the mass spectra measured in the meteorite you find major similarities to the compounds formed by non-living processes in dust and other meteorites that have nothing to do with Mars. Gibson argues that this is not true: you do find some similarities, but the mass peaks are not in the same ratios as in other meteorites, and some peaks seen in these other samples are not seen here. The riddle at the present time is incomplete: if you look at the spectra one way you see peaks that argue for similarity, if you look at it another way you see data that suggest differences.

To sum up, the doubters argue that there is nothing seen in 84001 that requires life to have been present on Mars. The pro-life group is saying that, yes, all the observations could conceivably be explained away without invoking the presence of Martian life, but the *simplest* explanation is that they are all due to one phenomenon: the effects of living organisms. It's the difference between a civil and a criminal trial, as we all learned ad nauseam a year or two ago: the doubters say that the conclusion of life beyond Earth is so important that it demands the highest standards of proof, proof beyond any reasonable doubt. The prolifers argue that the preponderance of evidence suggests life. They are hanging their case on a basic part of the scientific method, Occam's razor, enunciated by the philosopher William of Occam in the fourteenth century: "Entities should not be multiplied unnecessarily," or in the present vernacular, "KISS" (Keep It Simple, Stupid).

The argument for simplicity is a compelling one. If it's not kept in mind, you can always explain anything with a false, complicated series of premises. For example, when the earliest philosophers were thinking about the structure of the universe, the simplest explanation for the observations (the sun, moon, and all the stars rise in the east and set in the west) was that everything revolved in a circle around the Earth. When the planets were discovered to move with a varying motion this theory was revised to make the planets move in circles upon circles, as if they were mounted on the rim of an invisible wheel which itself revolved around the Earth. As their motions were measured more accurately, it was found that in order to keep this theory one had to assume the planets were mounted on wheels upon wheels upon . . . The concept lost its basic simplicity, and this is what inspired people like Copernicus to set up an alternative theory, that the Earth revolved along with the planets around the sun. This much simpler theory turned out to be true.

And so it goes. Throughout the history of science, simpler explanations usually turn out to be better than more complex ones. Chris Romanek, David McKay, Everett Gibson, Kathie Thomas-Keprta, and their colleagues are in good historical company when they fight for the simplest explanation that will cover all the observations. On the other hand, there is another edge to Occam's razor, and we might perhaps heed the words of Albert Einstein: "Scientific explanations should always be made as simple as possible. But not simpler."

5 | The Oceans of Europa

The great depths of the oceans are entirely
unknown to us. What creatures might live in
those remote depths, we can scarcely conjecture.
JULES VERNE, *20,000 Leagues Under the Sea*

. . . relay this information to Earth . . . only
survivor . . . Please listen carefully. THERE IS
LIFE ON EUROPA. Repeat: THERE IS LIFE
ON EUROPA . . .
ARTHUR C. CLARKE, *2010: Odyssey Two*

1.

San Juan Capistrano is a beautiful name with an ugly town attached. It sits just inland from California's Pacific coast, between Laguna Beach and San Clemente, and between the Santa Ana Mountains and the ocean. At one time it was lovely but now the twentieth century has caught up to it, ravished it, turned it into a commuting way station on the road to Los Angeles, and left it bedraggled, bothered, and bewildered.

The morning and evening rush hours clog its narrow streets

with cars and its air with carbon monoxide. The natural geologic glory of the region has been smothered by people and automobiles, and in their wake is a town with strict standards for
its architecture but none at all for its uses. Thus the streets are
crowded with a homogeneous faux-Spanish ambience, and the
stores have cute names like *The Cat's Meow, Nutcracker Sweet,
Pet Agreed* (get it?), *Whimsey Hollow,* and *Daffy's Designer
Sandwiches.* Swallow's Inn ("Fine Food, Cocktails") celebrates
the famous swallows that return every year to the Mission. But
the Fine Food consists of hamburgers, cheeseburgers, or buf-
falo wings, and the swallows return to a crumbling, ill-kept
masonry and a million tourist cameras.

And yet, San Juan Capistrano may live on in the memory of
our civilization. Not for the neglected Mission, certainly not
for the local culture, but because on November 11, 1996, in a
small building on a dingy street called the Camino Real, the
San Juan Institute hosted the first conference dedicated to
finding life on one of the moons of Jupiter.

2.

The story begins with an interesting contradiction about the
life on Mars debate, which has to do with the PAH evidence.
On Earth these are usually indicative of life processes—but
here too there is argument. They are found, as we said, when
meat is burnt (as on a charcoal grill—these are the chemicals
which make cookouts carcinogenic). Clearly these PAHs are
related to life, for the meat was once alive. They are also found
in the exhaust of automobiles (again a carcinogenic factor)
and, as everyone knows, fossil fuels are just that—fuels made
of fossilized life forms, the remains of living organisms.

But, as it turns out, not quite everyone knows that. Tommy
Gold, professor emeritus at Cornell University, one of the
most irritating people in science and one of the all-time greats,
questions that. Gold is the innovator of the Steady State the-

ory of the creation of the universe, which has recently been proved wrong but which stimulated cosmological thinking for twenty-five years. He was the one who explained what pulsars are (spinning neutron stars) when no one else had any idea of what they could possibly be, and he was one of the dominant thinkers behind our lunar exploration strategies in the 1960s. When I (D. E. F.) first met him in 1962 he was director of Cornell's Center for Radiophysics and Space Research, which was housed in a modern building known to the students as Uncle Tom's Cabin. Then in his early forties, Tommy was at the height of his vigor; in the cellar of his house he kept a thick rope dangling from the ceiling, and delighted in challenging his guests to match him as he sat on the floor with his legs outstretched, grabbed the rope, and hauled himself up to the ceiling while keeping his legs stiff and horizontal. Aside from that ordeal, dinner at his home was an event to be looked forward to. Conversation would drift over astronomy and physics, politics and theater. We would settle such topics as how to utilize nuclear energy to heat highways and melt the snow, and what the hell we were doing in Vietnam. He has been known to bring religious guests to the edge of apoplexy with his casual but scathing irreverence toward the concept of God, and he frequently does the same to the more conventional-minded of our scientists when he analyzes and dismisses their fundamental beliefs.

One of which has to do with the origin of fossil fuels. The conventional wisdom, assumed by so many for so long that the concept comes perilously close to dogma, is that they're formed from the decay of living marine organisms buried under successive layers of sediment and transformed by increasing pressure and heat over millions of years. Gold argues instead that oil and gas are primordial remains of nonbiological hydrocarbons from deep within the mantle of the Earth. This idea is an old one that was supplanted early in the twentieth century when the occurrence of oil fields matched very well with the sites of ancient seabeds, and when common biological compounds were identified in every single oil field.

Gold argues that aside from its association with biocompounds there is nothing peculiarly biological about petroleum itself: the hydrocarbons that comprise it are common everywhere in the universe. The largest concentration of hydrocarbons in the solar system is in Jupiter's atmosphere, while Titan (Saturn's brightest moon) and faraway Pluto are known to be covered with methane ("natural gas") ice to a large degree, but no one is claiming there is any biology on any of these bodies.

Instead, he goes on, another datum is of more significance: helium is also found associated with oil fields, and there is no possible biological mechanism that could concentrate it since biology has no use for helium. He points out, however, that if the petroleum is a deep mantle product that has seeped up toward the surface—rather than being a near-surface biological product that has seeped down—it should have flushed out vast quantities of helium from the mantle and carried it along.

The scientific community has no answer to that but comes right back at him with two questions. If petroleum comes from deep in the Earth, why do all oil deposits contain biomolecules? And if petroleum has nothing to do with ancient marine life, why are oil deposits found precisely where ancient seabeds once existed?

Easy, Gold replies. The oil deposits contain biomolecules because the oil is the *food* of life, not its remains. If hordes of microbes live deep within the Earth and feed on the petroleum hydrocarbons, then of course the oil deposits we see always have biomolecules in them; but these biomolecules are incidental to the oil rather than indicating their origin. And the oil deposits are found where ancient seabeds existed because that's where we look! Our theory tells us that's where they are, and so we look there and we find them. But according to Gold the oil is everywhere under the ground, and could be found wherever there are cracks in the rocks to allow it to seep up and an impervious crown of rock to cap the reservoir and hold it in.

Not many people are convinced. But ten years ago Vattenfall, the State Power Board of Sweden, caught in an energy crisis, agreed to attempt digging the world's deepest well, in granite, where no one else in the world expected to find petroleum. By 1990 they had spent three years and forty million dollars, and all they had reached was the limit of their engineering capabilities. They found no petroleum; the hole was dry. End of story, end of theory.

But not quite. They stopped too soon, Gold insists. There were indications of a sealed reservoir 7.4 kilometers down, but the company ran out of money and enthusiasm at 6.8 kilometers, and there they quit. And the hole wasn't exactly empty. There was an assortment of hydrocarbons, and there was magnetite. The hydrocarbons were seepage from adjoining sediments or from the drill oil itself, say almost all the biogeologists who have seen the data; but Gold says these hydrocarbons are merely the tip of the iceberg that lies just another half kilometer down. And what about the magnetite? Tons of it were brought up: tiny, micrometer-sized grains, just like the stuff found in ALH84001 as evidence of bacterial life.

The Martian prolifers are right about that, Gold says, wrong about the PAHs, and right about their conclusions. The polycyclic aromatic hydrocarbons of Mars are no more the remains of life than are our own petroleum deposits; both of them are the food on which bacteria live. What McKay and Gibson and their group have found in the Martian meteorite is the food and the excrement of living organisms: the PAHs and the magnetite grains. Taken together, Gold claims, they are reasonable evidence of life on Mars. Just as he predicted six years ago.

In a 1992 paper titled "The Deep, Hot Biosphere," published in the *Proceedings of the National Academy of Science*, he gathered together evidence for a whole new type of biology: creatures living kilometers deep within the Earth. He claimed there would be more life found deep within the Earth than on its surface, that down there was where life started, and that life will be found in similar environments elsewhere in the solar system. Earth is unique only in its surface conditions: alone among the planets and moons of our solar system it has abundant sunshine and liquid water: rivers and streams, lakes and oceans. But if life originates deep within a planet, this unique circumstance is no longer important. "The other planetary bodies in our solar system do not have favorable circumstances for *surface* life," he acknowledged, "but with the possibility of *subsurface* life, the outlook is quite different. Many planetary bodies will have temperature and pressure regimes in their interiors that would allow liquid water to exist. . . . Meteorites are being collected at the present time that are thought to have derived from Mars. Can one find traces of biological substances in them?"

When that was written in 1992 almost everyone thought Gold was crazy. In 1998, it doesn't sound quite so nuts.

3.

Tommy Gold's thesis depends on the existence of flourishing life deep within the Earth. Until recently, this was a subject fit only for science fiction. The Earth's biosphere, defined as the sum of all living creatures, was envisioned as a thin film surrounding the surface of the Earth; it extended a few hundred feet up into the atmosphere and not much further down into the oceans. Above and below that, life quickly died out.

Or so we were taught. A couple of generations ago students at universities learned that the ocean floors are vast, undeveloped, uninteresting deserts. The typical picture portrayed them as smooth saucers lying under the oceans, unchanging geologically and devoid of life, which existed only in the surface layers of the oceans because of the intense pressure, the extreme cold, and the lack of sunlight in their depths.

The pressure seemed to be the greatest obstacle to life. Listen as M. Aronnax, the hero scientist of Jules Verne's *20,000 Leagues Under the Sea* (1870), explains it to Ned Land, the master harpoonist, while they search the seas for an elusive bottom-dwelling sea monster:

> "If such an animal is in existence, if it inhabits the depths of the ocean lying miles below the surface of the water, it must possess strength which would defy all comparison."
>
> "Why?" asks Ned.
>
> "First you must understand that the pressure of the atmosphere is represented by the weight of a column of water thirty-two feet high. When you dive, Ned, your body bears a pressure equal to that of the atmosphere, roughly 15 pounds for each square inch, for each 32 feet of water that there is above you. It follows then, that at 32 feet this pressure equals that of 10 atmospheres, at 3200 feet it's 100 atmospheres, and it reaches 1000 atmospheres at 32,000 ft, or about 6 miles, which is a typical ocean depth. This means that if you could reach this depth in the ocean, each square inch of your body would bear a pressure of 15,000 pounds. . . . If you are not crushed by the pressure of the air on the surface, it is because the air penetrates the interior of your body with equal pressure. Hence perfect equilibrium between the interior and exterior pressure, which thus neutralize each other. But in the water it is another thing."
>
> "Yes, I understand. Because the water surrounds me but does not penetrate."
>
> "Precisely, Ned. So at 32,000 feet the pressure on your whole body would be 39 million pounds; that is to say, that you would be flattened as if you had been drawn from the plates of a hydraulic machine!"

This explanation, though it reflects the general scientific opinion of the times, turns out to be wrong. It's true for such creatures as ourselves since our lung cavities, filled with air, would certainly be crushed by the water pressure. But this is because gases such as the air in our lungs are easily and almost infinitely compressible to liquids: if the gaseous air in a fully expanded pair of lungs was compressed into liquid air it would barely fill a medicine dropper. But liquids and solids undergo no such change of volume when pressure is applied, and so living bodies composed entirely of liquids and solids would be incompressible. They would still be flattened by millions of pounds of pressure, as M. Aronnax claims, except for the fact that deep in the ocean they would *not* be "in the plates of a hydraulic machine," which would apply pressure in the vertical direction only and allow the crushed body to flatten out in the horizontal plane. In the ocean the hydraulic pressure would be equal on all sides, and no "flattening" could occur.

The extreme cold (bottom ocean waters are about 2° Centigrade) also turned out not to be a problem, as we have found living creatures in any environment that allows the presence of liquid water. Even before Jules Verne wrote his deep-sea novel there was a school of scientists who believed that life could indeed exist deep in the ocean. Because conditions there were thought to be not only poor for life but also unchanging, they postulated that these creatures might be living fossils: ancient forms of life that had resisted evolutionary change and would exist today in their ancient forms, still perfectly suited to their unchanged environment. In the mid-1870s HMS *Challenger* sailed the seven seas in the world's first comprehensive expedition to explore the deep oceans. They found more than four thousand species of deep-sea life never before known, although none that could be classified as a living fossil.

It was realized, therefore, that life could exist in conditions of extreme pressure and cold. Yet the understanding was that any life at the bottom of the sea was an outcast from the world above, struggling to eke out a bare existence under terrible conditions, subsisting at starvation levels. The biggest problem such life forms faced was neither the temperature nor pressure but the lack of sunlight, since the sun's rays can't penetrate that depth of water, and without sunlight there can be no photosynthesis and without photosynthesis there can be no life, or so we thought. As recently as 1959 one of the twentieth century's outstanding life scientists and popular writers, Dr. Loren Eiseley, was telling us:

> Things that do not love the sun, that could walk through total darkness upon slender footholds over evil waters . . . had come down there by preference from above. It was in this way that the oceanic abyss was entered . . . Life did not arise on the bottom; the muds of the deep waters did not compound it. Instead . . . it has groped its way down into the dark.

His reasoning was based on the fact that up here on the Earth's surface virtually all living creatures either photosynthesize or eat creatures that do; one way or another, we are all dependent on each day's life-giving light. Down in the depths the base of the food chain could only be creatures from above, based on photosynthesis, who die and fall down to the bottom where the bottom-dwellers could eat them. And such a rain of food from above could not be more than a drizzle, as most of the flesh would dissolve in the corrosive water as it fell.

But we should have been smarter. We should have realized that there is nothing magical about sunlight; it's simply a form of energy, and energy can take many forms. We know from basic principles such as the second law of thermodynamics that energy is necessary for life. The second law, after some mathematical fiddling, takes various forms; one form states that in *an isolated system* the universe tends toward increasing disorder, and by an isolated system is meant a part of the universe into which no energy flows from outside the system. Life is a highly ordered system: it begins as isolated atoms and molecules distributed randomly throughout its environment, and it arranges itself into an efficient, working, *ordered* system. So if, for example, we were to find indigenous life on a chunk of rock far out in the emptiness of interstellar space where no starlight reached and no other form of energy was available—well, life simply could not exist in such circumstances. If we ever do find such a system we would have to call it a miracle; and miracles don't happen. Life needs energy to be able to order itself into the complex molecules of which it consists.

But many different sources of energy can do the trick; it doesn't have to be sunlight. Although nobody was paying much attention at the time, one scientist began to realize this way back in the 1920s. Edson S. Bastin, a University of Chicago geologist, was wondering why underground water in oil fields contained hydrogen sulfide (H_2S). He knew that anaerobic bacteria—those which live without the presence of oxygen—can obtain their energy by reducing sulfates ($SO_4^=$) to sulfides ($S^=$), and so he speculated that such bacteria might be living underground in these oil fields. He succeeded in culturing bacteria from these waters, but he didn't manage to convince anyone of his ideas: They all thought that the bacteria he cultured were only contamination picked up as the oil was brought to the surface, rather than deep-earth life forms, and there the idea rested for several decades undisturbed by the thoughts of men. Things began to change drastically after the Second World War, when a few men like F. G. Walton Smith came calling.

Walton, as he was known, was a British biologist who had been working on sponge fishery problems for the Colonial Office in the Bahamas and who decided he liked the warm weather better than the drizzly, cold fog back home. One day he happened to meet Dr. Bowman F. Ashe, the first president of the University of Miami (then known as Cardboard U. because it was housed in one building and the classrooms were separated by cardboard partitions). They got to talking and Walton told him he could build a world-class oceanographic institution in Miami and it wouldn't cost the university a cent.

That's the kind of talk a university president likes to hear, so he invited Walton Smith to sit down and tell him all about it. What Walton Smith said was that the war was over, and the navies of the world had learned that if they wanted to operate efficiently in the oceans they had to know more about them. Sonar operations, for example, depended on sound echoes to locate submarines; but sound waves traveled differently through the water depending on the little-known variations in physical parameters of different layers of the water, and various living organisms in the oceans emitted and absorbed sounds which disguised and interfered with the sonar signals.

The United States Navy, through its Office of Research and Development, was going to be spending a lot of money trying to find out more about the oceans, and since they didn't have enough scientists themselves they were going to be looking for universities willing to carry out the necessary research. Walton gestured out through the windows at the tropical sunshine: Miami was the perfect place for America's first institute for tropical marine biology.

A few other people at other places had similar ideas, and so in the years following World War Two, as the Navy's research money flowed in, Miami's Marine Laboratory was joined by expansion and development of the Woods Hole Oceanographic Institute on Cape Cod, the Lamont Geophysical Observatory in New York, and the Scripps Institution of Oceanography in California. And the oceans began to yield up their secrets.

The main secret was that the oceans are split down the middle by a continuous, snaking ridge of the highest mountains in the world, the Mid-Ocean Ridge system, formed as the sea floor splits open and molten magma from deep in the mantle wells up and spills out. In this process the ocean floor is continually growing, leading to the concept of seafloor spreading (which had originally been proposed by a Princeton geologist, Harry Hess, who thought the idea so fanciful he dubbed it "geopoetry"). By the 1960s, evidence was beginning to accumulate that the geopoetry was really geoscience. One of us (D. E. F.) moved to Miami in 1966 specifically to test the hypothesis by dating the seafloor. (The other of us had no choice.) I had been working on methods of measuring the age of iron meteorites, and in discussion with Miami scientists had worked out a procedure for testing the seafloor spreading hypothesis. We carried out the experiment in the summer of 1967 and showed that the theory was wrong. Rocks on the ridge itself, which according to the theory should have been exceedingly young, turned out to be the oldest ever measured, with ages of more than half a billion years. But when our group analyzed the entire geological picture, we realized that there was something wrong with the experiment which measured the age as a function of the buildup of argon gas, formed at a steady rate by the radioactive decay of potassium. The argon is normally flushed out of rocks when they erupt onto the surface of the Earth. This flushing effect sets the radioactive clock to zero, and the buildup and time are measured from that event. But the seafloor rocks were different in that the high hydrostatic pressures on the seafloor didn't allow the argon to flush out when they erupted, spoiling the calculation of their age and making it meaningless, and teaching us once again that experiments do not necessarily take precedence over theory. Many experiments since then have confirmed that the theory is correct, and have led us to use the trapping of argon and other gases as a probe into mantle conditions. It's easy to do experiments, but hard to . . .

Even earlier than this, just after the Second World War, Claude Zobell of the Scripps Institution in California had found evidence of microbes buried in sediments on the seafloor, but these were assumed by everyone to be surface microbes that managed to survive for a while after burial, and once again did not excite much interest. It wasn't for another twenty-five years that more serious concerns surfaced. By the late 1970s we were beginning to worry about the quality of our drinking water, which is largely stored and transported in and through underground aquifers. The United States Geological Survey and the Environmental Protection Agency began to realize that they didn't understand the processes which were going on underground, while the Department of Energy (DOE) was beginning to face up to the problems which had been (literally) buried during the pressures of the Cold War—including both the radioactive and chemical poisonous wastes from nuclear bomb-building facilities throughout the country.

Frank Wobber was a geologically trained administrator at DOE who realized that if the earlier hints of deep-earth microbes turned out to be true, they could have an important effect on his problems. Bacteria can carry out all sorts of chemical reactions; for instance, they might be useful in converting organic poisons into harmless compounds. On the other hand, they might degrade and crack open the buried nuclear disposal containers. Either way, Wobber reasoned, DOE had better find out if they existed and whether they could be utilized or destroyed, and so he initiated the Subsurface Science Program which by 1987 was bringing engineering investigators together with biologists and geologists to study the deep-earth environment.

What they found was the evidence quoted by Gold in his 1992 article, evidence for a whole new environment down there, teeming with life. As Gold put it, "every place there is room for life, there is life." About 3500 species of bacteria were known at the time they began their researches; today, although they still have not been identified and classified, the suspicion is that there are millions of bacterial species living under the surface of the Earth. At first these creatures were found living in the voluminous pore spaces of loosely packed sedimentary rocks, gaining their energy from the reduction of sulfates and their food from organic decomposition products of plants that had grown on the Earth's surface; so ultimately, these bacteria depended on sunlight and photosynthesis for their food supply. But as research progressed into the deepest parts of the seafloor we discovered life further down than anyone had ever suspected, existing totally independently of the surface of the Earth and of light from the sun.

4.

It has been said that we know more about the surface of the moon than about the bottom of the ocean, and that's not terribly surprising. All we have to do is look up and we can see the moon, and a lot can be learned by looking—especially if we look with a variety of instruments at different frequencies. To be sure, we've learned a lot about the ocean depths by remote sensing equipment, by bouncing sound waves through the water layers, and by dredging up rocks and sediments. But the only way to really find out what's down there is to go down beneath the obscuring layers of water and take a look.

There are four different ways of attempting this. The most ancient is the simplest: just fill your lungs with air and dive down as far as you can. This is fun, and dangerous, and can be profitable if you're looking for pearls in oysters, but as far as investigating the ocean depths it's useless. The deepest a man or woman can free-dive is a few hundred feet, and the floor of the ocean is many thousands of feet below that.

The first real attempt to delve into the depths was with a helmet fitted with a pipeline to the surface. The helmet fit over the head and rest on the shoulders, with or without a protective diving suit, and air was pumped down into it. The depth limit with this device was still only a few hundred feet, but at least it allowed the diver to stay down and look around. When he did he found a profusion of marine life, but nothing vastly different from that found higher up. The mysteries of the depths still beckoned.

The first man to answer that siren call was William Beebe,

curator of ornithology at the New York Zoological Gardens and later director of the department of tropical research at the Zoological Society. His interest in birds had drawn him to Bermuda and the tropical islands of the Caribbean, and the clear waters there drew him down into their depths. He went down in a helmet and described with lyrical writings the beauties he found there, attracting the attention of an American engineer named Otis Barton, who had designed a spherical steel vessel for use in undersea observation. It was to be provided with portholes and an air-line to the surface, and would be suspended by a steel cable from a boat. It sounded like a fine idea to Beebe, and together the two men raised the money and built the *bathysphere*. After a few unmanned trial dunkings to test it for leaks, Beebe and Barton crawled into the hollow ball, which was about four feet in diameter, screwed the sewerlike entrance hole shut behind them, and gave the signal to the crew to let them down into the water.

On June 11, 1930, they reached a depth of four hundred meters, or about thirteen hundred feet, and in 1932 they took it down to nine hundred meters, or about three thousand feet, deeper than any human had ever ventured before. They came back up safely, thrilled with the new world they had discovered. But, like Moses, they were granted only a distant vision of the promised land; the bathysphere wasn't able to take them down to the deepest floor of the ocean. It wasn't free to descend at will; it hung from its cable, and the weight of the thick cable limited its length. Moreover, it was totally dependent on that cable: a break would have meant its sinking into the depths without hope of recovery, bringing a terrible death to the two men inside. The cable was as strong as they could make it, but surface waves would rock the support ship, and with each surge the bathysphere would be jerked up and down; a fatal snap could have occurred at any time. They also flirted with death on several occasions as the bathysphere twisted in the deep currents, winding the suspension cable around the air-line and threatening to choke it off.

To efficiently investigate the ocean depths we needed a vessel that could operate independently of any connection to the surface. It would have to be able to maneuver on its own and to carry its own oxygen supply. We needed a submarine.

Luckily, it had already been invented. Unluckily, it had been invented for warfare rather than exploration, and consequently was designed to operate just under the surface instead of plunging to the depths. Cornelius Drebble, court engineer to James I of England, had built the first maneuverable submersible in the seventeenth century, but it was maneuvered by oars and turned out to be useless. In 1776 the American *Turtle*, invented by David Bushnell, tried to attach a bomb to the hull of a British ship outside New York harbor. The *Turtle* looked like an elongated clam, just big enough for one man. It had six glass windows, "each the size of a half-dollar piece," and a hand-driven crank that drove two short oars. With his other hand the pilot steered by means of a rudder. Depth was controlled by letting water in to sink, and pumping it out to rise. For navigation there was a pocket compass, and a cork floating in a tube connected to the outside monitored water pressure and therefore depth. It is no surprise that anyone smart enough to have invented this beast was too smart to go down under the water in it, so Sergeant Ezra Lee volunteered.

On a moonlit night in September he was towed out into the East River, just above its terminus, and cast off. By 1 a.m. he had managed to pull himself abreast of the British man-of-war *Eagle*, at which point he pressed on the water vent and sank down into the waters. He slipped under the *Eagle* and tried to attach his single bomb by means of drilling into the hull and hanging it by a hook, but he wasn't able to drill through the metal-covered hull. When he tried to maneuver the *Turtle* around to try another spot he lost his equilibrium and the submarine shot up to the surface, whereupon Sergeant Lee hastily beat a retreat, and no one ever tried the *Turtle* again.

Robert Fulton introduced the movable horizontal fin for depth control in 1800 and tried to sell his ship to Napoleon. It had a normal mast and sail for use on the surface, and one of the crew cranked a propeller to drive it when submerged. Despite some success in staged tests (the *Nautilus* successfully attacked a moored ship in harbor) it failed to convince the French admiralty that it could sink British ships. Without hesitating for a moment, Fulton then tried to sell it to the British for the purpose of sinking French ships, but they too declined and he returned home and settled for designing America's first commercially profitable steamboat.

The submarine disappeared from view for nearly a century. In the Civil War the Confederate *Hunley* sank the USS *Housatonic*, but after sending the agreed-on signal that they were safe—two blue lights blinking across the dark waters to watchers on land—the *Hunley* and its crew disappeared forever, or at least for a hundred and thirty years. In 1995 the sunken submarine was found off the coast of South Carolina. It has not yet been raised, and what happened to it is still conjectural: while waiting for the tide to turn so that its man-powered propellers could bring it back to shore, it was probably swamped by ships racing to rescue the *Housatonic* survivors. And that was that: the *Hunley* had previously sunk itself three times on demonstration and test voyages, and the Confederates decided enough was enough; they used no more submarines. By the 1880s both England and France had built truly maneuverable submarines, featuring battery-powered electric motors that could operate under water. But these had the same disadvantage as today's electric cars: their range was severely limited by the need to put in frequently at recharging stations. Then in 1900 the United States built the USS *Holland* with electric engines for submersible operation and gasoline engines for surface propulsion and recharging, and humanity had a truly terrible weapon.

In the two world wars that soon followed, the submarine nearly provided the turning point of a German victory, but never did these machines venture deeper than a few hundred feet. Shortly after the Second World War, however, a Swiss inventor named Auguste Piccard designed a workable combination of the bathysphere and the submarine which he called the bathyscaphe. After nearly ten years of construction, redesign, and reconstruction the newly named *Trieste* reached a world-record depth of 13,000 feet in 1954. The United States Navy bought a copy and redesigned it for even greater depths. In 1960 Jacques Piccard (Auguste's son) and Lieutenant Don Walsh of the U.S. Navy took it down to a record 10,916 meters (35,810 feet).

In 1963 the nuclear submarine USS *Thresher* sank with all hands, and the Navy accelerated its program of deep-sea recovery, building two new versions of Piccard's submersible. In the end, however, the two *Trieste* craft were used more for Cold War espionage and spying as the United States and Russia grappled throughout the 1970s and 1980s in an invisible and secret war deep beneath the oceans. The scientists needed something different; in a symposium held in Washington, D. C. on February 29, 1956, they argued for more mobility and vision, and better equipment for sample deployment and recovery.

The *Trieste* was not tethered to its mother ship, as the bathysphere was, but in truth it wasn't much more mobile. Not only was it slow, but more important it was exceedingly difficult to maneuver. If you saw something interesting, you had to hope it would turn around and swim up to you; you couldn't hope to chase it down. And the chances of seeing anything interesting were remote, since there were only two portholes and each of them was only wide enough to permit squinting with one eye open. As Piccard himself has written, the *Trieste* was "a blind and cumbersome prototype" of things to come. What the scientists wanted, what they needed, was a true submarine that could fly through the ocean like an airplane through the air, settle on the bottom like a crab, and look around to see what was there and where it wanted to go. It needed arms that could put out bottles for water collection or break off pieces of rock from the seafloor or snatch critters from the deep and bring them safely home again.

It turned out that the Reynolds Metals Company had designed such a craft, the *Aluminaut*, and was looking around for possible users. Charles B. "Swede" Momsen, Chief of Undersea Warfare at the Office of Naval Research, was willing to pay for it, and a group of scientists at the Woods Hole Oceanographic Institution were eager to operate it, but the head of the Reynolds company, J. Louis Reynolds, turned out to be a hard man to pin down. Every time the Woods Hole people got him to agree to revised specifications, every time Momsen got him to agree to a binding contract, Reynolds would change his mind and make himself scarce. Finally, frustrated and anxious, they decided to look around for an alternative design. They sent their specifications out for bids, and finally awarded the contract to General Mills, the makers of Wheaties, the Breakfast of Champions.

It sounds like an unlikely corporation for such a task, but General Mills was a diverse company that had built torpedo and gun parts for the Navy during the past war, and its Mechanical Division numbered thousands of employees, including an aeronautical engineer named Bud Froehlich. In 1962 he began work on a submarine that could dive more than a mile deep. In June 1964 the *Alvin** was commissioned at Woods Hole.

Designed to carry three men (or, after much argumentation, women) the *Alvin* is twenty-two feet long and eight feet wide, weighing sixteen and a half tons. It can cruise along at

---

*Members of the Woods Hole group thought that Alvin the chipmunk, a major entertainment celebrity in the late 1950s, bore a marked resemblance to Allyn Vine, one of the prime movers for the submersible. In the irreverent spirit of oceanographers everywhere, and against the vehement protests of the United States Navy (which wanted something more dignified), they insisted on this name.

four kilometers per hour, carrying a crew of one pilot and two scientists in a titanium sphere two meters in diameter, and its dozen or so lights can illuminate the black waters out to fifteen meters. Brought to its oceanic destinations on a mother ship, it is gently lifted up and lowered into the water where it begins its daylong work: a typical dive takes two hours to reach bottom, four or more hours at work there, and another two hours to climb back up again. To explore the depths it has two six-foot-long arms with mechanical fingers on the ends, one window forward for the pilot and one on each side for the scientists, and can use two automatically operating forward-looking thirty-five-millimeter cameras and four television cameras. Below the lights and cameras is a rack typically filled with scientific gear: bottles for sampling, space for recovered samples, and instruments typically lashed down—in normal oceanography spit, string, and sealing wax technique—" by rubber bands hooked onto plastic milk-carton carriers, which in turn are connected to the metal rack."

To a Navy man, accustomed to submarines as sleek killers of the deep, it's a squat, ugly little bugger. To oceanographers who until now had to explore the ocean depths by remote instrumentation, it is a thing of beauty and a joy forever.

In 1977 it proved its worth when a geological expedition took it to a part of the midocean ridge known as the Galapagos Spreading Center, to look for a suspected but not yet observed phenomenon known as submarine hydrothermal systems. The idea was that hot molten magma was being thrust up along the ridge. As it came into contact with the cold ocean water it would freeze into solid rock, and as the frozen rock surface contracted over the hot interior it was expected to form deep cracks into which the ocean water would percolate. The water would be heated by the hot interior and should surge upward and out again as more cold water came in, setting up a continuous flow of water: cold in, hot out. A remotely operated, towed temperature probe had found a spike of warm water along the Galapagos Ridge, and so the group, led by John Corliss of Oregon State University, took the *Alvin* out there to dive down and take a close look at what should be there if we understood undersea tectonics correctly.

Barely out of graduate school in 1977, Corliss looked—and still does—like an overgrown hippie. A large bear of a man who gives the appearance of being even larger, with long hair and thick beard usually blowing about in the sea breezes, spectacled and sandaled, he looks as if he would be more at home in a Berkeley coffeehouse than in the cramped confines of *Alvin*. But he squeezed into the sub along with John Edmond, a geochemist from MIT, and Jack Donnelly, the pilot, and led her down into the depths, and there they found the hydrothermal system, all right. They also found a totally unexpected phenomenon: a dense jungle of living creatures, weird plants and animals that had never before been seen, wiping out in one blow our concept of a dead, desertlike sea floor.

5.

"We all started jumping up and down. We were dancing off the walls. It was chaos. It was so completely new and unexpected that everyone was fighting to dive. . . . It was like Columbus."

The initial impact was due to the profusion and variability of life clustered around the hydrothermal vents. They found not only the microbial forms anticipated in hostile environments, not only dandelions and anemones and limpets, but large animals: shrimps and crabs and lobsters, fish and clams and mussels, and worms as big as snakes. All of this down where—as the saying goes—the sun don't shine.

The clue to the abundance of life came in three forms. First, the water at the bottom of the world was warm. Instead of the usual deep ocean temperature of about 2° Centigrade, *Alvin's* thermometer showed a luxurious 17° Centigrade, or 63° Fahrenheit. Second, when the animals were captured by the submersible's tongs and brought up to the surface they stank of rotten eggs. That smell is the odor of hydrogen sulfide (H_2S), and its presence indicated a different source of energy for these creatures.

Not all forms of energy are biologically useful. Any object in motion, for example, has kinetic energy; but there is no known or conceptualized life form that can utilize this energy for its own purposes. We animals use chemical energy from molecular oxygen (O_2), which is produced from CO_2 and water by photosynthetic plants that get their energy from sunlight; in effect, the energy of the sun is radiated across a hundred million miles of empty space, captured by the chlorophyll in plants, converted into chemical energy and stored in oxygen, and later used by us. Oxygen is, of course, not the only chemical compound which can store energy. Hot seawater and basalt (the type of rock that constitutes the seafloor) can produce H_2S, which can be utilized by the bacteria living in and around the deep sea vents to release energy to drive their metabolism.

These bacteria were the third clue. They came flocculating forth from the hot vents in incredible numbers, forming a dense cloud of life in the water hundreds of meters high and kilometers across, and coagulating into thick mats on the ocean floor. At first it was thought that perhaps the eruptions may have pushed out not living bacteria but only chemical nutrients which attracted microbes living in the ambient seawater, but when the microbes at this and other soon-discovered sites were captured by *Alvin* and brought to the surface it was found that they grew and flourished at temperatures greater than 90° Centigrade, and died at temperatures less than 55° Centigrade, so they couldn't ever have lived in the cold ocean waters.

Known as thermophiles (heat lovers; or hyperthermophiles for those who *really* love the heat), these bacteria thrive in temperatures that had been thought to be fatal to life. Indeed, when the temperature of the vent fluids was tested, the results were astounding. In 1979, searching for similar communities, *Alvin* found thick black smoke pouring out of what looked like a smokestack sitting there on the ocean floor. It was the first of what have become known as "black smokers," and the temperature of that black cloud was measured as 91° Fahrenheit, or 33° Centigrade.

That was the highest temperature ever recorded on the seafloor, but when *Alvin* came up to the surface and the temperature probe was examined it was found to be faulty: its plastic tip had melted. When they looked up the melting point of the plastic, it turned out to be 180° Centigrade, higher than the normal boiling point of water. (The water didn't boil because of the high pressure down there.) Further investigations brought back temperatures reaching up to an incredible 350° Centigrade.

A few varieties of thermophilic bacteria had been known before the discovery of the deep-sea vents, but they were thought to be anomalous freaks. As the exploration continued, however, it became abundantly clear that they are not only ubiquitous* but quite possibly the basis of all life on Earth.

Traditional evolutionary theory envisages life beginning in a pleasant, warm little pond. The pond is full of organic chemicals that were either formed on Earth as various forms of energy flashed through the atmosphere, in much the manner of the Miller-Urey experiment, or that rained down on Earth entrapped in meteorites, cosmic dust, and comets. But we mentioned in Chapter 3 that evidence of ancient life forms has been discovered on Earth with an age of about 3.8 billion years, and we know from lunar cratering studies that the moon was bombarded by a rain of meteorites until about that time. The

---

*At the time of writing, eight separate submarine eruptive events have been found and studied during or soon after eruption, and each and every one has produced massive effusions of thermophilic microbes.

Earth's gravity, being larger than the moon's, would have cast a larger net and drawn in even more of these crater-forming meteorites.

There is no guarantee that the oldest fossils we have are actually the oldest that exist; they provide only a lower limit to the origin of life on Earth. It is therefore possible that life formed on Earth before the end of the cataclysmic bombardment and, so the argument presented by John Corliss and coworkers goes, no pleasant, undisturbed, warm little ponds could have existed on the surface of the Earth during this period.

Envisage the Earth as it forms, whirling through the disk of rubble that surrounds the early sun, reaching out with its gravity to suck in the dust, rocks, and clumps of metal, bringing them screaming and burning through the atmosphere to explode with the force of atomic bombs on the ground. The Earth is continually rocked and shaken and cracked open by this furious bombardment. Molten lava streams forth as the interior is heated by vastly more radioactivity than we have today. (Our volcanos and earthquakes are fed by the heat released mainly by four radioactive isotopes: uranium-238, uranium-235, thorium-230, and potassium-40. These have all been fading away since their creation, according to their own individual half-lives. Four and a half billion years ago there was only a little more thorium-230, but there was twice as much uranium-238, almost ten times as much potassium-40, and more than fifty times as much uranium-235.)

The Earth's surface is a boiling, roiling cauldron of flaming lava and exploding meteorites. But far below, in the cracks among the rocks, perhaps there will be places where hot water flows gently, fed by the deep radioactivity but not ravaged by it; quiet oases hidden from the sun and the stars and the cosmic bombardment. Perhaps, Corliss and his coworkers argue, life formed deep in the ocean floor, oblivious to the turmoil above.

6.

One of the most fascinating aspects of science is the way in which ideas and experimental results reverberate through the community. At its most exciting, it's similar to the whispering game children play at parties: everyone sits in a line, and the first person whispers a secret to the one sitting next to her, and she whispers to the next, and so on down the line, until the last person gets the word and announces what it is. And everyone laughs loudly because the secret at the end is usually so different from what it was at the beginning.

When that happens in science it's a joyful and wonderful thing, and that's what we're talking about here. In fact, in science it's even more complex; it's as if we had three or four different lines of people, and there are several whispered secrets that come in from the ends of each of the·lines and meet somewhere in the middle. The first line began with John Corliss and his coworkers who went down to the bottom of the sea in *Alvin* and brought up the word that there was life down there. A second line began at the NASA Ames Research Center in California, when a group led by Ray Reynolds—a theoretical planetary physicist, and one of the nicest guys you could ever hope to meet—began to think about a subject that had nothing to do with oceans or life.

In the March 2, 1979, issue of *Science*, Reynolds and two coworkers published a series of equations describing the gravitational effect of the total Jovian satellite/planetary system on Io, the Galilean moon closest to Jupiter. The effect of the giant planet plus all its moons forces Io into an eccentric (elongated) orbit, and they predicted that this forced eccentricity would lead to tidal heating effects as Io approaches and recedes from Jupiter, and this in turn would be sufficient to melt a major fraction of its mass. They predicted that a likely consequence of such internal melting would be violent volcanism.

Moons do not have volcanos—not because they are different in kind from volcanically active planets but because they are smaller. Being smaller, they have a higher surface-to-volume ratio and so they cool off more quickly, and so any moon that might have been originally hot enough to generate volcanos is no longer so. A similar example is seen on Mars, where volcanos were once active (as we discussed in Chapter 4 in relation to the SNC meteorites), but since Mars is so much smaller than Earth, it has long since cooled off and volcanism is no longer seen. No moon in the solar system is large enough to have held its heat sufficiently to remain volcanic for the four and a half billion years since it was created. But the mechanism discovered by Reynolds and his friends is a continuing, present-day one, and so they predicted on March 2, 1979, that when the *Voyager* spacecraft, which was on its way to Jupiter, reached its target and began to take pictures of the satellite system, its cameras might return "images of Io that may reveal evidence for a planetary structure and history dramatically different from any previously observed."

And just one week later, on March 8, 1979, *Voyager* came to within a few million miles of Io and took a picture that showed a cloud on the surface—but Io has no atmosphere. Further pictures and thermal emission studies showed that what they were seeing was a volcanic eruption bigger than any on Earth, and one by one another six volcanos were spotted in the act of erupting. The average surface temperature on Io is *minus* 145° Centigrade, but the tremendous Jovian tidal forces keep the interior molten and roiling, resulting in continual production of heat and eternal massive volcanism.

The third line of evidence came in when *Voyager* turned its cameras on the other moons of Jupiter. The second one out, Europa, had never been anyone's idea of unique or exciting, but when the *Voyager* pictures came back to Earth the scientists began to look like Charlie Chaplin doing a classic double take. "Here's Io, look at that volcano! Isn't that exciting? And here's

Europa, nothing much going on there. And the next one is Ganymede, and then this next one is—Whoa! What's that on Europa? Let's take another look!"

What was on Europa was, at first glance, nothing. And like Sherlock Holmes's dog that didn't bark, *that* was what was so exciting. Where were all the craters that pockmark every other moon in the solar system? The surface of Europa was, at first glance, singularly bland; on closer inspection it did reveal markings, but of a most peculiar kind. There were long—very long—scratches cutting the moon; to some they were reminiscent of Percy Lowell's Martian canals, to others they looked like satellite photographs of the region around Earth's North Pole. They weren't straight like Lowell's imaginary canals, although they were as great in extent; instead they curved in smooth lines, sometimes even swirling around in near-circles. They did appear, the more one looked at them, to resemble the great cracks observed in polar ice.

And the moon was intrinsically much brighter than our own moon, which meant it was reflecting more sunlight. What could make it so bright, so reflecting? Looking at Mars, the ice caps stand out as exceedingly bright, as do the polar caps of Earth as seen from satellites. No wonder: snow and ice reflect nearly a 100 percent of the sunlight that strikes them, as we all know from the blinding glare of a fresh snow-covered landscape in bright sunlight. The conclusion was obvious: Europa was covered with ice.

That same year Ray Reynolds and his coworkers took the story a bit further. The nearly total absence of craters indicated to them that not only was there ice covering the entire surface of Europa, but that it had to be warm ice. That is, the ice had to be warm enough so that its viscosity was low enough so that the meteorite impacts which had to have occurred on Europa did not leave long-lasting craters. The line of reasoning was that since every solid body in the solar system whose surface we have seen—including our moon, Mars, Mercury, and all the Jovian satellites—has extensive craters, Europa should too. But it doesn't. Earth's craters, we know, have been largely washed away by erosion, sedimentation, and tectonic activity. What could have erased Europa's?

The only reasonable mechanism is for the icy surface to be warm enough that the ice slowly flows into and fills up the craters. This implies that underneath the surface ice is a body of liquid water. This idea was supported by an extension of the thermal calculations just done for Io. The tidal effect on Io is enough to cause the interior of the moon to melt and burst through the surface with active volcanos. The tidal effect on Europa won't be as great, since it's further away from Jupiter, but will still be strong. Combined with heat from radioactivity, the calculations indicated, there would be enough heat inside the moon to melt the ice deep below the surface, where the ice meets the silicate interior; as they put it in their paper, the tidal plus radiogenic heat must have led "to formation of a liquid water ocean overlain by a thin crust of ice." And perhaps it did even more: perhaps there was enough heat to melt the silicate mantle. If it did, the resulting magma flows could lead to the formation of hydrothermal systems at the floor of the Europan ocean. And since we find life on Earth in our own hydrothermal systems, there might well be similar life on Europa. For the first time since Richard Adams Locke's 1835 hoax, described in Chapter 1, we visualized the possibility of life on a moon, and indeed a moon so far from the sun that there could be no liquid water on its surface and little hope of photosynthesis.

That is, *if* the tidal energy is enough to cause melting of the ice and formation of a subsurface ocean, and *if* there are hydrothermal systems on Europa and, finally, *if* life will indeed form wherever such hydrothermal systems exist.

7.

By 1983 Reynolds and his coworkers had published a paper titled "On the Habitability of Europa," which examined these questions and concluded that there was every reason for optimism.* However, their arguments were speculative since no further experiments could be done from Earth to test the hypothesis, and so for a few years the idea lay dormant. Then in 1989 another spacecraft was sent off to take a closer look at Jupiter and its satellites.

Galileo was a working symbol of the 1990s' determination for international partnerships in science (born more of the necessity to convince Congress that we weren't paying for everything alone, than of a sincere desire to share the knowledge and excitement of research). It was launched by NASA in October 1989, in collaboration with Germany, Canada, France, and England. Instead of heading directly out to Jupiter it flew by Venus, taking measurements and photographs as it passed, then used Venus's gravity to whip around back to Earth, which again whipped it around before flinging it off into the further reaches of the solar system. It took the first close-up pictures of asteroids as it passed by Ida and Gaspra, saw the great collision as comet Shoemaker-Levy 9 hit Jupiter, found clouds resembling thunderstorms in the Great Red Spot, and sent a probe down through Jupiter's thick atmosphere to measure conditions at different altitudes. It was programmed to then turn its attention to the Jovian moons, orbiting and taking pictures of each of them in turn, utilizing each moon's gravity rather than its own fuel for the course changes so that it could spend years out there making observations. It would take data continuously for a week or two, then send the data back to Earth while speeding to the next moon.

*Arthur C. Clarke beat them to it. His novel *2010: Odyssey Two*, with the quote that begins this chapter, was written in 1982.

As *Galileo* was on its way, the project manager, Torrence Johnson, went to Seattle to give a talk at the University of Washington. John Delaney is an oceanographer there who has pioneered undersea exploration and was one of the main investigators of the novel life forms seen around hydrothermal vents on the sea floor. He was immediately struck by the possibilities of similar life beneath the hypothesized Europan ocean. He got in touch with one of Ray Reynolds's original team, Steve Squyres, now a professor at Cornell University, and invited him out to Seattle to talk about Europa.

Squyres couldn't take him to Europa, but Delaney could take Squyres down to the bottom of the ocean. The two of them went down to the depths in *Alvin*. Or at least, they started out for the depths, but they ran into technical problems when *Alvin*'s support systems gave out, and the dive was aborted at 600 feet. They came back to the surface, but the ship wasn't ready for recovery, and so they bobbed about in twelve-foot seas for an hour with nothing to do but think either about getting seasick or about the strange life thousands of feet below them and the possibly even stranger life four hundred million miles above them. As Squyres puts it, they sat there bullshitting about oceans and life and Europa, and each fed on the other's enthusiasm for his own specialty, and they both got more and more interested in the possibilities of life on Europa.

The problem was that the astronomers studying Europa didn't know much about the chemistry and biology of the sea floor, and the oceanographers didn't know anything about planetary science. They decided that if they were to make any progress, that was the first situation that had to be remedied. So when they got back to land they decided to organize a conference which would bring these separate groups together. Delaney went to his funding agency, the National Science Foundation, and Squyres went to his, the National Aeronautics and Space Administration, and each of them told their agency that the other agency was interested, and they got joint support for the San Juan Capistrano conference in the fall of 1996.

Intellectual soulmates, the two scientists present a study in physical contrasts. Both are vibrant, young, and impressive in their own ways. John Delaney is a tall, strongly built giant with a leonine head of thick black hair and a trim white beard. He appeared before the assembled congress snazzily attired in a dark blue blazer with white turtleneck, looking more like John Barrymore than Albert Einstein, and took command with his first beaming, confident words. Steve Squyres, smaller and slimmer, slouched to the podium in rumpled jeans and open-collar shirt. If Delaney is a movie star, Squyres is a cowboy. But like Delaney, when he speaks everyone perks up and listens. Each is an expert in his own field, and by now has learned enough about the other's so that they are ideal leaders for this conference and the work to come.

At the time of the conference *Galileo* had not yet reached the Jovian system but had taken its first pictures of Europa from a distance of 155,000 kilometers. Its itinerary would include three close encounters with Europa, and it was expected to send back pictures with much more detail than we had ever seen before. *Voyager*'s closest approach had brought it to a distance of two hundred thousand kilometers from the moon, while *Galileo* would come to within one thousand. Its first close pass was to be in December of 1996. In February 1997 it would come back for another close look, then zoom away for some of the other moons, returning to Europa in November of 1997.

As a consequence, the conference in November of 1996 was idea-rich but data-poor. It served its purpose, however, which was to bring the two scientific communities together, teach each what the other had learned, and spark waves of interest. It also sparked waves of frustration, as we sat there on the precipice of perhaps the greatest discovery in our history, peering over the side, unable to see into the depths of an ocean hundreds of millions of miles away—indeed, one that might not even exist. The questions that bobbed out at us from every speaker were as frustrating as they were intriguing. Is there an ocean under the ice of Europa? Is there heat bubbling up from the interior of the moon? Are there living creatures feasting on that heat, unseen and unknown in that eternal darkness?

From questions of ultimate truth we moved on to questions of technique. Would *Galileo* be able to find a "smoking gun"? Did it have the capability to provide definitive evidence for life on Europa? These were the "real" questions that circulated at the conference, the questions to which reasonable answers could be expected, and the predicted answer to all of them was "No." The spacecraft hadn't been designed to do all that and it didn't have the capabilities to do it. What it might do was find a *lack* of evidence, which would diminish the hopes for life. But as it turned out, it didn't do that, either. Every photo sent back since the conference has only raised more excitement, upping the odds for life out there in the cold and dark depths of space.

Those first pictures which came in during August and were discussed at the conference were greeted by the imaging team as "tantalizing." The long cracks in the surface had been recognized early on as consisting of two dark bands with a white stripe down the middle. These "triple bands" stretch for thousands of miles, dwarfing any feature seen on Earth. The model suggested to explain them had visualized the ice crust being pulled apart by stresses from below, with fresh white ice bulging up and forming the central stripe. This would give sharp cracks, and the edges would be sharp. But the new photos showed that the edge of the bands are diffuse rather than sharp. A new theory was needed to explain the fuzzy edges, and it wasn't long in coming.

In this model the dark fuzzy stuff at the edge of the bands is formed by a line of dirty geysers erupting and throwing out a mixture of ice and silicates from below. This initial eruptive event is then followed by more gentle flowing of ice, similar to rock magma on Earth, which fills in the crack with cleaner, whiter material. These postulated eruptions imply a large supply of heat under the ice, which in turn implies that under the icy crust is a liquid ocean. Or so we thought. But we had to be cautious: the imaging team found that the "new images can offer no solid evidence of such an ocean."

Other details were equally tantalizing. "In some areas the ice is broken up into large pieces that have shifted away from one another," Dr. Ronald Greeley of Arizona State University and a member of the imaging team explained. "This shows the ice crust has been or still is being lubricated from below by warm ice or maybe even liquid water."

Well, that's the rub, isn't it? To be or not to be depends on whether it's warm ice—which would still be below the freezing point of water and would not be a reasonable environment for life—or actual liquid water down there. But that was as much information as those first pictures could give us, and everyone eagerly awaited the first really close pass, which came in December 1996.

When those photos came in, the chances of life on Europa went from "tantalizing" to a "high potential," according to the imaging team. One picture showed a bulging surface feature that was interpreted as evidence for a rising plume of warm icy mush. Other pictures showed cracks in the surface that might have come when a similar rising plume actually broke through the surface in what might be called an ice volcano. This still doesn't prove there is water beneath the ice, and even if there is water at those spots marked by the rising plumes it doesn't mean that an ocean exists; the plumes could be isolated hot spots in a solid ice surface. Or it could be that an ocean once existed on Europa but has since frozen over, leaving behind a few scattered remnants of liquid water oozing out and freezing.

In February 1997, *Galileo* slipped in to within 363 miles of

PERCIVAL LOWELL. [Lowell Observatory Photograph]

Martian "canals" as seen by Lowell.

[Lowell Observatory Photograph]

The first picture from Mars, 1976,
showing the *Viking* lander's foot.

[JPL/NANA]

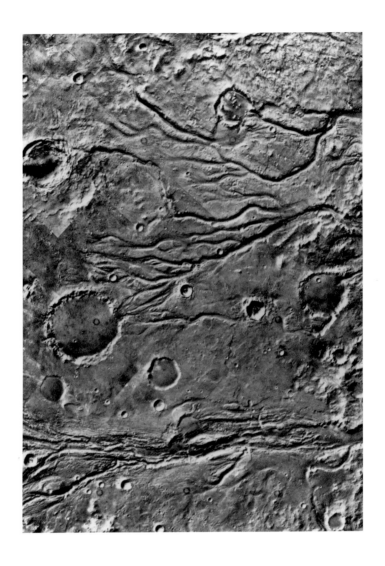

Surface of Mars as seen by *Viking* from orbit,
showing watercarved channels.

[JPL/NANA]

Norman Horowitz.

A typical day during the search for meteorites on the antarctic continent. [*Ancient Life on Mars???* (Lunar and Planetary Institute)]

Meteorite ALH 84001.

[*Ancient Life on Mars???* (Lunar and Planetary Institute)]

Chris Romanek.

DAVID MCKAY (left), KATHIE THOMAS-KEPRTA,
and EVERETT GIBSON.

[NASA]

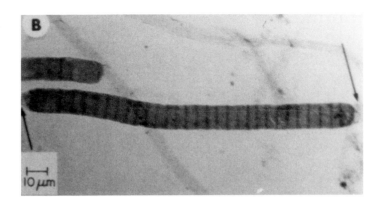

Fossil bacteria, about 950 Myr old, found in Siberia.

[W. Schopf]

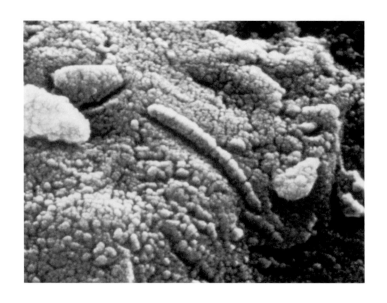

"The First Martian To Be Seen on Earth." See page 119.

[*Ancient Life on Mars???* Lunar and Planetary Institute]

Harry McSween, Jr.

Ed Stolper.

STEVE SQUYRES, at the Capistrano conference.

[D. Nash, San Juan Institute]

JOHN DELANEY, at the Capistrano conference with David Fisher.

[D. Nash, San Juan Institute]

Tommy Gold.

Geoff Marcy (left) and Paul Butler.

PAUL HOROWITZ.

[Mike Blake]

Europa. The pictures were as clear as we get of Earth itself from satellites, and the chances for water and life continued to improve. Icebergs and frozen puddles were seen, and NASA was jubilant. "It's becoming more and more clear that there is a subsurface ocean on Europa, that the ocean is driven by heat, and that there is abundant organic material in the water," they announced. The frozen puddles—smooth, crater-free patches of ice—indicate that the ice crust is thin enough to allow water from below to break through and coat the surface, and the absence of craters on the puddles indicates that they formed recently. There is a lot of argument about what "recently" means, however, with some scientists arguing that they could still be a billion years old, and that therefore the crust may not be thin today. (By "thin" is meant less than about a mile thick. From density and *Galileo*'s gravity measurements we know that the total water layer on the surface is about a hundred miles thick, so an ice layer this thin leaves the oceanic depth unchanged at roughly a hundred miles, compared to an average depth of two or three miles on Earth. The Europan ocean—if it really exists—is massive compared to our own. We may have gone from being the only planet with an ocean to being merely second-best.)

There is still more circumstantial evidence to be found in the pictures from *Galileo*. Some of the icebergs appear to have been twisted loose from the ice pack and rotated, which strongly suggests that they are floating on water. And since there is no atmosphere on Europa, there are no winds to move the icebergs around; the only thing that could do it is a rising convection current in the water below, fed by heat from the rocky mantle.

"We're intrigued by these blocks of ice," Dr. Greeley says. One would think so! NASA immediately announced that it was providing an extra burst of 50 million dollars to extend the *Galileo* mission for another two years. (The spacecraft, by cleverly using the gravity of the Jovian moons to steer itself, has been conserving its fuel better than a Japanese car, and has enough to last several more years. The limiting factor for an extended mission was radiation damage—we don't know how much she can take without giving up the ghost—and money. There's no way to pump money into the spacecraft itself, of course, but that 50 million dollars is necessary to keep the support and receiving systems operating here on Earth.) The hope is if we keep it flying, close passes similar to those already accomplished but coming a couple of years later will show these same icebergs in different positions, proving conclusively that they are moving at the present time rather than millions or billions of years ago, and thus that an ocean exists today under the ice.

The final answer might come sooner. There is always the possibility that during its extended mission *Galileo* might actually see a geyser erupting, just as *Voyager* saw a volcano on its first try at Io. Many of the investigators don't feel it's necessary to wait for that; they feel the present pictures are enough to answer the question.* "These are really mind-blowing," Richard Terrile of JPL said at the press conference which announced the results. "How often is an ocean discovered? The last one was the Pacific by Balboa." Torrence Johnson, the mission's chief scientist, went right past the ocean question to that of life, pointing out that the pictures so far provide "no evidence directly bearing on life" on Europa. But John Delaney was just bursting with optimism. "The answer is, I'm sure there's life there," he announced.

Yes. Well, maybe. Delaney's argument is that on Earth we

*In the fall of 1997 another piece of data was reported to the American Geophysical Union: *Galileo* found traces of a magnetic field in Europa. It is too weak to be accounted for by a molten iron core, as in the Earth. What is needed instead is a less-conducting liquid, and a model proposed independently by David Stephenson of Cal Tech and Fritz Neubauer of the University of Cologne suggests that a salty ocean might be just the thing. Experimentalists now claim an "impressive fit" between the predictions of the model and the actual data sent back by *Galileo.*

have found life every single place we've looked that had a suitable environment, and now with strong evidence for water and heat on Europa the environment looks suitable. But there is still the question of whether life formed on our own planet as an event unique in the universe, or whether life forms wherever and whenever it can. This is really the crux of the whole matter.

8.

To our mind, the most important point to come out of all these studies so far was best expressed by Tommy Gold: As long as life was thought to need photosynthesis, the Earth was unique among all the bodies of the solar system. It was (and is) the only body with liquid water on its surface, and therefore it was a uniquely suitable habitat for life. This meant that we were unlikely to answer the foremost question in evolutionary theory, expressed in this diagram:

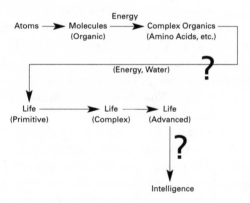

The diagram shows the total chain of evolution. Simple atoms combine into molecules. Given a source of energy and water as a solvent, organic molecules suitable for life will form.

Given enough time, and the continued presence of water and a suitable energy flux—one neither too traumatic nor discontinuous—these molecules will form a living system. This first, simple living organism will mutate into more complex forms in response to environmental pressures, and eventually will form conscious, intelligent creatures.

That is the theory in a nutshell. Most of it has been tested and has passed the tests successfully. We know that atoms will combine into molecules, we know that organic molecules of suitable complexity can and have been formed in the laboratory, in a water solution with various forms of energy passing through the solution. We know from the fossil record that life on Earth began as simple creatures and evolved progressively into more complex forms.

All this we know beyond reasonable doubt. But there is a large question mark over the stage in which life forms, reflecting our ignorance of this basic point. Is life a miracle, a once-in-a-universe happening? Or, given the right conditions, will life form spontaneously?

As long as the right conditions were to be found nowhere else in the solar system except our Earth, there was little chance of testing this point. We could always produce the physical and chemical conditions in the laboratory but we could never give the experiment enough time, for it is likely that millions of years are necessary. Now, however, we know that life can exist without photosynthesis, which means that it is no longer necessary to consider only the surfaces of planets as suitable locations. And that in turn means that Earth is no longer unique: other planets and even their moons may have liquid water trapped beneath their surfaces, together with the requisite chemical species and sufficient thermal energy to breathe life into those chemicals.

This does not mean that life exists throughout the solar system. It does mean that if it does not, we are going to have to re-think everything we think we know about evolutionary processes. It opens up a century of exciting science: first to explore the other planets and moons to ascertain which of them have the proper conditions for life, and then to search and either find life there or convince ourselves that it is truly absent. Whichever way the facts fall, we will learn more about the nature of life than any other series of experiments or philosophical arguments could possibly tell us.

Finally, there is one other question mark in that diagram. It points to the evolution of consciousness, of intelligence, and brings us to what for many people are the most intriguing questions of all: Does intelligent life exist elsewhere (or anywhere) in the universe? Is it a natural consequence of the evolution of living systems, or is it unique? If it exists elsewhere, how can we make contact with it? Do we *want* to make contact with it?

It sounds like science fiction, but it is not. At this very moment, at various spots around the world, serious scientists are listening for messages from extraterrestrial supercivilizations. Their methods are based not on hunches or intuition, but on high-level astrophysics and mathematics.

It all began with the Wizard of Ozma.

TWO

Distant Stars

The known is finite; the unknown infinite.
THOMAS HENRY HUXLEY

6 | The Princess of Oz

I think it's part of the nature of man to start with romance and build to a reality.
RAY BRADBURY

1.

On April 8, 1960, Frank Drake set his alarm for 3:00 a.m. He probably didn't need it. Just turned thirty, the young astrophysicist was about to fulfill a lifelong dream. He walked across the National Radio Astronomy Observatory campus where he had lived for two years, through the cold fog of the West Virginia night to the eighty-five-foot bowl that was the main telescope. He was about to listen for communications from alien civilizations.

With two student assistants watching, Drake rode up the equivalent of five stories in an open-air elevator called a telescoping tower to tune the special amplifying equipment housed in a small metal canister suspended over the dish. "I must have spent at least forty-five minutes inside this glorified garbage can," he remembered decades later, "lying on my side, tilted at the same odd angle as the dish, and twiddling the controls."

By 5:00 a.m. Drake and his assistants were in the control room getting ready, preparing the signal processing equpiment: one simple spectrometer with a moving pen that would translate the signal picked up by the telescope into a series of peaks and valleys corresponding to signal strength. When they had everything checked out, they pointed the enormous dish in the direction of the star Tau Ceti, one of our relative neighbors in space, twelve light-years away. Drake had selected Tau Ceti and another neighbor, Epsilon Eridani, as targets because of their proximity, their similarity to our sun, and their likelihood for possessing planetary systems (though at that time there was no direct evidence of planets around any star except our own).

"And then," he said, "there was nothing to do but wait."

2.

Seven months before, in September 1959, the British scientific journal *Nature* had published the first serious article about the possibilities of listening for extraterrestrial communication. In "Searching for Interstellar Communications," Giuseppi Cocconi and Philip Morrison, two Cornell University professors, revived enthusiasm for Marconi and Tesla's idea that we might communicate with extraterrestrial civilizations via radio waves.

Just a few months earlier, communication with aliens had been the farthest thing from their minds. Morrison was a theoretical physicist who had studied under J. Robert Oppenheimer at Berkeley and during the war years worked on the Manhattan Project at Los Alamos, New Mexico. In 1945 he had ridden in the back seat of the car carrying the atomic bomb's plutonium core into the desert for its first successful test. At Cornell in 1959 he was working with the Italian physicist

Cocconi on the possibilities of gamma-ray astronomy, when one day Cocconi walked into his office. Cocconi and his wife, also a Cornell physicist, had been musing about the possibility of using gamma rays for communication across space. What do you think, he asked Morrison, about the idea of civilizations communicating across the galaxy* using gamma rays or other radiation? Soon after, they rendezvoused in Geneva, where Cocconi was on sabbatical, and quickly hammered out the paper that would become the seminal work in a new field.

Their thinking began with the simple question: assuming that other beings exist elsewhere in the galaxy, what would be the best way to communicate with them? Travel by spaceship would be too slow and expensive. Signals sent by visible light would be overwhelmed by light from stars. After going "systematically through all the possibilities," they concluded that radio signals were the most logical medium of interstellar communication. Such signals travel at the speed of light, which Einstein established as the universal speed limit—nothing can move faster.†

Electromagnetic radiation is a natural phenomenon throughout the universe, and so any advanced civilization would undoubtedly have developed a means of propagating radio waves just as we have. Our own capabilities in radio astronomy had by then developed to the point where it would be possible to detect radio transmissions from civilizations that possessed even our own admittedly rudimentary level of equipment. And the cost, or energy required, to send such transmis-

*We are about 30,000 light-years from the center of our galaxy, where most of its hundred billion or so stars are concentrated. Other galaxies are typically millions to billions of light-years away, so although intelligent life might exist anywhere in the universe it makes sense to think first about our own galaxy.

†Radio waves are one part of the electromagnetic spectrum, distinguished from gamma rays, x-rays, and visible light only by their longer wavelengths (or lower frequencies).

sions out across the universe is infinitesimally small in comparison to that of sending actual spaceships between stars.

At first, though, the task of tuning in might seem impossible: which of the infinite numbers of different frequencies* should one investigate? Cocconi and Morrison went so far as to recommend a channel for optimal reception. To begin with, they ruled out very short frequencies, as these would be absorbed by the interstellar medium and the Earth's (or a similar planet's) ionosphere. On the other hand, water vapor molecules tend to absorb radio waves of much higher frequencies, and so our own atmosphere—and possibly that of any inhabited planet—would black out these frequencies. They concluded that we should be looking at frequencies lower than about 10,000 MHz but higher than than about 20 MHz. (Frequencies are measured in oscillations of the radio wave per second, and the unit is called a hertz, after the nineteenth-century German physicist, Heinrich Hertz. A megahertz, or MHz, is a million oscillations per second. Visible light has a range of frequencies just below a billion MHz; your AM radio dial spans from 0.540 MHz to 1.6 MHz, and FM goes from 88 MHz to 107 MHz.

But there are other reasons besides absorption to concentrate on this part of the radio spectrum. All the matter in the universe radiates energy; the frequency is determined by the molecular or atomic makeup. This background noise—both from the distant universe and our own atmosphere—makes it almost impossible to search for signals at certain frequencies. Any civilization advanced enough to be trying to contact other civilizations would realize that choosing a frequency in the noisy part of the spectrum would be like setting up cellular phone channels in the middle of the commercial radio band.

*Or wavelengths. Electromagnetic radiation frequencies are related to their wavelengths by a simple equation: *frequency* times *wavelength* equals the speed of light (3×10^{10} cm/sec).

If we plot the amount of each type of background radiation at each frequency, we can see where it is noisiest and where it is quietest.

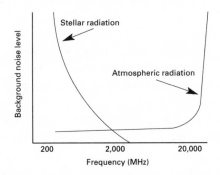

The area where both terrestrial and galactic background noise are quietest happens to correspond roughly to the window of frequencies that get through our atmosphere without being absorbed. Considering both criteria yields a reasonable spectrum for interstellar communication: from about 1000 to 5000 MHz. It makes perfect sense from both points of view—broadcaster and listener—to use this chunk of the spectrum, but Cocconi and Morrison were even more specific. (They needed to be: even between 1000 and 5000 MHz, there are about four billion broadcast channels, assuming 1-Hz-wide channels.)

To choose a particular frequency they turned to a clever bit of physics published in the 1940s by a young Russian physicist, Iosef Shklovskii, who predicted the existence of a "hydrogen line" at 1420 MHz: a background hum of radiation corresponding to a previously unobserved ground-state transition of hydrogen, the most abundant element in the universe (details are given in the Notes). Cocconi and Morrison suggested that any civilization as advanced as our own must have noticed this natural "landmark." Perhaps, they reasoned, 1420 MHz is the standard broadcasting channel for interstellar communication. Other civilizations may have been chatting on this frequency for billions of years. All we need do is tune in to join the party.

On his own, Frank Drake had already come to the same conclusions as Cocconi and Morrison and was planning an actual search, something they had no intention of doing. Drake was still a child when he began to be obsessed with thoughts of alien civilizations. Like so many people throughout the millennia, Drake "could see no reason to think that humankind was the only example of civilization, unique in all the universe." As an undergraduate at Cornell (the university that would become the breeding ground for extraterrestrial intelligent-life studies), he was inspired by a series of guest lectures by the great astrophysicist Otto Struve.

Struve had just worked out a technique of using the measured Doppler shift of individual stars to determine their rates of spin. If a star is spinning so that Earth is in the plane of its equator, half of it is spinning away from us while the other half is spinning toward us. As seen from above:

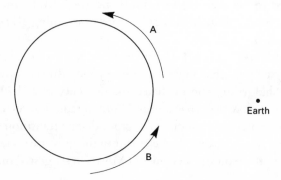

The half of the star represented by the curved arrow *A* is moving away from the Earth, while that represented by *B* is moving toward us. When light is emitted by a shining object such as a star, it comes out in a series of discrete frequencies known as a "line spectrum," with each frequency observed as a single line. If the source of light is moving with respect to the observer, the frequencies are shifted: when the source is moving away from the observer (arrow *A*) the light is shifted toward longer frequencies, and when it is moving toward the observer (arrow *B*) it is shifted toward shorter frequencies, with the magnitude of the shift proportional to the velocity of approach or recession. Struve suggested that by measuring this "Doppler shift," one could accurately determine how fast the star is spinning.

A series of such measurements gave a curious result: most stars could be grouped into two categories of spin velocity. Very large stars, much larger than the sun, spin very rapidly; smaller stars spin more slowly. The two groups, he claimed, are clearly delineated; very few stars fall in the middle.

Struve's explanation, with which he enthralled the young Drake, was that the smaller, slower-spinning stars had planets. Stars are formed from enormous cloudy maelstroms of gas and dust that slowly congeal due to their own gravitational pull. According to a universal law, the angular momentum, or "total spin," of each protostellar cloud must be preserved. This is the same law that allows a skater to spin faster by pulling her arms in close. In the same way, if a star forms with planets circling it in a wide orbit, it will spin slower, while if it forms without planets—with all its mass pulled in close to its center—it will spin faster.

There was no proof that Struve was right, but it seemed to make sense. "In the space of a few moments in the lecture hall," remembered Drake, "Struve had raised the number of planets in the galaxy from the nine we knew to more than ninety-nine billion." He went on to speak the very words that

Drake was thinking: the prevalence of planets means the prevalence of life. After all, the organic molecules found on Earth are present throughout the universe. If, then, there are billions of planets orbiting around stars similar to the sun, how could it be that these molecules would develop into life only here on Earth, a typical planet around a typical star located out in an unremarkable spot in the Milky Way?

A few years later, as a graduate student at Harvard, Drake first thought about using radio telescopes to search for alien signals. Sitting alone at the controls of Harvard's Agassiz Station telescope out near the town of Harvard, Massachusetts, twenty-six miles west of the university, Drake had plenty of time to daydream. For his doctoral work, he was using the sixty-foot radio telescope to measure the radio spectrum of the Pleiades star cluster. While waiting for his data to collect, however, he would make idle calculations about how sensitive a telescope would be needed to detect an alien transmission of what strength and where might be the most likely direction to point one's telescope.

One night in 1956, his fanciful ruminations were interrupted by a heart-stopping aberration in his Pleiades data. There, in the midst of otherwise routine signals, as if in answer to his daydreams, was "a strikingly regular signal—too regular, in fact, to be of natural origin." He had already thought of the 1420 MHz hydrogen frequency as a promising channel for extraterrestrial communication, and this signal was right in that region. What's more, the source of the signal was moving at the same velocity, with the same red shift, as the Pleiades. Drake was stunned:

> I was twenty-six years old that night. I'm past sixty now, and I still can't adequately describe my emotions at that moment. I could barely breathe from excitement. . . . What I felt was not a normal emotion. It was probably the sensation people have when they see what to them is a miracle:

you know that the world is going to be quite a different place—and you are the only one who knows.

Hands trembling, Drake proceeded to move the telescope to point in a different direction, just to make sure the signal really was from the Pleiades. The signal continued. It wasn't from outer space after all. It must have been a terrestrial source, probably military, since 1420 MHz was a frequency never assigned a civilian license due to its great importance in radio astronomy.

Drake dropped into his chair in disappointment, "sweating and shaking." He would not be the last pioneer in this field to be fooled by zeal, naivete, and a suspicious radio signal.

3.

Two years later he was at Green Bank as one of an inaugural staff of three radio astronomers at the new National Radio Astronomy Observatory. While he helped oversee the construction of three radio telescopes, he discreetly continued to plot how he might use the facility to further his search for extraterrestrials.

Much in the same way as Cocconi and Morrison were thinking at Cornell, Drake calculated how powerful a receiver he would need to detect a transmitter of the same size located in another solar system. In fact, an alien civilization transmitting signals would almost certainly be far more advanced than we, since our level of technology is the bare minimum necessary. The civilization we hear could be anywhere from our own level to—literally—billions of years ahead of us. Since their transmitters might very well be orders of magnitude more powerful than ours, it might be far easier to hear them than we thought. But to be safe, it seemed reasonable to assume equal technologies.

The twenty-five-foot telescope Drake had used at Harvard

to observe the Pleiades failed that test: it could have detected
its twin's emissions at a distance of only two light-years, too
weak to reach the stars. At Green Bank, however, an eighty-
five-foot dish was going up immediately (in advance of larger
dishes of 140- and 600-foot diameters), and this put the theoretical radius of communication over the "escape" threshold.
The eighty-five-foot dish could hear signals propagated by a
similar dish up to twelve light-years away, and within that distance lay several stars similar to the sun, which might possibly
have planets of their own.

At a small country diner in a neighboring town, Drake finally sprung his plans on the observatory's interim director,
Lloyd Berkner. To Drake's surprise, Berkner jumped at the opportunity and authorized him to proceed. Project Ozma
(named for the princess in L. Frank Baum's *Wizard of Oz* series) was off the ground.

Drake designed receiving equipment to analyze signals on
the 1420 MHz hydrogen frequency for two reasons. First, as
mentioned before, it was an obvious possibility for a universal
communication channel. Second, it was an important frequency for other radioastronomy research, and so Drake could
argue that the equipment he was building—even at the paltry
cost of $2,000—wouldn't go to waste if Ozma proved fruitless.

He received another lucky break when none other than Otto
Struve, the astronomer who had so excited Drake at Cornell
with his declaration of the probability of extraterrestrial civilizations, was appointed as new director at Green Bank. With
a true believer in charge of the observatory, Drake sped ahead
with his preparations. So unprecedented was this type of project, though, that he remained apprehensive about how it
would be received by the scientific community. Consequently,
he kept it under wraps.

Then, in September 1959, months into his work on Ozma,
Drake picked up the new issue of *Nature* and read the
Cocconi-Morrison paper. Completely independently, they

had made the same calculations as he, come to the same conclusions, and even settled on the same 1420 MHz channel as a
good starting point. Drake had picked it mainly because of its
uses in other radioastronomy projects; he was "being frugal,"
as he put it. Cocconi and Morrison, however, were speaking
theoretically, not practically. For them, the 1420 MHz line was
a poetical signpost, a landmark, or "cosmic water hole," where
all sorts of intelligent beings throughout the universe might
agree to meet.* Drake seized on the coincidence, yet another
lucky break. Many newspapers carried the story that scientists
were ready to search for signals from aliens, and when they
turned to look at his work, it would look all the better that he
was on the right frequency.

Otto Struve made sure that Ozma would be seen. More
hardened than Drake by a long career of competitive science,
he was incensed that Cornell had beaten Green Bank to the
punch. To minimize the damage, he devoted an entire lecture
at MIT soon after the *Nature* paper to "the search for extraterrestrial intelligence as it was proceeding at the National Radio
Astronomy Observatory."

This put the pressure on Drake, but he proved up to the
task. With the help of a few colleagues he designed the receiver, which would look at a single [100-Hz-wide] channel at
a time. After hearing of Struve's announcement of Ozma,
Dana Atchley, Jr., president of an electronics firm near Boston,
called up to donate a state-of-the-art amplifier that would increase the telescope's sensitivity manifold.

It was this amplifier that Drake had to ascend five stories
to tune on the inaugural morning of Project Ozma. Then, as
described earlier, he pointed the telescope at Tau Ceti and
waited. The telescope had an equatorial mount, so it could

---

*In fact, the term "water hole" is often used to refer to the region around
1420 MHz, specifically 1420–1700 MHz. Not just for the oasis metaphor,
but because this is the chunk of spectrum bordered by the frequencies of
hydrogen and the hydroxyl radical (OH), which together form H_2O.

counter the Earth's rotation and remain fixed on a star as the Earth turned. By the time Tau Ceti set in the west around noon, Drake and his team had heard nothing.

The next night, with somewhat less bated breath, they moved the telescope to Epsilon Eridani, their second target. Within five minutes Drake was experiencing the same rush of excitement he'd felt at the Harvard observatory four years before. This time, though, he had several colleagues bouncing around the room with him. The spectrometer needle was jumping like a Geiger counter over Hiroshima, and the loud-speaker Drake had hooked up to auralize the signal was sounding off like a fire alarm. The pulse was powerful, rapid—eight beats a second—and unlike anything normally encountered in astronomy. It was just what one might have expected from an intelligent broadcaster.

They moved the telescope away from Epsilon Eridani, and, unlike Drake's Harvard experience, the signal immediately stopped. It must be coming from that star.

But then they turned back to Epsilon, and the signal failed to return. All day they tried desperately to find it, searching that exact channel and scanning around it in 100-Hz increments, but they were met with silence. After several nights of this, the initial thrill had worn off. "I would find myself thinking almost fondly at times of retuning the parametric amplifier," wrote Drake, "as it would get me outdoors and give me something to do."

Then, on day ten, the signal returned. This time, however, there were no whoops and shouts of joy and astonishment. For one of the ways they had killed time waiting for it to return was to install a screening device in the form of a regular terrestrial antenna connected to a horn. And now their "alien" signal had returned as a double blast, sounding eight shrieks a second through both the telescope's horn and the terrestrial one. The signal was from life on Earth, probably a nearby aircraft.

Project Ozma ran for another couple of weeks, took off a

month to free the telescope for other work, and then ran for another month. It detected no other signals. But Ozma was far from a failure. It had taken the first step, set the stage for future searches. And though it hadn't found extraterrestrial life, it was not without results.

For one thing, the Ozma experience showed the necessity of multiple antennas for verification. Another antenna to check for ordinary terrestrial signals would become routine equipment in future searches. The idea of "piggybacking"— using equipment that has other radioastronomical uses— would also become standard procedure. Furthermore, Drake learned a valuable personal lesson about his new endeavor. After the first few hours of a new search, the excitement quickly faded. Aside from keeping the equipment running, there was nothing to do. A career of eternally waiting for a message that might never come could drive a person mad. He realized very soon, as others would, that an astronomer would have to combine the search for extraterrestrial signals with other work that provided more immediate gratification.

Ozma also taught a valuable lesson about false alarms. Frank Drake had been fooled twice now; he wouldn't let his eagerness overcome him again. Other researchers would suffer similiar false triumphs, but not he. He was learning that if we wanted to communicate across the galaxies, we were in for the long haul; success wasn't going to come in the first fraction of the first step of the marathon. Even back then it must have been apparent how rudimentary Ozma was. Of the billions of channels extraterrestrials might use, they had looked at only one. A supercivilization on Epsilon Eradani might very well have been broadcasting a welcoming message to Earth, their neighbor whose radio and television broadcasts they had been detecting for decades. But if they were broadcasting at 1500 MHz, or any other frequency except 1420 MHz, their message would have bounced right off the Green Bank valley along with the sunlight and the moonlight. Or if the eighty-five-foot telescope was just a tad too weak, the message would have fallen on hard-of-hearing ears. After the equivalent of twenty-four Earth years had elapsed, enough time for their signal and a reply to have traversed the twelve light-years between them and Earth, they would have had to conclude that the Earthlings were still plodding along the road of progress, stuck somewhere between *I Love Lucy* and interstellar communication.

Project Ozma, however, was a crucial beginning. Simply by making an attempt, it showed that an attempt could be made. As Cocconi and Morrison had written, "The probability of success is difficult to estimate, but if we never search, the chance of success is zero." The publicity that it generated attracted other astronomers who together would engender a new subfield of astronomy.

4.

The very next year, a group of nine men convened in Green Bank for the first official conference in this new field. Drake had received a call in the summer of 1961 from J. Peter Pearman, an officer at the Space Science Board of the National Academy of Sciences (NAS), who suggested that they bring together all the scientists who had been working on extraterrestrial life for a brainstorming session. As it turned out, there were only ten such people to invite. The group that arrived in Green Bank that Halloween formed one of the smallest yet most diverse scientific conferences ever. Of course there was Pearman, an erudite Englishman, and the thirty-year-old Drake. The youngest member was twenty-seven-year-old Carl Sagan. Sagan had written to Drake a couple of years before when he was doing his doctoral work on the atmosphere of Venus, which Drake was also studying. But Venus was just a side interest for Sagan, as it was for Drake. Sagan's true passion was the consideration of life on other worlds, and at twenty-seven he was already a recognized expert in the field—a member of the Space Science Board's Committee on Exobiology and also of the NAS's Panel on Extraterrestrial Life. Sagan was currently at Berkeley, working with famed chemist Melvin Calvin on studying the origins of life on Earth. Calvin was rumored to be a top contender for that year's Nobel Prize in Chemistry for his work on photosynthesis in plants. Pearman and Drake invited him to Green Bank, too, for good measure.

There was John Lilly, an M.D. neuroscientist who looked like a movie star and had made his name by studying a known, supposedly intelligent, nonhuman species: dolphins. There was Barney Oliver, the imposing, garrulous vice president of Hewlett-Packard. An accomplished inventor and physicist with a lifelong love of science fiction, Oliver had received a Ph.D. from Cal Tech in 1940 at the age of twenty-four and worked on radar for Bell Labs during the Second World War; he had also written a brief paper on the possible use of radio to contact other civilizations. During Project Ozma, Oliver had flown down to Green Bank in a colleague's private plane to see the proceedings for himself.

And then there was Dana Atchley, the electronics entrepreneur who looked like an electronics entrepreneur (and who had loaned the parametric amplifier for Project Ozma). Morrison and Cocconi were on the list, but only Morrison could attend (Cocconi was busy in Europe). Otto Struve, Drake's boss, rounded out the group, and he asked that they also invite his former student, Su Shu Huang, of the new National Aeronautics and Space Administration (NASA), with whom he had worked on identifying stars that might have planetary systems.

When the three-day conference opened on the morning of November 1, Drake stepped up to the blackboard and wrote an equation:

$$N = R f_p n_e f_l f_i f_c L$$

It was the product of his musings over the past few weeks, and it represented the agenda of the meeting. It would later come to be known as the Drake equation and would be presented in any work dealing with intelligent life in the universe. Today, in the very spot where that blackboard hung, there is a plaque with the Drake equation to commemorate the moment it was first presented.

The N on the left side of the equation was the answer to the puzzle: the number of detectable civilizations in the universe. On the right side were seven factors relating to it: the average rate of star formation R, the fraction f_p of stars that form planets, the number n_e of planets in each system with conditions favorable to life, the fraction f_l of those planets that actually do develop life, the fraction f_i of planets with life where intelligent beings evolve, the fraction f_c of those intelligent civilizations that develop means of interstellar communication, and the average lifetime L of such a civilization. If the conferees could determine the value of each factor, they would know just how many hopeful correspondents were out there waiting for us to join the club.

There is one little problem with the Drake equation: it is impossible to solve. Each factor carries a wide range of probabilities and is open to a great range of interpretations. The men sitting around the conference table in Green Bank, however, were more qualified than anyone alive to interpret them. They sunk their teeth into the equation with relish.

For the first factor, R, the astrophysicists in the room came to a quick agreement. They felt confident that there are about ten billion stars in our galaxy,* and that the average lifetime of these stars is about ten billion years, so the average star formation rate was about one star per year.

From there on it got trickier. In fact, each succeeding factor in the Drake equation is more difficult than the last to assign a value to. Beginning with the second one, the percentage of

*Current extimates are closer to 100 billion.

stars with planets, the panel found themselves disagreeing. Struve, who had invested so much time in studying the spin velocities of stars, declared that stars are rarely formed without some sort of celestial companion. He guessed that half the stars are binaries (pairs of stars), and that the other half have planets. The value of f_p, then, was ½. Morrison, though, pointed out that stars might have different sorts of bodies orbiting them, such as asteroids, too small to develop life. His guess for f_p: ⅕. The group decided that they would use these figures as limits, but they were clearly speculative. Sagan in 1966 adopted a value of one for f_p; many other scientists thought the value could turn out to be virtually zero.

For the next question, how many planets in a solar system might be suitable for life (n_e) Huang suggested a value of one. He and Struve felt that ours is a typical solar system, and we know that the value of n_e for our system is at least one. (As we know from our discussion of Mars and Europa, we may soon learn that our n_e is greater than one.) Sagan pointed out that a planet needn't necessarily orbit its star at what we would call a comfort zone—that is, neither too far away where it would be too cold, nor too close where the heat would be prohibitive to life. This is because of the greenhouse effect, in which a planet's atmosphere lets in the sun's ultraviolet radiation but then traps the energy re-emitted from the planet as infrared heat. For instance, a planet with an atmosphere rich in carbon dioxide (as many people know, CO_2 is an effective greenhouse gas which threatens to cause too much of a good thing here on Earth) might have a similar surface temperature as the Earth even though it had an orbit much further away from its sun. This widens the scope of possible "comfort zones," and so the Green Bankers decided on a value for n_e between one and five.

Sagan and Calvin led the discussion of f_l, the fraction of these planets where life actually developed. They had shown in the laboratory, they felt, that in an environment where life could arise, it inevitably would. The distribution of molecules in the early Earth was nothing unusual in the universe. The presence of ultraviolet radiation to provide the initial burst of energy would certainly be present on any of these planets orbiting a star. The fact that organic molecules were generated in the laboratory under these circumstances, coupled with the fact that life apparently arose on Earth almost immediately, led them to believe that the development of life was the most routine of processes. The group agreed and adopted a value of 100 percent, or $f_l = 1$.

Estimating the probability of intelligence evolving was an even more difficult task. It had occurred on Earth, of course, but if it hadn't there would be no one to debate the question. (The same reasoning applies equally well to the discussion of f_i.) Of course, there is a whole range of intelligence among animals on Earth, but in this context intelligence refers to the ability to communicate in a complex way and to build a civilization. John Lilly regaled the group with tales of his work with dolphins, arguably another example of intelligent life on Earth. He played tapes he had recorded of dolphins communicating with each other by whistling through their blowholes in a great range of frequencies (some out of human hearing range). Dolphins could never develop technology with only flippers and their mouths to work with, but that capability had nothing to do with f_i. Considering, then, what a great advantage intelligence is in the process of natural selection (at least until a species develops nuclear weapons and the concomitant ability to destroy itself, as Sagan pointed out), and the fact that two intelligent species had developed on Earth,* the group concluded that the development of life would always lead to intelligence, or $f_i = 1$. (Other workers, including Sagan, have

---

*The intelligence of dolphins has been overestimated. Although at the time the group was "quite enthusiastic" about it, according to Phil Morrison, "within a few years the subject had pretty much dissipated, and Lilly's work was not found to be reliable."

argued that this is unrealistically optimistic and have suggested lower values, somewere around $^1/_{10}$; still others feel it might be zero.)

That was as far as the conference got on the first day, and in the early morning of the second day the news came from Stockholm that Melvin Calvin had won the Nobel Prize. Drake had been forewarned by Pearman, and champagne was waiting in the cellar. It was before five in the morning, but everyone got out of bed to drink a celebratory toast to their colleague. "This may sound crazy," wrote Drake, "but sharing that moment with Calvin was almost like sharing the prize itself. I felt somehow that our group and the topic we were discussing had suddenly been raised to a higher credibility level" just by their association with the Nobel Prize winner. All the papers the next day would report that Calvin had been reached for comment from Green Bank, where he was attending a meeting on the prospects for intelligent life elsewhere in the universe.

When the meeting got back under way later in the day, in between phone calls from the media, the group tackled f_c, the question of how likely an intelligent species was to develop the means and the desire to communicate with other worlds—quite a different matter from determining the birth rate of stars or evaluating the probability of life emerging from organic molecules. Here the committee was in the realm of fantasy and conjecture. The greatest drawback in evaluating any of the equation's elements is the tendency to extrapolate from the example of life on Earth. In trying to determine the likelihood of life elsewhere, our first clue is to observe that life arose on Earth. But of course it did. Any being in the universe tackling the same question would begin with the same observation; where life did not arise, there is no one to wonder about these things. In predicting the probability of intelligence, we can only look at ourselves for inspiration. To think one could come up with an accurate number seems quixotic. Yet the Green Bank group had boldly accepted a figure of 100 percent: all worlds with life would produce an intelligent species.

Even more recherché is the value for f_c. We know of one planet that produced a race with the technological means to communicate across the cosmos, but we have no more examples of planets that did or did not. The best assumption seems to be that, given enough time, most intelligent life-forms in the universe would discover the same basic physical principles that we have, and that this knowledge would eventually lead to the ability to propagate radio waves as a form of communication. On the other hand, humans have just developed this capability in the last one hundred years, after millions of years of evolution, in the last few billionths of geological time. The calculation of f_c is a hazardous guess at best. The great minds assembled at Green Bank, though, felt that a conservative estimate was that 10 to 20 percent of intelligent species would develop radio technology and care to use it to probe the cosmos. Their value for f_c, then, was $1/10$ to $1/5$.

And then there was L, the final and perhaps most speculative item in the equation. It was also the most ominous. For L represents the average lifetime of a technological civilization. And if we are once again to look to ourselves as an example, as we must, then we come face to face with the specter of self-annihilation. Presumably, as in our case, a civilization creates the means for such total suicide around the same time it develops a means of electromagnetic propagation. It is not impossible that, as a general rule, self-destruction occurs soon after it becomes possible. On the other hand, life as we know it has shown nothing if not a remarkable tendency to survive and adapt. So it is also a considerable possibility that civilizations generally find a way to avoid nuclear holocaust—that they develop peace as a survival mechanism.

Sagan saw the value of L as one of these two extremes. "If [a technological civilization] survives more than 100 years," he wrote, "it will be unlikely to destroy itself afterwards. In the latter case, its lifetime may be measured on a stellar evolutionary timescale (L much greater than 10^8 years)." The other causes of annihilation—collisions with asteroids or other celestial bodies, and the eventual death of the home star—are not likely to occur, on average, for millions of years. And the longer a civilization survives, the more likely it is to develop the means to survive such catastrophes. If Sagan was right, then L is by far the most influential factor in the Drake equation, altering the final result by more than six orders of magnitude.

The Green Bank conferees came to the same conclusion. The other factors all had ranges between $1/10$ and 5; they approximately canceled each other out. Drake wrote a new equation on the blackboard: $N = L$. Taking our current technological lifetime—the fifty years since we developed the atomic bomb—as a lower limit and 10^8 as a conservative upper limit would lead to the conclusion that there are between 50 and 100,000,000 advanced civilizations in our galaxy. Not quite ready to believe that our own civilization might come to an end the next day, though, they plugged in a rough number of one thousand years for the lower limit of L and thus suggested that there are at least a thousand alien civilizations.

It was a stunning conclusion. Astronomers had previously mused to themselves about how there must be other beings out there in the inconceivably vast universe. But now there was a number to the dream, arrived at by cold scientific calculation, however much estimated. The spread in that final number is terribly large, but so is the number itself, and so the Green Bank conference had established at least one hard fact: that the search for such civilizations was a reasonable scientific undertaking. How exactly to go about it was another question. But with such a sober technical analysis to back them up, they would never have to worry about charges of whimsy, fanciful notions, or of wasting money on what belonged in the realm of science fiction. Or so they thought.

Of course, the longer we survive as a species, the more we raise the lower limit for L. As Sagan put it, "For the evaluation of L there is—fortunately for us, but unfortunately for the discussion—not even one known terrestrial example." The ending of the Cold War and subsequent diminishment of the threat of nuclear war that hung over our heads for 45 years appear to have significantly increased the chances for a large L. Another fifty years and we'll have passed Sagan's critical age. So far we are justifying the optimism of the 1961 Green Bankers.

The conferees spent the remaining hours after the "solving" of the Drake equation immersed in specific practicalities and grandiose daydreams. They discussed what sort of frequencies should be monitored, which stars should be targeted first, and how it might best be done—with telescopes dedicated to the purpose or with a larger number of "piggybacking" experiments on general-purpose telescopes. They also imagined together the nature of the interstellar communication that might ensue.

First, say, the recognition of an intelligent signal from a thousand light-years away. A quick reply from Earth—"Message received!"—and then two thousand years to try to decipher just what the first message was saying, what stores of unimaginable information it might contain. A whole field of science would open up dedicated to the task. They called it CETI: communication with extraterrestrial intelligence. And when the practitioners of that science received the next reply, they would be ready. It would be a project in which the whole world might participate, a dialogue requiring one unified communiqué from Earth each century, or each millennium, depending on the distance involved.

The Green Bank conference was over. To commemorate the occasion, the nine men whimsically declared themselves members of "the Order of the Dolphin." With the one remaining bottle from the Nobel Prize celebration, they drank a champagne toast, led by Otto Struve: "To the value of L. May it prove to be a very large number."

5.

In 1965, five years after the fleeting moment of excitement at Project Ozma, the scene was reenacted with even more hoopla halfway around the world. At the Sternberg Astronomical Institute in Moscow, leading Soviet astronomers held a press conference to announce the discovery of an alien civilization.

In the 1960s and 1970s, the Soviet Union took up the cause of searching for extraterrestrials more enthusiastically than did the United States. Perhaps a government with no ties to—and even an antipathy for—organized religion was more amenable to the idea of humanity's ordinariness. Or maybe the Soviets just saw the search as an arena in which they could outdo the Americans. Whatever the reason, where American scientists had to piggyback their searches onto other astronomical projects and convince their funders that the work had benefits in other areas of astronomy, their Russian counterparts had no problem procuring funds to search for signals.

One of the Soviet Union's leading astrophysicists was Iosif Shklovskii. Shklovskii was the son of a rabbi-turned-construction worker in a small Ukrainian town. In the early 1930s, when he was forced to stop school at fourteen and earn a living, he became a railroad construction worker in Siberia, where he happened upon a magazine article about the discovery of the neutron. Inspired, he worked himself through university and eventually received his doctorate from the Sternberg Astronomical Institute. In the 1940s he made his name in astronomy by predicting the existence of the 1420 MHz hydrogen line. He was not thinking about a universal frequency for communication at that point; the hydrogen line was an important concept simply because of the fundamental role of hydrogen in the universe.

When the Cocconi-Morrison paper in 1959 proposed a bold new use for his hydrogen line, Shklovskii saw his holy grail. He jump-started Soviet interest in extraterrestrial life with his 1962 book, *Universe, Life, Intelligence*; he was already such a famous figure that his ruminations on the likelihood of alien life and the feasibility of searching for it had immediate credibility. (In 1966 his book was published in the United States as *Intelligent Life in the Universe*, with Carl Sagan as coauthor. Sagan brought it up to date and added about as much material as had been in the original, and *Intelligent Life* became a seminal text.) Known for his wit and effusive personality—"I thought he'd make a great maître d'," said Drake—Shklovskii also initiated many students into the nascent field.

One of these was Nikolai Kardashev, Shklovskii's star pupil at Sternberg, who spearheaded the first Soviet search in 1963. In a landmark paper published in 1964, Kardashev made bold predictions of the types of life that might exist throughout the universe, in civilizations billions of years more advanced than our own. He divided civilizations into three types. We fit into his most primitive class—Type I. These simpletons have figured out only how to make use of the energy of their home planet. Type II civilizations have developed a means of capturing all the energy of their home star. A Type III civilization, on the other hand, would have learned to harness the energy of all the stars in its galaxy, an amount of power lending mind-boggling capabilities. The last two groups were of particular significance, for if there were societies out there harnessing such enormous amounts of energy, it might not be necessary for them to be sending a beacon specifically to us for us to detect them. We could merely look for unusual sources of significant radiation. Or, as the Princeton astrophysicist Freeman Dyson speculated around the same time, we could look in the infrared region for their "exhaust" energy. (Any use of energy inevitably leads to a certain fraction being wasted, or exhausted. This usually creates radiation in the infrared.)

Kardashev put together a network of telescopes across the Soviet Union, from Vladivostok in the far east to Murmansk on the Finnish border, but he was unable to detect any Type II or III exhaust heat. Then in 1965, when he was examining cosmic radio sources at the Crimea Deep Space Station with his colleague Yevgeny Sholomitskii, his quarry revealed itself to him. A clearly extraterrestrial source of radio waves of a sort never seen before was pulsating in the vicinity of cosmic radio source CTA-102 (the 102nd object listed in the Caltech Catalog A of stellar radio sources). It wasn't the sort of signal that one would expect from an alien civilization: it was broadband instead of narrow,* and its frequency of 1000 MHz was just on the edge of the quiet microwave part of the spectrum—why would they have chosen 1000 MHz? But when Sholomitskii showed Kardashev a graph of his observations of CTA-102, the man who conceived of Type II and III civilizations could only stare. The graph looked like a perfect sine wave: the radio source was transmitting its signal with a steady pulse, peaking every one hundred days. For months, the two men tracked the signal, and it rewarded them with an unvarying pattern. It was as regular as a metronome, and its source was so powerful that Kardashev could not help but imagine a Type III civilization thriving out there near CTA-102.

At the Sternberg Institute, over a hundred foreign cars—the luxurious mark of foreign correspondents—swarmed the parking lot and courtyard. Sholomitskii and Kardashev had called a press conference to announce the first discovery of an alien civilization. They presented their data regarding the pulsing signal, expressed their confidence that it was an artificial source, and speculated on its meaning. Perhaps, they said, the one-hundred-day pulse was simply a beacon announcing their existence, and there was a second signal of shorter period for conveying detailed messages. (For complex mathematical reasons involving Fourier analysis and information theory, this is

---

*Naturally luminous objects, like the sun, emit radiation over the whole spectrum of frequencies. An artificial beacon would be more likely to emit over a narrow range, in order to concentrate all its energy to reach out to greater distances.

a likely method for extraterrestrials to use. A simple beacon with little information will be easier to detect, while a signal with much modulated information is more difficult to differentiate from noise.) Or, on the other hand, perhaps each hundred-day oscillation contained a single word or similar amount of information, and a message might be decoded over a few centuries (possibly a manageable time span for other life forms). Shklovskii, the elder statesman of Soviet astronomy, was skeptical from the outset and initially tried to postpone the press conference until the results could be verified and further studied, but pressure from the government was too great—they were finally going to trump the Americans—and he was swept along with the proceedings.

The announcement from Moscow was heard around the world. Drake wired his congratulations and a request for more detailed information. From the California Institute of Technology, though, came another telegram; this one contained news that showed Kardashev should have listened to his old teacher. Unbelievably, the Russians were unaware of work at Caltech showing CTA-102 to be a quasar.

Quasars appear at first to be stars (the name derives from "quasi-stellar radio sources"), but in fact they are something totally different: each one's energy output is equivalent to that of hundreds of billions of stars. We now believe that they are galaxies at the outer edge of the universe, receding from us at enormous velocities, with black holes at their centers that are swallowing up the matter in the galaxy and causing intense radio emissions. The fact that CTA-102 is a quasar did not necessarily prove that the signal was not intelligent in origin, but it certainly provided another reasonable explanation. Sholomitskii had detected the first pulsating emission from a quasar (many others have since been found), but he had not discovered aliens. The Soviet "breakthrough" dissolved in embarrassment.

———

Although the Russians deserved a bit of shame for having made such extravagant public claims without checking around the scientific world first, such false alarms were commonplace in the early years of CETI. After all, no one had ever listened for intelligent signals before; for all they knew, outer space was as congested with interstellar communication as a modern city's air is with FM, AM, and television broadcasts, cellular phone calls, shortwave radio, and other electromagnetic traffic. So when an astronomer first pointed his radiotelescope at a target and saw what looked like it might be intelligence, his adrenal glands couldn't help but go haywire.

If it weren't for the CTA-102 incident, and perhaps also Drake's false alarm with Project Ozma, there might have been another such misunderstanding in Cambridge, England in 1967, when a graduate student named Jocelyn Bell accidentally noticed a radio source that was emitting pulses of radiation with phenomenal regularity. Every 1.3 seconds it was launching a shot of radiation lasting one hundredth of a second. When Bell reported this to her colleagues, everyone's first thought was that she had found alien signals. How else to possibly explain such awesome precision, signals they later likened to Morse code?

The Cambridge group, however, did not make the same mistake as Kardashev and Sholomitskii; before publishing, they analyzed the data thoroughly. The signals were too strong and spread out over too great a range of frequencies to be the work of living beings. It would have been a wasteful squandering of energy to broadcast over a large frequency range instead of concentrating the beam into a narrow frequency; anyone intelligent enough to build the apparatus would hardly make such a mistake. In addition, by the time they published their paper in *Nature*, in February 1968, they had found several more of these pulsating radio stars and concluded that they were more likely to be the result of a natural mechanism at work, so they merely alluded to their initial suspicions.

Though most scientists agreed with them, some picked up on the early alien hypothesis, and for a while the objects were commonly referred to as "Morse code stars" or "LGMs," short for "little green men." Frank Drake, for his part, wasn't so sure Cambridge's first impression was wrong. If you were creative it wasn't too hard to explain the wideband frequencies. A Type III civilization, for instance, with the energy of an entire galaxy at its disposal, would laugh at the suggestion that they conserve their resources by transmitting only over a narrow band. And if they weren't actually creating the bursts themselves, Drake reasoned, they might have developed a means for switching a natural broadband source on and off.

Still on the Cornell faculty, Drake was now living with his family in northern Puerto Rico, where he was director of the university's 1,000-foot radio telescope. Built under the direction of Tommy Gold, it was the largest radio telescope in the world. Its twenty-acre dish lies ingeniously nestled in the hills surrounding the town of Arecibo, formerly best known as a prominent stop on the cockfighting circuit. Suspended above the dish is a six-hundred-ton steel platform—an equilateral triangle with 210-foot sides carrying the transmitting and receiving equipment. When scientists or engineers need to work on that equipment, a cable car carries them fifty stories into the air to a covered walkway leading to the platform.

Intrigued by the strangely pulsating beacons, Drake led an effort to use the powerful Arecibo dish to investigate them. Strangely, it required him to go to the local Sears Roebuck and buy a large but ordinary television antenna—the frequencies observed at Cambridge were not only much lower than had been expected from an intelligent source, they were even lower than the antenna feeds on the big dish could receive. The Sears antenna worked like a charm, and with the thousand-foot dish beneath it, Arecibo became humankind's greatest tool to unearth the secret of the blinking radio sources.

Try as he could, Drake was unable to separate any sort of intelligent message from the signals. The best he could do was to coin a term for them—"rapidly pulsating radio sources" was getting tiresome in the vocal debates resounding around the scientific world. In an article in *Science*, Drake and two graduate students first used the term "pulsar," and it stuck.

But what were they? Most scientists seemed to agree that, as the name suggested, they were stars that pulsated for some reason, until Tommy Gold came out with another of his wildly original ideas. What if, he said, a star was spinning rapidly, and its radiation was emanating in a concentrated beam? The effect would be like that of a lighthouse—to us it would look like a blinking light (or, in this case, a blinking radio source). As for what was causing the radiation, he pointed to work done by Franco Pacini, an Italian astronomer in residence at Cornell before the pulsars were discovered. Pacini had suggested that when massive stars explode in supernovas, the end result is a phenomenally dense spinning object with a strong magnetic field, known as a neutron star. The catastrophic implosion taking place would cause the neutron star to spew great amounts of radiation, and the magnetic field would focus it like a searchlight. Gold felt that these must be what pulsars were. Most people agreed that if it could be shown that pulsars were slowing down, like any spinning object, Gold's theory would take the prize. When David Richards, a graduate student at Arecibo, showed that the rate of blinking of a particular pulsar was slowing down—by 36 billionths of a second each day—the debate ended.

A Nobel Prize in Physics was awarded in 1974 for the discovery of pulsars, but not to Jocelyn Bell, who discovered them, or to Tommy Gold, who explained them. Instead, Bell's advisor, Anthony Hewish, who had overseen her work, was honored. Bell, seemingly unperturbed, went on to a fine career in x-ray astronomy, and Gold, as we have seen, just keeps rolling along. But the award remains a controversy to this day. In any case, the scientific world had an exciting new astronomical phenomenon to study, but still no company in the universe.

6.

In the 1970s, the search for extraterrestrial intelligence moved into a new phase. Gone was the giddy period when every unidentified signal raised blood pressures and brought newspaper headlines. Yet the concept that there were probably advanced civilizations out there and that we had hardly begun to search made for a period of rapid development.

One major catalyst of this efflorescence was Project Cyclops. John Billingham of the NASA Ames Research Center in Moffet Field, California, was an English gentleman, an Oxford-educated physician with a long-standing interest in the physiology of flight and space travel. He had been with NASA since 1963, and had, among other contributions, designed the water-cooled space suits worn by the *Apollo II* astronauts who had walked on the moon. Billingham had become one of the early converts to the belief that alien civilizations must exist and that we should be looking for them. Every summer he organized an engineering design workshop at Ames, at which scientists from all over the country worked as a team to solve some space-related construction problem. In 1970, Billingham decided the group should devote themselves to the problem of detecting intelligent extraterrestrial signals.

Billingham recruited Barney Oliver, the engineer, inventor, and Hewlett Packard executive who had been part of the Order of the Dolphin, to codirect the workshop, and together they brought a number of experts to give lectures to the group on everything they needed to know about astrophysics, radio astronomy, and possibilities regarding exobiology.

By the end of the summer the workshop had produced a design incorporating a new concept called *long baseline interferometry*. They envisioned an array of fifteen hundred dishes, each one three hundred feet wide, forming a great desert mosaic in the southwestern United States. Moving slowly in perfect unison like three hundred synchronized swimmers, they would track the sky with a sensitivity and power that would make Arecibo look like a transistor radio. Oliver christened the beast Cyclops, after the mythological giant with one huge eye in his forehead. Argus, his first choice, would have been more appropriate—the watchman with a hundred eyes. But Argus was the brand name of a cheap camera, so he settled for Cyclops.

The giant never had a chance. In 1971, $10 billion was a bit more than Congress could be expected to allocate for alien-searching. Its proponents were doomed by their own exuberance, with which they proposed a mosaic of ultimate proportions—the most lavish undertaking they could imagine. After all, the point of the workshop was to dream up the best design imaginable, not the most practical. A smaller-scale Cyclops, which was in the workshop's final report, might have been considered, but the popular image of fifteen hundred telescopes was what caught the public's attention. Some feel that in giving Congress the idea that CETI required billions of dollars, Cyclops set the field back ten years.

Nevertheless, Cyclops did succeed in fomenting CETI enthusiasm at NASA and around the world. NASA formed an Exobiology Division at Ames, with Billingham in charge. And once again the human race began a serious attempt to find other civilizations.

The first international CETI meeting took place in the Soviet Union, for two reasons. The Soviets, led by Kardashev, coorganizer of the conference, were plunging ahead full speed at CETI, with several searches going on simultaneously, while in the United States there had been no formal search since Ozma ten years before. And it was much more difficult for Russian scientists to travel outside the Soviet Union than it was for foreigners to visit there. So in September 1971, the group met at the Byurakan Astrophysical Observatory in Armenia.

The meeting was coorganized and cochaired by Carl Sagan,

who managed to bring together an international cadre of astrophysicists, together with people as diverse as Francis Crick, codiscoverer of DNA; Marvin Minsky, a pioneer in artificial intelligence; and Richard Lee, an anthropologist who had lived among the bushmen of the Kalahari Desert.

Shortly after the Byurakan conference, the great search began again on American soil, tentatively at first, then with a slowly building acceleration of enthusiasm. Gerritt Verschuur, working with the 140-foot and 300-foot telescopes at Green Bank, led Project Ozpa (a paternal counterpoint to Ozma, as he named it) in 1971. Ozpa ran for only thirteen hours, but searched nine stars as opposed to Ozma's two, and was able to analyze 384 channels in the waterhole region where Ozma had managed only one. Then, from 1972 to 1976, another Green Bank project, Ozma II, spent five hundred hours looking at 674 stars at the hydrogen frequency, using the same 384-channel system.

Needless to say, no alien signals were identified. But that doesn't mean that these projects were failures. Even with their modest advances over Ozma, the leaders of these searches were well aware that they were taking only the very first, rudimentary steps of a journey that might last centuries or more. Looking at a few hundred stars out of hundreds of billions was like beginning to empty the Atlantic Ocean by dipping a teaspoon into it. By now everyone knew that success would not come so easily. What was important was that the attempt was finally being made.

Frank Drake was not going to miss the new wave of projects, either. As director of Arecibo, he was busy from 1972 to 1974 overseeing a massive renovation of the telescope, a project that required the production of 277 miles of aluminum belt right on the observatory grounds, the laying of 40 miles of cable, and fifteen hundred concrete anchors. The result was a new apparatus ten times more powerful and accurate than the old.

It was enough to attempt a search of the sort Drake had suggested in his talks at Byurakan. At that meeting, he had noted that it did not necessarily make sense to look at the nearest stars first. After all, the stars that we see in the night sky with our naked eyes are not the nearest stars; they are the brightest. The natural distribution of stars puts very few of the bright ones in our vicinity. If you were undertaking to search for natural-light emissions throughout the universe, therefore, our closest neighbors would be among the worst targets.

In the same way, the distribution of intelligent signals throughout the universe may well place the strongest signals in the far reaches of the universe. A strong signal from a million light years away might be more detectable than a weak one from a hundred light years, and certainly more rewarding than none at all from nearby. There would be the drawback of unfathomably long intervals between communications if we desired a dialogue, but the important first step is simply receiving an intelligent signal. What's more, as a result of the geometry of space, with one directional search you could cover billions of stars instead of just one (since a given solid angle subtends more and more volume as the distance increases).

Ever since Drake had proposed an intergalactic search at Byurakan, Carl Sagan had been urging that they do it themselves at Arecibo. Now, with the renovations complete, he continued to pester Drake, who was so exhausted by the years of overseeing the construction that he could barely remember why they had done it. Finally, they wrote their proposal and were granted some observing time with the new apparatus, beginning in 1975.

As Drake described it, he and Sagan would rise in the middle of the night in their beach hotel (visiting quarters at the observatory were full) and navigate the dark winding roads toward the telescope, Drake driving and Sagan "propped up in the front seat, eyes closed, munching laboriously on scraps of dried-out garlic bread rescued from the previous night's dinner." When M-33,* the Great Nebula in Triangulum, rose in the dawn sky, they were ready in the control room. It must have been a heady experience for Drake, after a fifteen-year hiatus from active searching. In the first hundredth of a second of observing with the Arecibo telescope, they accomplished more searching than he had in the entire Project Ozma.

In their hundred hours of observing time at Arecibo, Drake and Sagan observed many billions of stars, and detected nothing. But even with the most powerful telescope in the world and the most advanced electronics then in existence, they knew that their search was paltry in comparison with what must come in the future. They had been able to look at only a few frequencies, and even then devote only a minute or so to each star. The odds of a signal coming at just the right time and at their exact frequency were infinitesimal.

Realizing this, they were not terribly disappointed, merely determined to improve their methods. They realized more than ever, though, that this was not a business to devote one's career to. The disappointment and ennui of long fruitless searches would be unbearable. It was an odyssey that would need to be undertaken by as many astronomers as possible, each carrying on other, more easily gratified, work at the same time.

The period of giddy hopefulness, however, was not entirely over. As recently as 1973, a Soviet radio-astronomer named Yury Pariisky felt sure he had detected intelligent alien signals. At a meeting of the International Astronomical Union in Sydney, Australia, he invited Drake and Sagan for a walk. Apparently afraid of being caught discussing his work with Americans, he spoke quietly as they strolled down the busy city streets. He and his colleagues had been tracking an extraterrestrial radio source for months. It was a wide-band signal, but it repeated itself every day, for a few hours, sending out clearly patterned pulses in groups of one, two, seven, and nine. He

*The thirty-third object listed in the catalog of fuzzy stellar objects produced by Charles Messier in the eighteenth century.

had checked with Soviet authorities to see whether he might be picking up a satellite, but they assured him that the description fit no satellite, Soviet or American. This was the real thing!

His American friends were polite but could not share his excitement. They had seen too many similar disappointments. Sagan doubted whether the Soviets would reveal their knowledge of American satellites, even to a comrade like Pariisky. He was right; soon after returning home, Sagan was able to send Pariisky word of an American reconnaissance satellite called Big Bird, whose signal and location matched the Russian's measurements exactly. The aliens, it turned out, were clothed in blue denim and ate cheeseburgers and cokes.

And then there was the "Wow!" signal, which was not so easy to explain.

In 1956, radio astronomer John Kraus of Ohio State University began overseeing the construction of an unusual looking radio telescope. Lying flat on a cleared area in a wooded glade off a dirt road, it consists of a metal rectangle with an area of more than three football fields. In fact, it looks like an enormous football stadium with the sideline seats torn away, leaving only the cheap end-zone seats. These end-zone seats at each end are actually the collecting areas; the effective sensitivity of the telescope is that of a circular dish with a diameter of 175 feet. Kraus named his brainchild "Big Ear."

Between 1965 and 1972, Kraus and his graduate student Robert Dixon used Big Ear to conduct the most comprehensive survey of the radio sky to date, discovering twenty thousand new radio sources. In 1972, funding for the project ran out, and Big Ear was a dish in need of a job. Dixon was a junior faculty member by then, and had caught the CETI bug when he participated in the Cyclops workshop at NASA Ames in the summer of 1971. "It did not take too much arguing," he said, to convince Kraus that Big Ear, with one of the most sensitive 1420 MHz receivers in the world, was the perfect facility for a major CETI search. Kraus managed some funding, and Big Ear CETI went on line in 1973.

Big Ear initially used an eight-channel receiving system, but in 1975 upgraded to fifty channels. Its state-of-the-art IBM 1130 computer analyzed the fifty channels simultaneously, looking for unusual spikes of a narrow-band type, and recording such "interesting" signals on punched computer cards. For four years the great metal plate scanned the sky, the Earth's rotation turning it slowly around the heavens. A number of "interesting" cards were recorded, but a reasonable explanation was always found. The telescope received two signal beams at all times, slightly offset from each other, and subtracted the two to weed out terrestrial interference; more often than not, the distinctive signals turned out to be of earthly origin.

On the night of August 15, 1977, however, a signal came in that defied such explanation. Jerry Ehman of nearby Franklin University, who had worked earlier on Big Ear's sky survey and who helped program the receiving computer, was going over the reception data when he saw an enormous spike in intensity right near the hydrogen line, over a narrow-band frequency, coming from far out in space. What's more, it turned itself on and off like a beacon right while Big Ear was listening to it. It was like nothing Big Ear had heard before. Ehman circled the incongruous numbers on the printout, and while he didn't call a press conference, he did write "Wow!" in the margin.

He and Dixon and the rest of the Ohio State team immediately tried to get another such signal from the same point in space, but were met with silence. Over weeks and months, they failed to find it again. They also failed to explain what it was. No known artificial satellite could have produced such a signal. It was almost exactly what one would expect from an alien civilization, except that it occurred only once, for a brief instant.

To this day, no one knows what the "Wow!" signal was. While it seems unlikely that a civilization would broadcast for only one moment, it is possible that for complex reasons that was the only moment that we were able to perceive it. "The 'Wow!' signal is a beautiful example of exactly what an extraterrestrial radio signal could look like," said John Kraus, "but we only observed it once, and so we can't say any more about it." It will probably remain a question mark forever. It may very well, Drake felt, have been an unextraordinary piece of cosmic din:

> Space could be full of [such] signals, falling on our planet like raindrops, one after another, each one making a brief, barely detectable splash before it disappears. My own observing experience, when being especially attentive for any receiver output that might be of intelligent origin, is that there seem to be weak signals just popping up out of the noise all the time. Looking at the data, I often have the same feeling I get when I am deep in a forest and now and then sense brief, vague reminders of the civilized world around me—the drone of a jet plane far overhead, a distant auto horn, the faint shout of one person calling to another. Could it be that in our radio observations we are hearing, all the time, the murmurs of countless other civilizations blowing past the Earth as on the wind?

The nascency of CETI was over. The search had begun, the first tremors of excitement had passed, the necessary starts of emotion over an unusual signal had calmed down, much as a child looking for precious minerals in the backyard soon learns not to get excited at every shiny stone.

A new era was beginning, marked by quotidian persistence on the scientific end and constant battle on the political side. As if to mark the scientific sea change, CETI's name changed. In 1975, scientists in the field decided they needed a new acronym. They realized that the goal of "communication" with other civilizations was germane but not immediate. First we needed to find someone out there; then, over the centuries or millennia, perhaps a dialogue might be established. The detection of a beacon, or even of incidental intelligent emissions, analogous to our television programs already blasting through space toward unwitting receptors, would in itself be the greatest discovery of all time. Accordingly, astronomers began referring modestly to SETI: the search for extraterrestrial intelligence.

The SETI pioneers settled down for the long haul. More advanced computers would be needed to process enough channels to give SETI a fighting chance. New scientists, too, were needed to carry on the crusade. And newer, bigger telescopes—maybe even radio telscopes in space, to escape the smothering blanket of the Earth's atmosphere. All of this would require public support—in short, funding. And that was where the battles would come, as the new science struggled for survival against the forces of skepticism, indifference, and ignorance.

7 | Guaranteed To Fail

> *To consider the Earth the only populated world*
> *in infinite space is as absurd as to assert that in*
> *an entire field sown with millet only one grain*
> *will grow.*
>
> METRODOROS THE EPICUREAN, ca. 300 B.C.

1.

By early 1978, SETI was gaining momentum. Though not without opposition, more and more scientists were beginning to accept it as a viable and worthwhile scientific enterprise. Ever since the Cyclops project, John Billingham had been working to bring together the NASA Ames Research Center and the Jet Propulsion Laboratory in Pasadena, California, to launch a joint SETI campaign which would make the earlier forays look like lone fishermen with bamboo poles next to an industrial-size North Sea trawler. Under Billingham's articulate spokesmanship, SETI was drawing a nourishing supply of attention and funding.

Then, on February 16, 1978, Senator William Proxmire of Wisconsin named SETI the recipient of his "Golden Fleece of the Month" award. The Golden Fleece was a clever political gimmick: in the post–Great Society era, where government prodigality was becoming a universal punching bag, Proxmire each month selected a federally funded project that he deemed not just undeserving of public funds, but full of poppycock.

Sometimes the recipients did seem worthy of ridicule. The Army, for instance won a Golden Fleece for spending $6,000 to prepare a seventeen-page document that told the federal government how to buy a bottle of Worcestershire sauce, and another award went to the federal funding of a $792 designer doormat. Often, though, particularly in scientific matters, Proxmire had little understanding of the work he denigrated, and serious research sometimes took a fall. Of course, that neither lessened the publicity he received nor saved his victims from public opprobrium. The case of SETI was a perfect example of this. In running through the long list of projects up for federal funding, his eyes must have caught on the words "search for extraterrestrials." Sounds like some scientist flakes must have enjoyed *Star Wars* a bit too much, he may have thought. And now they're wasting millions of the taxpayers' dollars to snoop around for little green men! Talk about a natural for the Golden Fleece award: this was even dumber than the $27,000 study to determine why inmates want to escape from prison.

The announcement of the award caught the SETI community by surprise, and made them downright nervous. Members of Congress were not known for being overly attentive to the needs of scientists or for spending much time studying the complex issues involved in various scientific projects. They were much more likely to jump on the Proxmire bandwagon and avoid being labeled government fat cats. The Golden Fleece award was comic entertainment in the newspaper columns, but it was a serious blow to SETI.

When newspaper reporters asked Frank Drake to comment

on the award, he swallowed his anger and tried to make light of the subject: he said that he was arranging an honorary membership for Senator Proxmire in the Flat Earth Society, an actual organization whose members refuse to accept the idea of a spherical Earth. In a ludicrous move that even Proxmire's most vehement opponents could not have scripted, one of Proxmire's assistants called Drake to inquire, in all innocence, whether the senator would be receiving a certificate or plaque for his office.

But it was nothing to laugh about. With the publicity casting a black cloud over SETI, it became nearly impossible to get Congress to openly approve funding. Billingham, however, managed to keep some money coming over the next few years by cleverly hiding SETI projects within other, more conventional NASA programs. This was working well enough, and he was accruing so much support within NASA, that he ventured back out from underground, placing SETI in plain view on NASA's section of the federal budget proposal for the fiscal year 1982.

Senator Proxmire did not take to this kindly. Now the chairman of the Senate Appropriations Committee, he could hardly believe his eyes. Golden Fleece winners were expected to scurry off with their tails between their legs, ashamed of themselves for leeching the taxpayers' hard-earned money. And here was SETI, not only having survived for four more years, but actually requesting two million dollars a year! Never mind that this was one tenth of one percent of NASA's budget, and one thousandth of a percent of the defense budget; Proxmire paraded this two million in front of the Senate like proof of treason. "I have always thought if they were going to look for intelligence, they ought to start right here in Washington," he proclaimed. He went on to propose an amendment to the appropriations bill stipulating that "none of [NASA's] funds shall be used to support the definition and development of techniques to analyze extraterrestrial radio signals for patterns that may be generated by intelligent sources . . . a project that is almost guaranteed to fail."

The Proxmire amendment passed. By July of 1981 NASA was forbidden to spend money on SETI. After all the progress since Project Ozma, the search was suddenly in worse shape than ever. No one took it too seriously in 1960, but then no one was proscribing it either. It was outrageous: government bureaucracy was deciding which scientific research was worthwhile and which was not.

Every SETI scientist must have been thinking the same thing: If only I could get Proxmire alone in a room for an hour. . . . Not to throttle him, but to teach him the basic ideas behind the search for extraterrestrial intelligence. (Or perhaps first the one and then the other.) The ban on funding was based not on a well-informed opposition, but completely on ignorance.

As with the weather, people complained about the foolishness of senators, but no one did anything about it. Instead, NASA ordered Billingham to draw up a detailed plan to dismantle SETI, reassigning equipment and personnel. And scientists gave up hope for the great NASA SETI project and resorted to cynical humor. At conferences, they introduced a new term into the Drake equation, f_g, which represented the fraction of planets with scientifically literate governments. If every civilization had a Proxmire, they reasoned, $f_g = 0$, and therefore $N = 0$. Actually, as it appeared then, the existence of our own Proxmire alone would render $N = 0$, for without funding we would never be able to detect other civilizations, no matter how many there might be.

Carl Sagan, however, decided to do something about it. He felt the senator would listen to reason, and he managed to secure a personal audience with him.

For an hour, Sagan calmly and assiduously gave Proxmire the SETI lecture: why it seems likely there are billions of other planets, why a certain percentage of those planets probably have conditions favorable to life, how it has been shown that life probably does tend to arise given the right conditions. Why, given the billions of years the universe existed before the development of humans, there are probably millions of civilizations far more advanced than ours. And why, considering the near impossibility of interstellar travel, radio waves constitute the most obvious way of communicating between civilizations. He also stressed what would perhaps be the most practical result of the search: that the discovery of an advanced alien civilization would provide proof that intelligent life can survive the invention of nuclear weaponry.

Proxmire, to his credit, listened well. Averting the threat of nuclear war was one of his pet priorities, as Sagan was well aware. "Well," Sagan reported at a SETI meeting a couple of years later, "he had never heard this! He had never heard *any* argument about the [distances between] the stars, about *any* aspect of this, but especially, he had never heard of the connection between the longevity of advanced civilizations and the question of how many of them there are. And before my eyes, I could see two neural nets, which were in different parts of his brain that had never met, be introduced to each other."

Proxmire quickly realized that he had been remiss four years before in awarding SETI the Golden Fleece award. He assured Sagan that SETI would no longer find congressional antipathy from his office. He didn't publicly admit his mistake and urge his colleagues to reinstate support, but that was perhaps too much to hope for from an elected official. And true to his word, he never again fought against funding for SETI.

Buoyed by the apparent new attitude in Congress, Billingham convinced his bosses at NASA to reinstate SETI in the budget proposal for 1983. With Proxmire turned to other perceived taxpayer-fleecers, the NASA budget went through Congress untouched, and SETI was back from the dead in one short year. Sagan and his colleagues celebrated, unaware that the vicissitudes of the recent past were just the beginning of the public-relations roller-coaster ride SETI would have to endure in the coming years.

2.

In the meantime, other SETI projects were springing up around the world. At Berkeley, in 1979, a search began with the curious name SERENDIP and with, once again, a Cornell connection.

Jill Tarter, a Berkeley graduate student in radio astronomy at the time, didn't get the SETI bug until she came out to California. But her maiden name was Cornell, and she was a descendant of Ezra Cornell, the university's founder. A tall, exuberant former high-school drum majorette, Jill had applied to Cornell for the tuition waver Ezra had decreed for all his male descendants. The university, sticking to the letter of the law above the spirit, refused, so Jill got her own scholarship—from Procter & Gamble—and went to Ithaca anyway. She continued her nonconforming ways by pursuing a doctorate in theoretical astrophysics—not a field teeming with women—and then by jumping into the SETI pool just as the federal government was draining its funding.

Stuart Bowyer, her mentor at Berkeley, had been inspired by the report of the Cyclops project, and he passed his enthusiasm on to Tarter. They had no money to spend, but together with a few other students they scrounged up all the old receiving and computing equipment they could find, and they redesigned and reprogrammed it to their needs.

In 1979, just as the Golden Fleece award was bringing aftershocks of skepticism from Washington, Tarter and Bowyer took their receiver out to the University of California's Hat Creek Radio Observatory near Mt. Shasta and connected it to the eighty-five-foot radio telescope. The "little black box" they hung unobtrusively would hear the same signals that were being observed for other purposes and redirect an amplified copy to Bowyer and Tarter's refurbished computer, which Tarter had programmed to analyze one hundred channels.

They named their project SERENDIP, after the Horace Walpole story about the three princes of Serendip who travel the world without a clue but continually make chance discoveries. (Along with Occam's razor, serendipity is a crucial addendum to the scientific method.) Tarter devised an acronym for their search, too: Search for Extraterrestrial Radio Emission from Nearby Developed Intelligent Populations. The work even felt serendipitous, for Tarter and Bowyer were unable to stay at Hat Creek and "direct" the signal reception; they had no official purpose there, and the staff had no interest in this strange piggybacking contraption. As a result, SERENDIP was left to fend for itself, to catch what it might catch. On their periodic checkup visits, Tarter and Bowyer would sometimes find that the equipment had malfunctioned and hadn't been working for days.

In nine years of searching at Hat Creek, SERENDIP found no aliens. But it helped set the standard for piggybacking which had been started by Frank Drake, and would become a model for other SETI searches.

At the same time, Tarter's contemporary, Paul Horowitz, was making his first contributions to the search. As a Harvard undergraduate in the late 1960s, Horowitz was inspired by his professor Edward Purcell, the recipient of the 1952 Nobel Prize in Physics for the development of nuclear magnetic resonance (NMR) measurements, who had used his new technique to discover the 1420 MHz hydrogen line that Shklovskii had predicted. Purcell gave a lecture on the likelihood of extraterrestrial life, and, Horowitz recalled, "he pointed out that communication through space by radio astronomy is extraordinarily efficient, for extraordinary small amounts of energy . . . whereas space travel to distant star systems deserves to be back where it came from, on cereal boxes." Later, Frank Drake came to Harvard to give that year's Loeb Lectures. "He did the Sagan and Shklovskii book for us in four or five lectures," Horowitz recalled, "and got me quite excited."

But it wasn't until 1978 that Horowitz, now a full professor of physics at Harvard, conducted his first SETI search, at Arecibo. His idea was to look for very narrow-band signals from a civilization that was specifically targeting our sun. It was a trade-off: the odds of such a signal existing were relatively low, but if it did exist the odds of finding it were relatively high.

After this initial effort, Horowitz turned his computing and electronics skills to devising better and better SETI processing receivers. He developed a machine that could process 128,000 channels simultaneously. It also fit into three boxes for easy transport to various telescopes and so acquired the name Suitcase SETI. Horowitz took his suitcases down to Arecibo in March 1982 and searched 250 cosmic radio sources but found nothing. His hardware, though, was impressive. As a test, he used it to plot a map of a particular radio source. The definition was so fine that if it had been printed out at two hundred dots per inch, it would have covered the entire thousand-foot dish.

Horowitz carried his apparatus back to Harvard and hooked it up to the eighty-four-foot telescope at the Oak Ridge Observatory—the same telescope where Drake had had his first false alarm while observing the Pleiades cluster. Harvard University was not interested in funding Horowitz's SETI work directly, but they didn't mind him indulging his curiosity on their time. He found funding from a new organization cofounded by Carl Sagan in 1981. The Planetary Society is a private non-profit group committed to the exploration of our solar system and beyond, and to SETI. Within five years it had grown to more than 100,000 members, making it the largest private pro-space group in the world. From its inception, its biggest project was Horowitz's Suitcase SETI. The suitcase, however, was now unpacked and permanently set up at Harvard, with a new name: Project Sentinel.

Sentinel had a very different mode of attack than Horowitz's Arecibo search. Since this telescope had a much wider beam, Horowitz modified his equipment for an all-sky survey. Each day, the telescope was tilted half a degree higher. As the Earth rotated, the beam would sweep the sky at that latitude. In about two hundred days, it would cover the entire northern sky.

Sentinel again searched for very narrow-band signals that had already compensated for the Doppler shift—that is, that were heliocentric. But Horowitz was well aware that the odds of success at this rate were infinitesimal. That we had guessed correctly at the magic frequency was improbable enough; that the broadcasters were customizing their Doppler shifts for every star they targeted seemed truly wishful thinking. It was much more likely that they either transmitted at the magic frequency and expected their audience to do the adjusting, or perhaps adjusted their frequency to some universal frame of reference. Perhaps all the supercivilizations of the galaxy chose the center of the galaxy as a frame of reference and shifted their frequencies so that they all looked like a 1420 MHz signal emanating from that center.

Clearly we needed to be able to process more channels. The more we could, the more likely we would stumble on paydirt. In a perfect scenario—if we could listen to every channel in every direction—we could answer the big question once and for all. Until that time, the more channels the better.

All Horowitz needed was a little money to reach the next electronic plateau. Enter *E.T.*

In the first big fundraising coup for the Planetary Society, Sagan had enlisted Steven Spielberg as a member of the board of directors. Spielberg's movie *E.T.* had recently become the most popular film in history, and although it probably encouraged the wrong kind of extraterrestrial interest—fascination with interstellar visitors to Earth, from which SETI scientists were trying so hard to dissassociate themselves—it led to at least one big boost for SETI. Spielberg donated $100,000 to the Planetary Society specifically for SETI, which meant specifically for Horowitz. He and his colleagues at Harvard immediately got to work designing a new generation of SETI processors. One hundred thousand dollars, a farthing to most scientific enterprises, was all they would need. For three years, they soldered half a million joints by hand, putting together 128 circuit boards to form a computing system as powerful as the best supercomputers of the time.

On September 29, 1985, a perfect New England fall afternoon, a dark limousine pulled up at the Oak Ridge Observatory. Out stepped Carl Sagan, his wife, Ann Druyan, and their three-year-old daughter, Sasha. Next appeared Steven Spielberg, actress Amy Irving, and their four-month-old son, Max. SETI had met Hollywood.

Sagan and Spielberg each spoke to an assemblage of press, dignitaries, and Planetary Society members. "This," said Sagan, "is a moment in which the most sophisticated search for extraterrestrial intelligence in the history of the planet Earth is initiated." Spielberg approached a big old switch that Horowitz had unearthed, reminiscent of Dr. Frankenstein's laboratory. It was a fake, not attached to anything, but as the director symbolically pulled the switch, a technician hit some real controls, and the telescope whirred into position and began its search.

META, or the Megachannel ExtraTerrestrial Assay, was sixty-four times as powerful as Sentinel and could process over eight million channels at once. This was enough to cover more narrower-band channels, in an area centered on hydrogen's 1420 MHz frequency, than any covered by Horowitz's previous searches; yet the area was wide enough, he figured, to allow for Doppler shifts based on the galactic center. With each new burst in technology comes a new surge of enthusiasm, and META seemed to promise the hope of some real results at last.

Close on META's trail was SERENDIP. The Berkeley group launched SERENDIP II in 1986, this time piggybacking on the three-hundred-foot telescope at Green Bank. The system had been upgraded to scan 64,000 channels, the same as Sentinel, but within a few years it would catch up to META. There seemed to be no limit to SETI's potential to cover *all* the universe's broadcast channels, given enough time.

And it appeared they would be given enough time. Ever since 1982, when John Billingham and NASA managed to get SETI back on the budget, momentum had been building steadily. In that same year the International Astronomical Union, the world governing body of astronomy (which names all celestial objects, for instance, and decides when to add a leap second to our clocks), approved SETI as an official world-wide goal of astronomy. Added to the IAU's previous fifty "commissions," SETI became "Commission 51: Bioastronomy: Search for Extraterrestrial Life." Of course this encompassed more than SETI: the search for planets, cosmochemistry, and the search for rudimentary life in our solar system were all included. But SETI was the inspiration for the commission and, inevitably, the ultimate goal.

Around the same time, a National Academy of Sciences panel, though including no active SETI scientists, published a report proposing that the government spend $20 million on SETI in the 1980s—even more than NASA had requested. Considering the billions that would be allocated over the same period for other scientific endeavors, not to mention the trillions overall in the budget, this seemed like a modest request. Yet if $100,000 was enough to fund the development of a million-channel SETI search, $20 million seemed like it would make the sky, or the cosmos, the limit.

With such numbers being tossed around, NASA in the 1980s went full speed ahead with their plans for a massive

NASA SETI project. NASA SETI promised to be everything that earlier searches might have dreamed of. In each previous project, a choice had to be made between searching the whole sky with low sensitivity or targeting particular stars with greater sensitivity; NASA SETI would do both. The program would utilize telescopes all over the globe, concentrating on Arecibo's mammoth dish but also using Green Bank and other dishes in California, France, and Australia. Half the project entailed a targeted search of a thousand sunlike stars within a hundred light years of Earth, with an unprecedented sensitivity that could detect a signal of a trillionth of a trillionth of ten milliwatts per square meter. State-of-the-art software would pick out pulsing signals, steady signals, and signals that drift off frequency. If we had neighbors in the galaxy, it seemed, NASA SETI would find them.

The other half was an all-sky survey, looking for distant but "bright" sources in the 99 percent of the sky not covered by the targeted search. It would use NASA's Deep Space Network, a series of thirty-four-meter dishes placed around the globe, to search the entire "quiet" region from roughly 1000 to 5000 MHz. The sensitivity would be only a thousandth of the targeted search, but still greater than any previous endeavor.

Granted "mission" status by NASA in 1990, the program had the same priority as the shuttle program or Mars probes and employed more than a hundred people. This included a rotating team of astronomers to be on call to respond to interesting signals as soon as they might be received. One of the major drawbacks to all previous projects was that signals were of necessity analyzed long after they were received. There simply wasn't enough computing power to do the job in real time. So even when an interesting signal was received—and META, for instance, recorded several such signals—it was often weeks or months later that it was recognized. When the apparatus was finally redirected back, the signal had disappeared. Because it was spurious? Or because the alien transmitter had moved on? Impossible to tell. Immediate recognition of anything interesting would enable the crew to lock onto it and investigate it further.

The twenty-eight-million-channel supercomputing power of NASA SETI also meant that the days of choosing a "magic" frequency in the dark and sticking to it were over. No matter how clever our astronomers might be in choosing the magic frequency, the odds of their logic coinciding with that of unimaginable alien minds remains small. NASA SETI would be able to investigate virtually every channel within the quiet part of the spectrum. It was still limited to that segment, but the quiet area as a whole is thousands of times more obvious than specific magic frequencies within it.

NASA SETI was everything Drake, Sagan, Morrison, Oliver, Billingham, and all the other SETI scientists had dreamed of. And as international support grew during the 1980s, it looked as though they were going to pull it off. By 1990, the NASA SETI team had already been working for years, spending millions of dollars preparing for the day when the dream search would finally go on line.

But they still had to contend with the Proxmire factor. Not the Wisconsin senator himself—he had recognized the ignorance of his former stand and had quietly turned to other crusades. His ilk, however, had since proliferated in the halls of Congress. Politicians had caught on to the act of ridiculing perceived governmental excess to gain points in the polls.

In the summer of 1990, the $200 billion savings and loans bailout was dominating the media, the federal budget deficit was a $160 billion albatross around Congress's neck, and "cut government spending" was more of a catchphrase than ever. In late June, as the House of Representatives Appropriations Committee was about to vote on the 1991 budget, two congressmen saw a golden opportunity. Representatives Ronald Machtley and Silvio Conte scrambled to put together an eleventh-hour amendment that would strike NASA SETI from the budget. NASA had renamed their project Microwave Observing Project (MOP) to hide it from skeptics of "intelligence" searches, but someone had alerted Machtley and Conte. It was the Proxmire amendment all over again; even the rhetoric sounded like reruns. "We are just beginning to realize the costs associated with the S & L bailout," Machtley thundered in a committee meeting. Instead of this multimillion-dollar "rip-off" to look for "little green men with misshapen heads," he asked smugly, "might we spend some of this NASA money to find where the absence of intelligence was that led to this failure?" Like a washed-up comedian falling back on the oldest jokes in the book, he suggested a substitute for SETI: SCOTI, "the search for congressional terrestrial intelligence."

His audience, though, apparently was new to the routine. Ignoring the reeducation of Proxmire eight years earlier, they bought the joke and passed the amendment. It was just a committee, but a powerful one. Once again, NASA SETI appeared to be in jeopardy.

Jill Tarter, who had been named the project scientist the year before, decided to go to Washington herself. She fought her way into the offices of key Senate personnel and explained all over again, as Sagan had done with Proxmire, the rationale behind SETI. When the Senate Appropriations Committee met, Tarter's persistence paid off: the committee voted to give MOP the full amount of requested funding.

Tarter and her colleagues were guardedly celebratory. They now realized that government support for SETI would never be a sure thing. It appeared that every year they would have to undertake a new campaign to give Congress a crash course in radio astronomy; every year they would have to battle the politicians who were looking for a financial scapegoat. For the moment, though, NASA SETI was a go. The teams at NASA Ames and the Jet Propulsion Laboratory got back to work.

The proposed startup date was only two years away: October 12, 1992, five hundred years to the day since Columbus sailed for America.

NASA met its goal, and on that historic day the computers switched on and began the greatest SETI in history. One one-hundredth of a second later, NASA SETI had done as much work as all of Project Ozma. One year later, though, although a number of "interesting" signals had been identified and studied, in each case the possibility of alien intelligence had been ruled out.

No one was surprised. For even with the mind-boggling capabilities of new computers and the world's greatest radio telescopes at its disposal, NASA SETI would be able to detect only a small fraction of the possible intelligent signals bouncing around the galaxy. Secondhand radiation—communications between two other stars, or "leakage" like our own television shows fading off into space—would still be almost impossible to catch with our earthbound dishes. And even if someone were sending a beacon directly our way, they would have to have agreed with us that 1000–5000 MHz is the logical frequency range in which to broadcast. As Barney Oliver said at the time, "We have stepped out of our house, but we're not even down the front steps yet."

Oliver was anticipating the critics who would cry prematurely that we were obviously alone and that the search was over. "If we don't [find something right off]," he said, "it might be taken as a sign that there is no extraterrestrial intelligence. That would be a mistake."

His instinct was correct, but his prolepsis was not enough. A year later, the sons of William Proxmire were back in the saddle again. The ringleader this time was Senator Richard Bryan, a Nevada Democrat who typically confused SETI with UFO hype. He told a House-Senate conference committee, and the public in a press release, that despite NASA's expenditure of millions of dollars on SETI, "not a single Martian has said, 'Take me to your leader,' and not a single flying saucer has applied for F.A.A. approval." For Tarter and her colleagues, dealing with Congress was like having a boss with Alzheimer's disease. Every year they had to explain all over again, from scratch, why they were there and why their work was worthy. This time, they couldn't overcome the blinding ignorance of the committee. After NASA had spent $50 million getting ready and was already one year into the ten-year project, Congress pulled the plug. Once again, NASA was prohibited from spending any money on SETI.

Tarter was disappointed but not surprised. The very week the project was kicking off, a year before, she had said, "We've been there [fighting Congress for funding] before, and every year it seems to be worse than the year before." The money allocated to the project amounted to less than one tenth of one percent of NASA's $14.6 billion budget—about one ten-thousandth of the federal budget. What's more, NASA had once more given the project an alias—the High Resolution Microwave Survey—to hide this pittance in their proposal from dubious senators. But it seemed that SETI, no matter how well hidden, set off warning lights for budget cutters in need of a scapegoat. Congress spoke, and SETI was gone; all that was left for it in the 1994 budget was one million dollars to cover shutdown costs.

Congress had had enough of SETI, and the SETIers had had enough of Congress. The annual crusade to educate politicians and fight for one more year of funding simply took too much of a toll in time and emotion. Tarter and her colleagues were spending too much time on political strategy when they should have been planning scientific strategy. It was no way to do science.

Of course, they had no intention of giving up the search. If the government considered the discovery of other life in the universe a useless endeavor, they would find other benefactors. In January 1994, just three months after the cancellation, the SETI Institute (a nonprofit group founded by Drake in 1984) announced that they already had private contributions of $4.4 million, more than half of what they needed to keep the targeted search going through mid-1995. NASA SETI would live as the principal project of the SETI Institute, risen from the ashes with yet a new moniker: Project Phoenix.

NASA, for their part, would use their final federal SETI million not to close shop but rather to help with the transformation to Phoenix. The SETI instruments and software that NASA had been developing for twenty years were of no use to them now, and they were more than happy to turn them over to the SETI Institute. The various subcontractors merely changed employers, Jill Tarter was named project director, and Phoenix hardly missed a stride.

In January of 1995, Phoenix went on line, beginning with a six-month survey of the southern sky from the 210-foot Parkes radio telescope in New South Wales, Australia. Over six months, it looked at two hundred sunlike stars in our "neighborhood," concentrating on the waterhole frequencies and searching 57 million channels simultaneously. Despite Congress's determined effort to waste an investment of twenty years and $58 million, SETI was not dead.

Nor was Phoenix its only program. The other two major enterprises—Harvard SETI and SERENDIP—had been making their own advances and were right with Phoenix, always pushing at the technological limit. In 1992, SERENDIP III had begun a four-year run at Arecibo with an updated system that could process 20 million channels. Unfortunately, they were tuned in at a frequency of 430 MHz—well below the quiet microwave region—but that was the drawback of their piggybacking strategy: they had to use what was available. Meanwhile, Paul Horowitz had been far from inactive with his much heralded META system at Harvard. First, he went down to Argentina and helped astronomers there install META II at the Instituto Argentino de Radioastronomia in Villa Elisa. Then, back home, in 1991, he and his team set to work building the next generation: BETA (*billion*-channel extraterrestrial assay).

Aside from the increased number of channels and gigantic improvements in computing power, BETA had another immense advantage over META. For the first time at Harvard, interesting signals could be rechecked immediately, in real time. On a bright October day in 1995, in a scene reminiscent of META's opening ceremony ten years earlier, the old knife switch at the Harvard/Smithsonian Observatory was pulled, and BETA, with the power of a trillion Project Ozmas, began to listen.

In a world where state-of-the-art home computers become obsolete after six months, however, the glamour of a "billion" channels (see Notes) quickly fades. Less than two years after BETA's opening, in June of 1997, SERENDIP IV went into operation, still—after twenty years—under the leadership of Stuart Bowyer. With the awesome sensitivity of Arecibo's newly renovated thousand-foot dish, it now had the computing power of two hundred of the world's largest supercomputers working together. What's more, this time it would be piggybacking on a study that happened to be using the 1420 MHz hydrogen frequency. For the next five years SERENDIP will check out the entire sky, looking at 168 million channels every 1.7 seconds.

"SERENDIP is probably the most sensitive search and also quite comprehensive in [frequency]," says Horowitz. "But the Phoenix guys will never admit that. And of course it depends how you count. Phoenix to its credit looks for every damn thing: pulses, chirped (Doppler-shifted) pulses, steady signals, and they do it with discipline and search the whole sky and actually get it done. No search comes close to Phoenix in that sense. But crazy guys do things like use Arecibo with SEREN-

DIP's enormous channel count which gives you a channel count times sensitivity that can't be beat even by Phoenix. Although Phoenix can relocate to Arecibo . . . ."

No sooner said than done. After several two-week runs at Green Bank between October 1996 and June 1997, Project Phoenix moved to Arecibo in early 1998 for a five-year continuation of its one-thousand-neighboring-star survey. But no single project can be said to be the absolute leader. Different programs with different strategies and different equipment are operating all over the world—at Ohio State, Italy, Argentina, and elsewhere. They are all part of the same grand vision, funded by the Planetary Society and many of the same private donors. Their scientists meet periodically to discuss the various problems and goals of SETI and readily share their data with each other. They realize that to have even the slightest realistic chance of success, they need to have telescopes all over the world on a constant vigil, tuned to as many frequencies in as many directions as much of the time as possible. The governments of the world have abandoned the journey, choosing to ignore the experts' confidence that the greatest discovery in history is just out there waiting to be found. But the private funding continues to trickle in, and these scientists, convinced of the import of their work, continue to search.

3.

Despite political drawbacks, other news coming through in the mid 1990s was cause for great optimism in the SETI community. In 1995 and 1996, the journals and e-mail lines were buzzing with word of discoveries that stirred the entire astronomical world. SETI researchers in particular were buoyed.

Thirty-five years earlier, the original members of the Order of the Dolphin tried to derive a value for the Drake equation. They did their best to estimate the various factors, but in the end they could give a reasonably accurate number only for R, the average number of stars formed each year in the Milky Way. To the best of their ability, they estimated a value for N, the number of detectable civilizations in the galaxy, that was very high—somewhere between 50 civilizations and 100 million. But they would have been the first to admit that the true figure could lie beyond even those wide limits. They really had no way of knowing the values for six of the seven factors. Their guesses were educated—more than anyone else's in the world—but they were still guesses.

Then came the announcements of 1995 and 1996, which promised to take the mystery out of another of the Drake equation factors: f_p was about to fall.

8 | Strange New Worlds

> *Nature does not wait for us to understand*
> *something before doing it. Thank goodness.*
> DAVID GRAY, astronomer

1.

The development of astronomical spectroscopy in the nine-teenth century made possible the dissection of starlight into its component frequencies, revealing that the sun and stars are composed of the same elements (overwhelmingly hydrogen and helium), and reinforcing earlier speculations that the sun is actually a star. We know now that it is a typical star in a typical location in a typical galaxy of 100 billion stars; there is nothing unique or even unusual about it.

With this knowledge it becomes natural to assume that many other stars have planets of their own. And indeed, if you asked a scientist of the eighteenth, nineteenth, or early twentieth century whether he believed that planets orbiting other stars exist, the answer would likely have been "yes." In fact, the great minds at the Order of the Dolphin SETI conference in 1961 estimated that one fifth to one half of all stars formed planets.

Yet no one could say for sure, for no one had ever seen an extrasolar planet—one outside our own solar system. This was understandable: for one thing, our sun's planets are incomparably closer to us than are other stars (and their putative planets). Secondly, the light that a planet reflects is on the order of a billionth as bright as the light coming directly out of its star. If extrasolar planets did exist, they would be hidden in the glare of the suns they circle.

So lack of observation was little reason to doubt their existence. By 1961 a pretty good theory was in place which explained how planets were created as a by-product of star formation. Enormous clouds of hydrogen slowly congealed under the power of gravity, spinning faster and faster and compressing more and more mass until the pressure fomented nuclear fusion reactions, transforming the center of the gaseous cloud into a gargantuan fireball. Not all the gas was captured in the star, however. Farther away from the center, gas and dust spun into a great disk around the star. Eventually discrete spheres formed, their centrifugal force counteracting the pull of gravity so that rather than being sucked into the maelstrom they were held in orbit. These balls didn't have enough mass to start nuclear reactions, and over millions of years they cooled into planets.

The theory was simple and elegant. It explained how our nine planets might have formed, and accounted for all their subtle characteristics—their chemical composition, atmospheres, moons, and/or rings were all functions of their distance from the sun. It seemed reasonable that the same process occurred throughout the universe, wherever stars were born. But the theory only showed what was possible. It didn't prove that such a process occurred.

In fact, there was another planet theory which envisaged stars forming without planets. In a rare, chance occurrence two stars might then come close enough together for one star's gravitational field to pull material off the other. Some of this material might fall back onto the star, some would be lost into space, and some could fall into orbit around the star and coagulate into planets as the initiating star continued on its way and disappeared into the vastness of space.

In this scenario, planet formation was unusual and, indeed, perhaps unique. Calculations indicated that such a process might have occurred just once, in which case our planetary system might be the only one in existence.

This was not a theory much beloved by astronomers. They argued that if you had two theories, one of which was a normal progression from what they knew—ordinary star formation—and the other was a bizarre concoction that depended on very unusual circumstances—two stars passing at just exactly the right distance from each other—then clearly the more ordinary explanation was to be preferred. But their argument was not valid until we knew if other planetary systems exist. If the galaxy is full of them, then certainly it must be an ordinary and usual process that produces them. But if in fact our planetary system turned out to be the only one, then just as certainly it took a unique concatenation of events to produce it.

Until we actually found other planets around other stars, we just wouldn't know.

Most of our sun's planets were discovered the old-fashioned way—they were "seen." Before written history began, the Babylonians saw that certain "stars" in the sky seemed to wander on their own wayward paths through the fixed heavens. The evening and morning stars were probably the first to be noticed, since they are the most starkly beautiful. As the sun dips below the horizon, while the sky is still too bright for most stars to be seen, the evening star shines brightly as it follows the sun. And just before dawn, as the sun's light begins to haze over the night sky, the bright morning star rises to herald the new day.

But not on the same night. The Babylonians noticed that

when the evening star shone, the morning star was absent, and vice versa. They also noticed that though nearly all the stars rose and revolved around the dome of the sky and set while always maintaining their positions relative to each other, the morning and evening stars changed nightly. The evening star would appear closer and closer to the sun each night, until it disappeared in its haze. Only after that would the morning star appear, close to the sun at first and then gradually moving further and further away until it reached its furthest distance. It would then turn around and, night by night, appear closer and closer to the sun until it too would disappear into the brightness of the rising sun, and then the evening star would once more appear to repeat the process.

It didn't take too much imagination to realize that the two stars were one and the same. The Romans named it Venus, for its beauty, and named the other wandering stars—or *planets*, as we call them (from the Greek *planetes*, wandering)—after their other gods: Mercury, Mars, Jupiter, and Saturn. And so things remained until the late eighteenth century, when William Herschel discovered *Georgium Sidus*, twice as distant from the sun as Saturn, and his colleagues renamed it for Uranus, who represented the heavens to the Greeks.

Uranus was not only the first planet to be discovered since ancient times; it was the last planet to be discovered by "seeing." The next one would have to be found the way extrasolar planets would one day be discovered, by deciphering the subtle dance steps of a less obscure partner.

In the decades after the discovery of Uranus, astronomers carefully plotted its orbit, drawing on new observations as well as old ones taken when it had been thought to be a star. Strangely, though, it failed to adhere strictly to a clean planetary orbit. Rather, it executed regular wobbles in its path, as though being tugged by a force in addition to the gravity of the sun and the other known planets. By the 1830s, many astronomers suspected that there was yet another planet out there beyond the orbit of Uranus, disrupting Uranus's orbit with its own gravity.

John Couch Adams, a young English mathematician, resolved while still a student at Cambridge University to discern the cause of this idiosyncrasy. In 1845, just two years out of college, he succeeded in calculating the mass and orbit of an eighth planet that would account for Uranus's motion. Unfortunately for him, England's astronomer royal, George Biddell Airy, was unconvinced and unimpressed with the young upstart. After all, Airy himself had written in 1837 that even if another planet were the culprit, "it will be nearly impossible to ever find out its place."

The next year, a French astronomer, Jean Joseph Leverrier, made similar calculations and predicted a planet in almost the same position as Adams's. His work convinced Airy to have someone look for the planet after all, but the man appointed, though he saw it twice, failed to recognize it as a planet. Meanwhile, Leverrier was having the same problems as Adams convincing his own countrymen. It took a German, Johann Gottfried Galle, to whom Leverrier had written, to get the job done. On September 23, 1846, he and a student began searching the sky where Leverrier had suggested, looking for a star that was not on the map. Within minutes they had found it, and Galle dashed off a letter to Leverrier: "The planet whose position you have pointed out *actually* exists."

Leverrier initially named the planet Neptune but then decided that "Leverrier" would be more appropriate. In a not particularly subtle move to support his case, he began referring to Uranus as "Herschel." He was unable to convince his peers that he belonged with the Roman gods, however: Neptune stuck. In the newspapers, England and France battled for the prestige of discovery; France won that one, since Adams's calculations had never been published, and the British press calumniated Airy for his stubbornness. Meanwhile, Adams and Leverrier became fast friends, and life went on in the suddenly more vast solar system.

Pluto, the last of our own planets, forms something of a bridge between methods of planetary discovery. It was originally predicted by none other than Percival Lowell, who, reeling from attacks on his Martian theories, decided to redeem himself and his Flagstaff Observatory by discovering the ninth planet. Neptune, it turns out, did not have quite enough mass to account for all of Uranus's orbital wobble. There was something else out there, people thought, invisible to the naked eye, pulling on the seventh planet. It must have had an even greater effect on Neptune, but that planet's 165-year orbit had not yet been completely observed. Between 1905 and 1915, Lowell calculated positions where the new planet, which he called Planet X, might be, and wired them from Boston to Flagstaff. By 1916, though, his astronomers had failed to detect his planet, and he gave up. One of his final telegrams to Flagstaff didn't mention Planet X: "Please see that wild flowers on [Mars] hill are never picked. Put up notices." Broken-hearted, he suffered a fatal stroke. One of his closest friends said that the failure to find Planet X "virtually killed him." He would never know that on the nights of March 19 and April 7, 1915, his telescope in Flagstaff had indeed photographed Pluto, a tiny dot wandering among the stars on film. No one noticed.

It wasn't until a farm boy from Kansas, Clyde Tombaugh, was hired at Flagstaff that Planet X came to light. Tombaugh, who built telescopes as a hobby, had sent his drawings of Mars and Jupiter to Flagstaff and had impressed its director. Operating a new thirteen-inch photographic telescope and utilizing a blink comparator, a machine which superimposed plates taken on successive nights to see if any points of light had moved, Tombaugh found his quarry in January 1930. The observatory waited until March 13—Lowell's 75th birthday and the 149th anniversary of the discovery of Uranus—to announce their discovery of the ninth planet. The usual scramble for names ensued (Lowell's widow, Constance, who had crippled the observatory with lawsuits and stalled the search, demanded the planet be named Constance), but the observatory staff voted unanimously on Pluto, god of the underworld, whose name began with Lowell's initials.

The final insult to Lowell's career came slowly to light: although Pluto had turned up where Lowell had calculated it should be, this turned out to be nothing but coincidence. By the 1970s it was clear that the new planet was far too slight—with less mass than our moon—and too far away to have had any observable effect on Uranus.* In the 1980s, some astronomers claimed that there must be a tenth planet out there—a new Planet X—causing Uranus to wobble, and some even announced its mass and location. In 1993, though, analyses of *Voyager 2*'s 1989 flyby of Neptune showed that its previously measured mass had been in error by 0.5 percent, and this difference accounted for Uranus's wobble. This mistake had spurred the search for a ninth planet, but in the end Pluto had been found with a good old-fashioned optical telescope. In this it truly marked the end of an era.

2.

Finding planets around other stars is a completely different ball game. For one thing, as we've said, the distances are incomparably greater. The very nearest stars are several light years away, more than ten thousand times farther from us than Pluto. At that distance any planet would seem to be so close to its star, which would be shining billions of times more brilliantly than the planet, that looking for it would be like trying

---

*In fact, many astronomers now feel that Pluto is too small to be considered a planet. Rather, they say, it is one of many small cometlike objects wandering about in the far reaches of the solar system. For a good discussion, see the *Atlantic Monthly*, February 1998 (www.theatlantic.com).

to see with the naked eye a firefly flitting through a five-alarm fire from a mile away.

Instead, work focused on the method by which Neptune's existence was predicted: detecting a planet's gravitational effect on another body. In the case of extrasolar planets, one is dealing with the effect of a molehill on a mountain; nevertheless, the mountain does move.

Any two bodies exert a gravitational effect on each other, and in the case of a planetary system, a planet and star actually both orbit the same point—their mutual center of mass. A star's mass dwarfs a planet's so much, though, that that center of mass actually lies inside the star, just slightly off its center. So it appears that the planet simply orbits the star. Still, the fact that the star is orbiting around a point off its center causes it to wobble slightly, just as Uranus wobbles due to Neptune's presence. If we were observing our sun from ten light-years away, the wobble caused by Jupiter would appear as a deviation of just a bit more than one milliarcsecond. For comparison, when we look at Saturn with the naked eye, it appears twenty thousand times wider than that. The amount of the sun's wobble in the above example is equivalent to the thickness of a human hair as seen from two miles away.

The situation is even worse than that. As it turns out, planets aren't the only things that can make a star wobble. The first detection of such an effect was due to the pull of one star on another. In 1844, astronomer Friedrich Wilhelm Bessel wrote that the motions of the stars Procyon and Sirius are "very sensibly altered." He proposed that this was due to unseen orbiting companions, but didn't live to see his theory confirmed. In 1862 and 1896, astronomers finally detected the sources of Bessel's observations: smaller and dimmer "twin" stars orbiting around Procyon and Sirius.

Bessel had pioneered the art of using astrometry to "see" invisible objects. Astrometry is the branch of astronomy that deals with measuring celestial bodies, in particular recording their movement through the sky. Simply by charting the paths of Sirius and Procyon through the night sky, Bessel had discovered, or at least predicted the existence of, their companions. Over a hundred years later, astronomers would use a similar technique to discover planets orbiting other stars. Confirmation, however, would not always come as easily.

The first announcements of astrometrical "discoveries" of extrasolar planets came in the 1940s, when wobbles were detected in the paths of several stars between eight and eleven light-years away. The calculated masses of the planets causing these disturbances, however, were twelve to fifteen times the mass of Jupiter. Today objects of such magnitude would probably be presumed to be brown-dwarf stars (stars that don't have quite enough mass to ever get their fusion reactions going) rather than planets. In any case, the incriminating evidence was later found to be false: the wobbles were due to observational errors.

Then, in 1963, Peter van de Kamp announced the existence of a planet around Barnard's Star. Van de Kamp, born in Holland in 1901, received his doctorate at Berkeley and joined the Swarthmore University faculty in 1937. Almost immediately, he began astrometrical measurements of Barnard's Star, a small "red dwarf" only six light-years away, discovered by Edward Emerson Barnard in 1916. Van de Kamp was a charismatic, entertaining professor and astronomer who made enemies almost as fast as he made friends, and was also a popular musician, directing Swarthmore's orchestra for ten years; on van de Kamp's seventieth birthday, Peter Schickele, of "P. D. Q. Bach" fame, performed a tune he called "The Easy Goin' P.V.-D.K. Ever Lovin' Rag." But van de Kamp will be best remembered for discovering a planet that never was.

After twenty-five years of carefully observing a wobble in Barnard's Star, van de Kamp reported the discovery of a planet 1.5 times the size of Jupiter, orbiting the star every twenty-four years. He died in 1995, in retirement in Holland, confident to the end that he was the first astronomer in history to discover an extrasolar planet.

Unfortunately, he was alone in that belief. In the 1970s, several younger colleagues, including one whom van de Kamp himself had hired at Swarthmore, proved that the Dutchman's measurements were in error. They found that van de Kamp's measurements were inconsistent across the decades, and traced the problem to periodic maintenance work on the telescope. Van de Kamp may also have overexposed many of his photographic plates, leading to more errors. Various groups have since tried to reproduce his results, with no success. Barnard's Star appears to be childless.

By the 1980s, with the dissolution of the Barnard's Star planet and the earlier reported planets of the 1940s, skywatchers were faced with the prospect of a universe in which planets had formed only around our own sun. This went against the grain: our sun was a typical star in a typical galaxy; it should not be unique in any way. Yet so far that was what observation had indicated. And of course, if there were no planets beyond ours, then there was no life out there—at least no life as we could possibly imagine it.

3.

In 1991, planet fever raged again, and the central figure was an old friend: the pulsar. Pulsars are the superdense collapsed stars that shoot out a spinning beacon of radiation like a lethal lighthouse. The effect as seen on Earth is a pulsating star, and their discovery in 1967, you'll remember, momentarily gave rise to suspicions of "little green men." The pulsating effect has other significance when it comes to orbiting bodies. If another object were orbiting a pulsar and causing a wobble in the pulsar's trajectory, we should be able to detect that wobble as an aberration in the otherwise near-perfect periodicity of the pulse. For as the pulsar is pulled closer to Earth, its next signal will arrive slightly sooner than it would have, and when it is pulled away from Earth, the next signal will be a bit late. In 1969 and 1979, astronomers reported just such findings and inferred the existence of planets around two pulsars, but their conclusions were quickly refuted; the disturbances were due to other natural effects.

Andrew Lyne of the University of Manchester's Jodrell Bank Observatory in England has been studying pulsars since their discovery, when he was a graduate student. In 1991, he and one of his own graduate students, Setnam Shemar, noticed a variation in pulsar B1829-10, in the obscure constellation Scutum the Shield, 35,000 light-years away. At first Lyne ascribed it to the same sort of phenomena that nullified the earlier "discoveries," but as Shemar put the data through various computer models for planetary gravitational effects, it passed time and time again with flying colors. The cover of the July 25 issue of *Nature* announced "First Planet Outside Our Solar System." Lyne and Shemar's planet had the mass of Uranus and orbited the pulsar once every six months.

Astronomers were either shocked or skeptical. The last place you'd expect to find a planet was around a pulsar. For one thing, pulsars are the remnants of red giants, stars so enormous that they would in their lifetime suck in and destroy any planets that might have formed as close as the one Lyne and Shemar had found. Even if planets did survive the red giant stage, they would surely be annihilated by the supernova explosion that precipitates the star's collapse into a pulsar. If this planet was real, it brought into question our entire theory of planet formation. New theories sprang across phone lines and internet connections: perhaps planets could form after the supernova, or perhaps pulsars could "kidnap" other stars' planets with their gravity. But no one, it seems, took much notice of one odd characteristic of the new planet, duly noted in Lyne's article but discounted by him as unimportant. The planet's orbital period was exactly six months, half an Earth year. Quite a coincidence for such a bizarre planetary system a couple of hundred-thousand trillion miles away.

Lyne was promptly scheduled to present his discovery at the next meeting of the American Astronomical Society (AAS), in Atlanta, Georgia, in January 1992. In the meantime, another astronomer was looking at another idiosyncratic pulsar. Alexander Wolszczan was a Polish astronomer who had left behind the difficulties of doing science under a communist regime to work in the relaxed tropical atmosphere of Arecibo, site of Cornell's great dish, where so much important SETI research had taken place. In 1990 he was observing a pulsar, PSR B1257 +12, that was only 1300 light-years from Earth and that spun once every 6.2 milliseconds. This "millisecond pulsar," however, wasn't quite fitting into the expected timing patterns. At first Wolszczan thought it must be another stellar body acting on the pulsar, but by June 1991 he was starting to think that maybe there were planets around his star. Then Lyne's paper came out, and he was both disappointed that he would not be the first and exhilarated at the confirmation of the possibility. A more precise position reading from a colleague at the National Radio Astronomy Observatory's Very Large Array in New Mexico (the grid of twenty-seven dishes used in the movie *Contact*), together with a fine-tuning of his computer models led Wolszczan to a firm conclusion: "I knew then that it could hardly be anything other than terrestrial-mass planets around that pulsar."

Wolszczan published his findings that December: PSR B1257 +12 had two planets just a few times more massive than Earth, with orbital periods of sixty-seven and ninety-eight days and orbital radii a bit smaller and larger than Mercury's, respectively. As the 1992 AAS meeting approached, all the buzz was about the new planets discovered by Wolszczan and Lyne. They would present their work side by side in Atlanta.

Just after New Year's, Lyne was doing more routine work on

his pulsar when he noticed a problem. The latest observations showed a slightly different position for his pulsar than had been used to compute the period of the supposed planet. He stared at the data and, as he said, "had a flash of insight. Unfortunately my insight was correct." With some trepidation, he quickly loaded the new coordinates into his computer model. As he told his colleague Ken Croswell, "Only a matter of three minutes later, the computer had reprocessed it, and the six-month periodicity went away. Within half a minute of seeing it disappear, I knew what had happened: there was no planet around this pulsar at all. For the next half hour or so, I just sat staring at my work station. I was completely numb and inert, figuring out all the consequences of all the rubbish that we had written about and spoken on in the previous few months."

The next week, over a thousand astronomers gathered in the hotel auditorium in Atlanta to applaud the work of Lyne and Wolszczan, discoverers of the first extrasolar planets. Their talks, scheduled right next to each other, formed the marquee event of the conference.

Lyne began by declaring calmly, "This talk is not the one I was originally proposing to give." He then described his initial observations and the sequence of rational steps that led to the inference that there was a planet orbiting the pulsar. The audience was already familiar with this information as it had been published, and they listened politely, giving their man his deserved moment in the spotlight. Then Lyne continued, with dignified humility: "The data can be accounted for either by a planet orbiting the star, or by a mistake. Unfortunately," he said, "it was a mistake." The wobble, it turned out, was due not to a planet but merely to the Earth's own orbit around the sun. As a few astronomers suspected when they heard that the period of the wobble was exactly six months, it was the Earth's changing position that accounted for the irregularity in the timing of the pulses. When he corrected for that, Lyne told his audience, "the planet just evaporated."

"I was sitting in the front of the audience," one observer said, as reported by Ken Croswell in his book, *Planet Quest.* "I remember him delivering the punch line, and feeling the intake of breath propagating back through the room."

Lyne had expected a somber, perhaps even reproachful, response, but instead his colleagues were completely taken with his honesty and accountability. They broke out into a rousing ovation.

Wolszczan faced the unenviable task of following such an emotional performance, but he succeeded in convincing his colleagues that no such error was responsible for his own results. And in fact, over the following months, as independent observers confirmed his findings, it became clear that *his* pulsar planets were for real. The fuss and excitement of the last year were not for naught: planets did exist around pulsars. Our own planets are not the only ones in the universe.

4.

As monumental as the discovery was, though, in and of itself it did little to improve the chances of extraterrestrial life. Pulsars are beacons of deadly radiation that would forever preclude the appearance of life on their planets. The main importance of Wolszczan's work was that it showed the feasibility of discovering planets by sifting through their gravitational fingerprints. Approximately Earth-sized planets had been found, simply by observing their pull on their central star. It would be harder with regular stars, without the handy pulse timing of pulsars, but it could be done.

The 1980s had brought a new method of searching for invisible planets. In the 1970s, two Canadian astronomers, Bruce Campbell and Gordon Walker, began to work on a technique previously used to look for double stars. Called Doppler-shift analysis, it detects the wobble arising from the gravitational pull of a neighboring body, but by a different method. In astrometry, you literally see the perturbation in the star's path across the sky. With the Doppler method, one detects the change in the star's velocity toward or away from the observer. When a planet (or companion star) is whipping a star toward the Earth, the star's light is shifted toward the blue, and when it swings back away from the Earth, there is a red shift.

The two methods have different strengths and weaknesses. Astrometry is better for finding planets with large orbits, which place the center of gravity farther from the star and cause a more noticeable wobble. Doppler analysis is better for finding planets closer to their star, for this causes a stronger whipping action and greater changes of velocity. The Doppler method has the advantage of being unaffected by the distance of the star from us; the change of velocity is the same no matter how far away the system, whereas with astrometry it is much easier to see wobbles in nearby stars. However, astrometry provides a much more accurate calculation of the mass of the planet. This is because with the Doppler method you never know just how much of the wobble you're detecting. If the star happens to be being pulled directly toward and away from you, then you have 100 percent accuracy. But if, as is more likely, it is being pulled in a direction somewhat tilted to your line of sight, then you are detecting only the component of the wobble that is in your direction. (As a limiting case, if the planet was orbiting in a plane perpendicular to your line of sight, you wouldn't see any Doppler shift at all.) So what you're really finding is the minimum red and blue shift, hence the minimum mass of the planet.

When Campbell and Walker first turned their attention to detecting planets, they ran up against the problem of precision. Their technique had a margin of error of about 1000 meters per second, yet a planet as massive as Jupiter induces a change in velocity of the sun of only 12.5 meters per second. By the 1980s, however, they had managed to improve their accuracy to 15 meters a second. Between 1980 and 1995, the Canadians observed twenty-one stars, hoping to find planets just a bit larger than Jupiter.

They found nothing. Nor did a number of other astronomers using similar methods. It was starting to look like there might not be so many Jupiter-sized (or Jovian) worlds out there after all. And if there were so few of that size, then why would Earthlike planets, which might harbor life, be any more prevalent?

Then, in October 1995, at a conference in Florence, Italy, a pair of Swiss astronomers made a stunning announcement: they had discovered a planet with 60 percent the mass of Jupiter orbiting a very sunlike star only fifty light-years away: 51 Pegasi.

Michel Mayor of the Geneva Observatory and his graduate student Didier Queloz had leapfrogged the Canadian team, getting their Doppler precision down to a margin of error of thirteen meters per second. In April 1994 they commenced their search of 142 stars. That December they noticed some discordant data coming out of 51 Pegasi—changes in velocity of about sixty meters per second. As this was far more than they expected even from Jovian planets, their first thought was that their instruments must need a tune-up. But after carefully observing 51 Pegasi for a week, they could make out a definite pattern. The star seemed to be swinging back and forth with a consistent period of four days, just as though a planet were causing the disturbance. In January they calculated the planet's characteristics. The magnitude of the change in velocity indicated a planet with a (minimum) mass about half that of Jupiter. But the incredibly short period—a four-day year!—meant that the planet was orbiting at a very tight radius. This enormous planet, 191 times as large as Earth, was even closer to 51 Pegasi than Mercury is to the sun.

Once again, Mayor and Queloz wondered about their instruments. If this planet were genuine, it would throw planet-formation theory into chaos, even more so than had the discovery of pulsar planets. Giant planets were supposed to form far away from their stars, where the temperature is so low—perhaps minus 150° Centigrade—that vast quantities of water ice and even hydrogen and helium can coagulate on top of the silicates to form a "gas giant" like Jupiter. Closer to the star the temperatures are obviously warmer and the ices and gases evaporate and are blown away. Since these form the vast majority of material in and surrounding the star, there simply isn't supposed to be enough material in the inner disk to form a Jovian planet.

Mayor and Queloz were well aware of the number of mistaken discoveries over the decades and were wary of making another. After all, the other groups had been searching for years with no results, and they had seen their whopper after only three months. Throughout January and February they made their observations and held their tongues.

At the end of February, they had to suspend their observations as 51 Pegasi passed behind the sun. For four months they went over their calculations again and again, checking for errors and making exact predictions of what 51 Peg's Doppler shift would be when it reemerged.

One night in early July, Mayor and Queloz entered the observatory, calculations in hand. Close to midnight, the Pegasus constellation rose in the east, and they trained the telescope on 51 Pegasi. A few minutes later, their computer had spat out the Doppler shift, and it was champagne time. The results matched their prediction exactly. They felt quite certain now that they had discovered the first planet orbiting a sunlike star.

In October, they traveled to Florence and made their announcement. They might as well have reported that they had found a star orbiting a planet. Their colleagues were in an uproar. How could a gas giant be so close to its star? It would be like Jupiter being four times as close to the sun as Mercury. Maybe it was really a brown dwarf star, they conjectured, not a planet. After all, the Doppler method provides only a minimum mass, not an exact one. But Mayor and Queloz had already thought of this and calculated the odds of the body being even four times as massive as Jupiter: less than 1 percent. Perhaps it wasn't an orbiting body at all causing the disturbance, but merely pulsations of the surface of the star (a reasonably common occurrence). Again, they had already ruled this out: they checked with optical astronomers, and 51 Peg's brightness did not vary. For every objection, they had an answer.

Still, many were skeptical. Surely Dr. Death would smite this planet from the realm of supposed existence.

Dr. Death is Geoff Marcy of San Francisco State University, who has earned his nickname by proving the nonexistence of a number of other supposed planets. His reaction when a colleague gave him the news from Florence of a giant planet with a four-day year orbiting 51 Pegasi at an eighth the radius of Mercury's: "Give me a break. Are we going to have to waste telescope time debunking yet another false detection?"

Marcy had long ago dismissed 51 Peg since it was classified as an old G2 subgiant in his 1982 catalog of bright stars. Old subgiants "puff out and their surface becomes frothy and gurgly," giving extra Doppler signals which obscure the expected signals. But rechecking in a newer edition, he now found 51 Peg had been misclassified. It's a main sequence, sunlike star— a viable candidate to have planets.

In 1994, with his graduate student Paul Butler, Marcy had developed a system that yielded an accuracy of three meters per second. This was far better than the Canadians or the Swiss could do, but it required enormous computer resources to analyze the data. So much so that Marcy and Butler's results were still lagging far behind their competitors' when the Swiss astronomers made their announcement.

With just one star to concentrate on, however, with the purpose of certain refutation, Dr. Death and his assistant could get a result quickly enough. Just five days after the Florence announcement, Marcy and Butler began a four-night run at the Lick Observatory on Mt. Hamilton in California. As it turned out, they needed only three nights—not to bury Mayor and Queloz but to praise them. Their results confirmed exactly the Swiss results. As strange as the planet was, it really was there. They put out a quiet press release announcing that they had confirmed the Swiss results.

"It was a wonderful drama depicting how science should be done," said Marcy. "One team makes a claim that is spectacular and clearly of historic value; another team, us, being skeptical and even cynical, comes in, and what happens? We simply confirm what they found. It was science at its finest."

What happened next was not. Mayor and Queloz had submitted their work to *Nature* for publication, and had agreed to abide by that journal's publication rule: no talking to the media until the article appears in press. But Marcy and Butler had not submitted anything for publication and so weren't bound to silence. Their press release unleashed a horde of interview requests, and the two Americans made the front page of newspapers around the globe. Then ABC's *Nightline* called. They were canceling their regular show to talk about the discovery. Marcy and Butler were filmed that afternoon, and Marcy appeared on the show live that night. ABC also invited Harvard astronomer David Latham, who took the opportunity to make an announcement of his own: he had discovered a second planet around 51 Pegasi, much farther out, which suggested that there might be other planets in a habitable zone. Latham's announcement was outrageously premature—he was searching with a system that had only a 500-meter-per-second accuracy—and, as it later turned out, wrong. The program left the viewing public with several false impressions: that the 51

Pegasi planet was the first extrasolar planet ever discovered (ignoring Wolszczan's pulsar planet), that Marcy and Latham were the discoverers, and that there were two planets around the star.

In any case, the scientific community clearly understood. Fifty light-years away, there was a real planet orbiting a star almost identical to our sun. The problem was how to account for its existence. As with the pulsar planets, new theories sprouted to explain the appearance of a gas giant so close to a star. A growing consensus left the formation theory alone but added a new twist: perhaps some planets wander from their birthplace. A gas giant that formed at about the same distance as Jupiter from the sun might subsequently be pulled in toward its star by the mass of the inner disk, and then (for some as-yet-unknown reason) stabilized at a close orbit rather than being pulled all the way into the star.

If you think that sounds a bit like grabbing for straws, Dr. Death agrees. "I think what's happening here," Marcy told Ken Croswell, "is that we scientists are doing what we always do: hanging on to the paradigm until the last possible moment, the paradigm being that Jupiter-like planets form at [Jupiter's distance from the sun]. If I had to put my money on the craps table, I would bet that we're going to learn later that giant planets can form closer in and that we've been deluded by our own solar system."

Despite all the news coverage Marcy and Butler received for confirming Mayor and Queloz's discovery, they knew as well as their colleagues that they had been beaten to the punch. While they had been accumulating tons of data that their computer would have to analyze later, the Swiss had done their analysis in real time and jumped right in front of them. So in November 1995, Marcy and Butler hoarded some computer time and crunched through all their old data. Within a month, the old ground yielded gold.

They found evidence for a Pegasus-type wobble around the star 70 Virginis, and then they found something even better.

Forty-six light years away another virtual twin of our sun, 47 Ursae Majoris, was wobbling. After careful analysis—always wary of a false announcement—they determined that the star had a planet with a probable mass of three Jupiters. And unlike the strange, tight orbit of 51 Peg's planet, this one orbited out at 2.1 astronomical units,* a distance between the orbits of Mars and Jupiter. So this planet really looked like one of ours. "This is a planet that we feel very strongly about," Marcy announced at the AAS meeting in January 1996, just a few weeks after making their discovery. "What's fantastic about this planet is that it is in close enough but not too close, so that the planet's surface will be warmed up to a temperature that's comfortable for almost all of us. It should have a surface temperature of about 80° Centigrade, about the temp of warm tea, so any water there would be in a liquid form."

Although it is presumably a gas giant like Jupiter, and therefore highly unlikely that it could harbor life, a planet like this might have a terrestrial-sized moon that might have liquid water which might mean life . . .

Back to reality. Later in 1996, Marcy and Butler continued to harvest discoveries from their accumulated data, announcing three more new planets, around the stars Rho¹ Cancri A, Tau Boötis A, and Upsilon Andromedae. All three were of the type that came to be known as "Pegasian." That is, like 51 Pegasi's planet, they were gas giants (Jupiter types) in orbits very close to their stars.

For a while Marcy and Butler's Pegasian finds seemed even more important than was first thought, for the very term *Pegasian* looked as if it might be a misnomer when, in February of 1997, a voice of dissension rose from Canada. David Gray of the University of Western Ontario presented evidence in a *Nature* article disputing the existence of 51 Pegasi's planet. He had been observing the star since 1989, and had recorded a

*An astronomical unit (AU) is the mean Earth-sun distance: roughly 93 million miles.

variation in the absorption lines of its spectrum with the exact same 4.23-day period as Mayor and Queloz had found. "When only the variation of line position was known," he wrote, "it seemed reasonable to interpret this variation as the 'reflex' motion of an orbiting planet pulling on the star. But with the new information, this planet hypothesis is no longer acceptable; a planet cannot alter the shapes of the spectral lines." Instead, he proposed that the variation was due to the oscillation of the star's surface—a tidal swell not unlike the sloshing of water in a tub.

Gray posted his paper on his web site, and even before it was published in *Nature* Marcy, Butler, Mayor, and Queloz had responded with their own internet rebuttal. For one thing, they said, such oscillations should result in changes in brightness, and no such variations have been seen. Also, Gray's objection applied only to 51 Pegasi, not to the other Pegasian planets: why would the method of ascertaining these planets' existence be wrong in only one out of four cases? Finally, they suggested that Gray's data were simply in error. "A planet explanation naturally explains all the data," they wrote, while Gray's oscillation theory was "far more extraordinary, far more unexplainable."

Gray dismissed the objections. His computer models accounted for the lack of observable brightness variation, and surface oscillation is known to be a fickle phenomenon, occurring on some stars but not on others. As for the oddity of the hypothesis: "Nature simply does not wait for us to understand something before doing it. Thank goodness."

At the July 1997 meeting of the AAS's Division for Planetary Sciences on the MIT campus in Cambridge, Massachusetts, the opposing forces met in an informal debate which was the highlight of the conference. Gray, the veteran star man, appeared casual and relaxed in front of an auditorium full of planet scientists. Though he avowed not to be terribly interested in planet discovery (or its refutation), he did joke that for the first time in thirty-one years at Western Ontario, he had received congratulations from his dean.

The result of the debate, though, was an agreement to disagree. William Cochran of the University of Texas, representing the planet seekers, assured his audience of Gray's reputation for fastidious collection of data. Marcy et al.'s charge of error notwithstanding, the scientific community must take his data seriously. Gray and Cochran agreed that more data were needed to settle the question. In a few months, when 51 Pegasi roamed back into view, five groups would seek to settle the question. "The verdict," said Gray, "will come in the fall."

And so it did. Four different groups, ranging from Texas to Paris, now agree: 51 Peg indeed has a planet. As Gray had pointed out, there was a less than 1 percent chance that his data challenging the planet were due to a spurious signal showing up strictly by chance; but "less than 1 percent" is not zero, and now it appears that that is what happened. "Nature played a dirty trick on him," as a member of one of the other groups said.

Gray's challenge to the planet's existence serves as a note of caution. Until we can detect extrasolar planets directly, there will always be the possibility of some as yet unknown phenomenon causing a star apparently to wobble. As Marcy himself said, we must be wary of hanging on to old explanations too long. Still, all the data taken to date indicate that what we are seeing are indeed real planets (or perhaps brown dwarfs; this distinction will be discussed in Chapter 10). The amazing proliferation of discoveries since 1995 by a variety of groups from all over the world, using different equipment and techniques, argues strongly for their reality. The latest catalog of extrasolar planets, as we go to press in 1998, includes nearly two dozen confirmed, with another dozen listed as preliminary or not yet confirmed (see http://www.obspm.fr/planets).

The appearance of so many planets after such a short period of time promises an abundance of planets to be discovered over the coming decades and centuries. Especially when one considers the primitive means at our disposal today, relative to the near future, the potential bounty seems limitless.

The extrasolar planets discovered so far have one common characteristic—the inability to harbor life. The most Earthlike of them are in orbit around deadly pulsars, and the others are gas giants like Jupiter, most of which are too close to their stars anyway, with surface temperatures of thousands of degrees. This lifelessness is no cause for disappointment among life-seekers, though. We simply don't have the precision yet to detect Earth-sized moons or planets. Around solar-type stars, we can only "see" Jupiter-sized planets at this point—and we found several of them almost immediately. There is every reason to think that as our technology advances, we will start fishing out planets of all sizes and types from the celestial sea. We can't know this until we do it, of course, but the big question has already been answered: planets do form as a matter of course in the development of stars.

9 | Where Are They?

Say first, of God above, or man below,
What can we reason, but from what we know?
Of man, what see we but this station here,
From which to reason, or to which refer?
Through worlds unnumbered though the God be known,
'Tis ours to trace him only in our own.

He, who through vast immensity can pierce,
See worlds on worlds compose one universe,
Observe how system into system runs,
What other planets circle other suns,
What varied Being peoples every star,
May tell why Heaven has made us as we are.

ALEXANDER POPE, *Essay on Man*

1.

The most optimistic estimates for f_p in the Drake equation, the fraction of stars that have planets, now appeared to be reasonably accurate. The news of Marcy and Butler's discoveries rang through the SETI newsletters and web pages with the thrill of corroboration. There *were* planets out there, just as they had always maintained. The arguments for the probability of intelligent life were stronger than ever. Perhaps now even the government would get back into the act.

Still, there were the naysayers. Even among the scientific community, there had never been a unanimous consensus for SETI. In fact, there remains a small group of astronomers who actively maintain that there are probably no other civilizations in the galaxy. The public opposition began in 1975, with the appearance of an article by Michael Hart, of Trinity College, Cambridge, entitled "An Explanation for the Absence of Extraterrestrials on Earth." Hart began by reviving the question known as the Fermi paradox.

Enrico Fermi had established one of the world's premier physics programs in Italy during the 1930s. When he won the 1938 Nobel Prize in Physics for the discovery of new radioactive elements and the role of slow neutrons in their production, he went to Stockholm for the ceremony and took the money and ran—not back to fascist Italy but to the United States. After helping his adopted country develop the atomic bomb, Fermi remained at Los Alamos after the war and began thinking about the possibilities for life elsewhere in the universe. Probably going through an algorithm similar to the Drake equation—the number of stars, fraction of stars with planets, and so on—he came to the conclusion that there must be many other civilizations out there, and that many of them would be on worlds millions of years older than Earth and thus should be unfathomably more advanced than we. If they, like us, have an irresistible impulse to explore their surroundings— and it's hard to conceive of any intelligence that would be so uncurious as to stay contentedly at home—they should have long ago developed a means of interstellar travel. And they should have found us.

"Where are they?" he asked his colleagues. "Where are these superior, advanced beings?" No one was able to give him a good answer at the time except Leo Szilard, the Hungarian-born physicist who was one of the earliest theoreticians to work on the bomb and Fermi's colleague on the Manhattan Project. "They are among us," he reportedly said, "but they call themselves Hungarians."

Fermi himself never concluded that his paradox meant that we were alone in the universe. He simply resigned himself to the fact that not enough was known—about the universe and about life and about intelligence—to tackle the question properly. Even to this day, though, opponents of SETI love to bring up Fermi in their arguments. In 1975, Hart claimed that there was still no good answer to the Fermi paradox aside from the conclusion that there were no other civilizations. He went on to make his own analysis of the conditions necessary to life. If the Earth were slightly closer to or further from the sun, he argued, life here would have been impossible. The odds of a planet with the proper elements forming at just the exact orbit that would permit the development of life were so infinitesimal, he said, that it had almost certainly not happened elsewhere in the galaxy. He had even done rudimentary computer simulations to show that planets even slightly different from Earth could never support life.

Later work, as we have seen, envisages life arising *within* a planet, where the distance from the sun is of little importance. Such life could not reasonably be expected to be sending intelligent signals through space, but more sophisticaated computer calculations which take into account the greenhouse effect and the moderating influence of clouds indicate that the "comfort zone" is larger than Hart thought, and that the odds of finding an Earthlike planet are perhaps small but far from infinitesimal. If only one star in a billion has a planet with life on it, that would give about a hundred such planets in our galaxy alone.

However, Hart found company five years later with the publication of two more dissenting arguments. The first was given in a series of three papers by Frank Tipler, a mathematical physicist then at Berkeley, now at Tulane. To begin with, Tipler invoked a theory known as the anthropic principle, first presented by the cosmologist Brandon Carter in 1974. It says, in effect, that it is not surprising that intelligent life has arisen on Earth, for if it had not, we would not be here to discuss it. The existence of other intelligent life, therefore, can not be assumed simply because we exist.

Tipler's second argument was similar to Hart's but drew on an idea first put forth by John von Neumann, one of the earliest computer scientists and one of this century's greatest mathematicians. Von Neumann thought it inevitable that one day we would develop robots that could reproduce themselves. Tipler insisted that if there were civilizations millions of years more advanced than we, they would have dispatched such robotic emissaries about the galaxy long ago. These mechanical pilgrims would visit other planets, use the raw materials there to reproduce, send their new generations off to new stars, and so on until the entire galaxy was colonized. Figuring an average of 100,000 years for each interstellar flight, 1,000 years to mine a new planet, reproduce, and build new starships, and an estimated six billion years for a habitable planet to develop a species capable of building these machines (based on our own example), Tipler estimated that it would take only 6.3 billion years to colonize the entire galaxy—a manageable span considering the galaxy's age of about 10 billion years. In fact, over half of the galaxy's stars (50 billion stars) are at least that old. Echoing Fermi yet again, he asked, "Where are they?" And he answered himself, "They don't exist because their supposed creators—the superior alien civilizations—don't exist."

Around the same time, in 1981, Robert Rood and James Trefil published a book, *Are We Alone?*, that treated SETI seriously but came to the conclusion that there were probably no other intelligent species in the galaxy. They were hardly vociferous opponents of SETI, though; in fact, Rood later became a SETI researcher himself. He felt that the search was essential in order to prove or disprove his theory that there were no extraterrestrials.

Despite the appearance of these scientific rebuttals, most scientists continue to believe in the possibility of intelligent life in the universe. It is true that it takes only one intelligent species for that species to contemplate the universe; but this does not preclude the existence of billions of other such species. As Isaac Asimov told author Frank White:

> In order for a raindrop to form, you have to have some concatenation of cloud and wind patterns and temperature change that will involve a large section of the Earth. But one raindrop isn't the only one that forms; *billions* of raindrops form, and I think if the universe had to be large enough for one world to develop life, it might well mean that many worlds could then develop.

The Fermi paradox— "Why aren't they here?"—is problematic only until you consider the prohibitive obstacles to interstellar travel. The very closest stars would require many years to visit, even traveling at the speed of light, which is impossible according to Einstein's theory of relativity. Today's fastest spaceships would require 200,000 years to travel to Alpha Centauri, our closest bright star. The energy required to send a hundred colonists to another star, as Frank Drake has pointed out, would be enough to meet the energy needs of the entire United States over a human lifetime. And these estimates are regarding nearby stars. When we consider the distances across the entire galaxy, and between galaxies, interstellar travel seems absolutely untenable.

Of course, many will point out, civilizations millions of years more advanced than ours will have developed technologies that we cannot even imagine. They may have spaceships that do travel near the speed of light, or perhaps they have even discovered the means of transgressing Einstein's law and traveling faster than the speed of light. Kip Thorne, an astrophysi-cist at Cal Tech, has suggested the possibilities of "wormholes" in space that allow one to disappear in one part of the universe and appear in another instantaneously. And then there are the von Neumann machines, that don't mind giving their lives for galactic exploration, and would colonize the galaxy over 300 million years.

All these Trekian wonders, and others that we cannot even imagine, are possible. But they are hardly probable. When we compare the enormous hurdles to interstellar travel with the relative ease of electromagnetic communication, it is almost mandatory to conclude that instead of spending monstrous amounts of energy and millions of years colonizing the uni-verse, a civilization would first send out beams of radio waves to communicate with their fellow beings. The cost is trivial, the potential rewards life-transforming. Clearly the place to expect the presence of extraterrestrials is in our telescope dishes, not on our soil.

2.

The search, then, continues. There is only one government-supported project in the world—META II in Argentina. Fed by the spirit of discovery, however, parabolic ears listen in Italy, Australia, and Ohio, at Green Bank and Harvard, and of course in the valley near Arecibo. Amateur astronomers are also keep-ing their ears open: the SETI League, an international group of radio astronomy enthusiasts, is devoted to building a net-work of people using their personal computers to listen for sig-nals at home. Directed by aerospace engineer Paul Shuch of the Pennsylvania College of Technology, the league is develop-ing their own Project Argus, an array of 5,000 amateur satellite dishes around the world that they hope to have up by 2001 (as of 1997, only 27 of their 468 members had dishes and software operating). All you need to join the search is a home satellite dish, a 486 PC or better, and special signal-processing software that the league provides. All the information you need is at their web site: seti1.setileague.org.

Even easier to join is the SETI@home program, a clever offshoot of the University of California's SERENDIP search, in which the power of normal personal computers will be summed and logged in with the Arecibo radio telescope. It uses a program that you can load onto your own computer. Like a screensaver, it activates when you stop using your computer and shuts off automatically when you return to work. The difference is that instead of producing pretty fish on your screen, it will be automatically analyzing data sent to it from Arecibo.

The project aims to sign up 50,000 people—that is, 50,000 individual PCs—and already has nearly 40,000 people on its mailing list, waiting to join in. The power of all those machines added together will provide a program as sophisticated as any other SETI project. Conceived in 1995, with prototype versions of the software developed during 1996, the program is currently on hold until more money can be raised. If you want to join, visit www.bigscience.com/setiathome.html.

The power of SETI searches has developed at a rate unparalleled in the history of human exploration, doubling every 235 days. In comparison to today's machines, the early attempts to detect intelligent signals seem quixotic indeed. Jean Heidmann of the Paris Observatory, France's leading SETI scientist, likens them to the "cast[ing] [of] nets a few times into a vast ocean of interstellar signals, searching for a minute bottle that may, perhaps, contain a tiny piece of paper." And no doubt the next generation of computers will make BETA and SERENDIP IV look like hopeless stabs into the abyss. As Heidmann points out, even Project Phoenix is looking at only one thousandth of the galaxy's diameter and examining only one hundred-millionth of its volume. What's more, each targeted star will be examined at each channel for only thirty seconds in ten years. If a planet's beacon is turned the wrong way when we happen to listen to it on the proper frequency, we'll miss it.

There are even more fundamental questions. The key to SETI is to guess the type of communication that an alien society would use. The best guesses so far have been that they would use radio waves, and that they would choose a frequency based on "universal" knowledge—for instance, the 1420 MHz hydrogen frequency. But these are assumptions formulated by the human brain. Who knows what sort of logic a superadvanced nonhuman life form might use? Even if they did use electromagnetic radiation, other frequencies might seem more obvious to them—multiple harmonics of 1420 MHz, for example, which are even quieter, without the hum of ubiquitous hydrogen in the background. With this in mind, Paul Horowitz focused his Suitcase SETI at Arecibo in 1982 at double the hydrogen frequency. But perhaps there is some other, more obvious universal magic frequency that we can not yet fathom.

At an international SETI meeting in Estonia, in the Soviet Union, in 1981, Drake pointed out that the hydrogen frequency, which he himself had championed for decades, might be all wrong. Radio signals, he reminded his audience, have a way of "widening" when they encounter clouds of free electrons in interstellar space. Even the most narrow-band signals, which had always been considered the earmark of intelligence, would be spread out over a wider band by the time they reached the Earth. This added a new complication to the effort to recognize signals. Unfortunately, the quiet microwave section of the spectrum, where the hydrogen line lies, is particularly susceptible to this broadening effect. Higher frequency signals are much less vulnerable. A compromise between this factor and the still crucial consideration of quiet background noise, according to Drake, suggested a frequency of about

70,000 MHz instead of hydrogen's 1420. Perhaps, then, advanced civilizations used a universal frequency of hydrogen-times-fifty, or 71,000 MHz.

There was just one problem with that, Drake informed his audience. These high frequencies are almost completely absorbed by the oxygen in our atmosphere. To receive them, we would have to place our radio telescopes in orbit around the Earth. "Everyone at the meeting," said Drake, "suddenly went for coffee."

Or perhaps there is some mode of communication other than radio waves, far better suited to interstellar contact, of which we cannot yet conceive. Just 150 years ago, an eyeblink in history, radio waves themselves were inconceivable, and we were thinking of lighting fires to signal the Martians. A few centuries from now, our descendants may laugh at our attempts to listen for other civilizations with radio dishes.

For the moment, though, we have no choice but to plumb these waters to the best of our ability, based on the full extent of our present knowledge. And that means using our most advanced radio telescopes to scan the radio spectrum. After all, we still have a long way to go along that one path before we exhaust it. When we develop the technology and the public support to make a truly thorough search of the galaxy, then we can begin to rule out certain scenarios. Until then, positive knowledge can come only with actual contact with another civilization.

What have we learned so far? As Paul Horowitz says, "It doesn't seem that someone is trying to get through to us in an intense way." That may seem obvious, but it was far from so before Project Ozma and its successors. In fact, Drake and the other early pioneers couldn't help but lean forward in anticipation each time the receivers were turned on. For all they knew, they were casting their bait into a stream swarming with fish.

Today, Drake and his new generation of colleagues know better. "No signals survived further scrutiny," is the routine lab-notebook entry of SETI scientists. To scientific skeptics and political opponents, such negative results are a cause to shut down the entire operation. In Carl Sagan's SETI novel, *Contact*, an astronomer opposed to continued funding for Project Argus (based on NASA's Project Cyclops proposal, with the name they should have used) delivers an argument so convincing that it gives even the director of Argus pause:

> No, Ellie, this is endless. After a dozen years you'll find no sign of anything. You'll argue that another Argus facility has to be built at a cost of hundreds of millions of dollars in Australia or Argentina to observe the southern sky. And when that fails, you'll talk about building some paraboloid with a free-flying feed in Earth orbit so you can get millimeter waves. You'll always be able to think of some kind of observation that hasn't been done. You'll always invent some explanation about why the extra-terrestrials like to broadcast where we haven't looked.

The rebuttal is in the bare nascency of the journey. When NASA SETI was canceled, according to one staff member, "it was as if the *Niña*, *Pinta*, and *Santa Maria* had all been called back and mothballed within moments of pulling away from the docks." Even with the resurrection of Phoenix and the progress of other projects, with all the advances and listening hours in the five years since then, our ships are still within sight of shore. No conclusions can possibly be drawn until a significant portion of the galaxy has been searched. And if the discovery of exoplanets is any indication, the best estimate for N in the Drake equation is going to go up, not down. Philip Morrison has likened SETI researchers to "all the people like Columbus who had his plan but didn't get it financed, didn't send the three ships to the Indies. . . . [SETI] hasn't properly begun yet, so we're in the position of these people we've forgotten about."

As we realize the full potential for the reception of electromagnetic radiation, a velvet curtain will draw back from in front of our eyes, and a whole new universe will emerge from obscurity. How do we realize this potential? Aside from predictable computing advances, there are telescope designs already thought of, some presently within our grasp and others fanciful treasures of the future, that will make even Arecibo seem deaf, dumb, and blind.

To begin with, there's the idea of a radio telescope in orbit, a companion to the optical Hubble space telescope. It would not only hear far clearer than terrestrial dishes, but it would also pick up frequencies that never reach the Earth's surface, such as the 70,000 MHz frequency that is absorbed by oxygen. An even more advantageous position would be on the surface of the moon. Drake envisions an Arecibo-like dish built on the far side of the moon, where it would forever be clear of radio interference from Earth. The low gravity and lack of wind to disrupt the hanging instruments would make possible a dish in a crater thirty miles or so wide—150 times the diameter of Arecibo—with several thousand times more collecting area.

Then, of course, there is the Cyclops design, an array of hundreds of dishes spread out across the desert, acting in concert as one impossibly large receptor. Combine that with the space designs, and you have Space Cyclops, a series of Arecibos on the moon, providing enough collecting power to transform the science of radio astronomy, aside from its SETI possibilities.

These designs are all feasible right now, with current technology. They require only funding and time to build them. Astronomers, however, have also looked beyond present means to envision even better ways to see. One of these is the gravitational lens.

The most stunning conclusion of Einstein's general theory of relativity is that objects, in exerting a gravitational pull, bend the space around them. One can imagine the skepticism in 1911 greeting such an outrageous suggestion based solely on equations scribbled on paper. But the proposed curvature of space could be tested, Einstein suggested, by observing a star as its light passed close to the sun.

Since starlight can't ordinarily be seen in daytime, the experiment could be done only during a total solar eclipse, which Frank Dyson, Britain's Astronomer Royal, pointed out would be observable from the eastern part of South America and the western tip of Africa in 1919. So while English and German soldiers were killing each other in 1917, he and another British astronomer made plans to test the German's theory. Two years later, Arthur Eddington, braving the tsetse fly, led a group from Cambridge to the small island of Principe, off the western coast of equatorial Africa, while Dyson's Greenwich team sailed to Brazil. On May 29, 1919, they had their equipment set up and waiting to photograph the stars near the sun during the solar eclipse, when the moon would obscure the sun entirely and make the stars around it visible.

Eddington almost cried when the skies above his telescope clouded over—"From [the time we arrived] no rain fell, except on the morning of the eclipse"—but at the last moment they cleared and he was able to photograph the positions of the stars near the blacked-out sun. On the other side of the Atlantic, in the small Brazilian village of Sobral, the skies were clear and the Greenwich expedition was making similar measurements. Hastily the two groups compared their pictures to photographs of the same portion of the sky taken months earlier at night, when the sun was not there, and to their great joy found that the stars whose light was passing close to the sun showed altered positions. As their light had skimmed past the sun, it had been bent by the sun's gravitational warping of space, and the stars appeared to have moved: Einstein's prediction of the bending of starlight was matched precisely by the measurements.

It took some time to check and verify the data, but finally on November 6, 1919, the results were reported to the Royal Society. The data were conclusive, and Einstein was right. The starlight had bent, and space was curved. The theory of relativity was, as the president of the Society proclaimed, "one of the greatest—perhaps the greatest—of achievements in the history of human thought." All of space was a twisty, curved maelstrom, an unimaginably complex conglomeration of invisible bends and turns, determined by the gravitational masses of all the matter in the universe.

Decades after Einstein's prediction was proved, scientists began to realize that the space-bending properties of gravity made each star into a potential telescope lens. The light from any star behind it would bend around and be focused onto an infinite number of focal points (the closer the light passed to the star, the closer would be its focal point). If we could send a telescope out to one of its focal points, we could use our own sun as a telescope a million times more powerful than the greatest terrestrial telescope ever imagined. We'd be able to see not only distant stars close-up, but also their planets, if they had them. Even particular structures on the planets, such as canyons or rivers, would be visible.

Leave it to Frank Drake to think about how this phenomenon might be harnessed in the service of SETI. As he pointed out at a conference more than ten years ago, not only light but also all other forms of electromagnetic radiation are affected by the curvature of space. A solar gravitational telescope would be not just an optical telescope but also an infrared telescope, a gamma-ray telescope, an x-ray telescope, and, of course, a radio telescope. Even the utopian Cyclops on the moon couldn't compete with a simple radio receiving dish only a few feet wide, placed at the focal point. Such an orbiting dish could detect the equivalent of a normal radio broadcast from across the galaxy.

Unfortunately, the nearest focal point is 51 billion miles away, 550 times as far from the sun as the Earth is. This is more than ten times as far away as Pluto, about a third of a light-year, way beyond the scope of any space mission so far. And in fact the resolution at that point would be far from optimal, for the radiation focused there would be that which passes so close to the sun that it would be distorted by the sun's corona and atmosphere. To get optimal resolution, we'd want to pick a point about twice as far away, to look at radiation that curved around the sun without getting too close.

This feat is not so far from our present capability as one might think. The Jet Propulsion Laboratory has already contemplated a mission to that distance to study the Oort Cloud of comets. Drake's dream, if that mission ever materializes, is to send a radio dish along to tune into the universe's ultimate telescope.

3.

It's one thing to speculate about the tremendous advances in telescope design and computational muscle that could arise in the next century. It's another thing to imagine what they might find. "Since we haven't found signals yet," says Paul Horowitz, "and aren't likely to in any given year, there's plenty of time to think about such things." While such speculation moves beyond the sphere of cold scientific calculation and logical inference and slips partway into the realm of science fiction, SETI astronomers have long contemplated the nature of the beings they might contact. Not surprisingly, their conclusions are often contradictory. Frank Drake is on record as saying, "They won't be too much different from us. . . . There are reasons for our basic anatomy being the way it is . . . A large fraction will have such an anatomy that if you saw them from a distance of a hundred yards in the twilight you might think they were human." Paul Horowitz disagrees: "[Alien civilizations] are guaranteed not to look like us, i.e. upright bipeds with stereo-scopic vision."

The prevailing opinion is that the great variety of possible planetary environments—from thin atmospheres with very little sunlight to red-hot, heavy atmospheres, to high gravity planets like Jupiter—combined with the unthinkable possible vagaries of evolution would produce life forms that would be nothing like humans, and perhaps not even imaginable by humans. Thirty years ago, Carl Sagan conjectured:

> There seems no reason for extraterrestrial organisms to have any particular number of legs—or, for that matter, any legs at all. . . . The greater the gravity of the planet, the smaller will be the largest animals. On planets with low gravity, there may be organisms which, from our point of view, would be long and spindly. The same applies to architecture . . . High-gravity worlds should have short and squat structures . . . On other worlds respiration may not be required . . . The usefulness of a sense of hearing depends on the composition, and temperature, of the atmosphere, which determines the velocity of sound . . .

But vision, he wrote, most probably would be a feature of the extraterrestrial body. Almost all stars that might support planetary systems emit most of their radiation in the visible part of the spectrum, and those frequencies too are least likely to be absorbed by atmospheres. A creature that "sees" infrared or ultraviolet radiation would be near-blind on most planets, because those frequenciess are mostly blocked by the atmosphere. Radio waves do get through, of course, but a creature would need enormous "eyeballs" to collect as much information from radio waves as we do from light.

In his book *Cosmos* Sagan describes beings that he and his colleague Ed Salpeter envisioned, creatures that could theoretically thrive on Jupiter—or on exoplanets similar to Jupiter. Enormous balloons the size of cities, with organic membranes that "might eat preformed organic molecules, or make their own from sunlight and air," these "floaters" would live in the middle layers of Jupiter's atmosphere, where the temperature is a comfortable 20° Centigrade.

"Nature always has more imagination than we have," says Freeman Dyson, the great theoretical physicist. Born and educated in England, and professor at Princeton's Institute for Advanced Study since 1953, Dyson is best known in SETI circles for his admittedly fanciful concept of superadvanced civilizations building shells around their solar systems to harness all the energy of their sun. To his amazement, both science fiction fans and serious astrophysicists picked up on these "Dyson spheres" as a serious possibility for advanced life.

Dyson, meanwhile, has become something of an astronomical seer. Among other things, he has put out the idea that advanced life need not necessarily be planet-bound. Pointing out that "we are messing the planet up very seriously, and before long we will undoubtedly move elsewhere," he has suggested that intelligent life may inhabit nonplanetary environs, such as comets. Intelligent life may even exist in a form completely incompatible with our concept of "beings." Perhaps clouds of organic gas swirling out in interstellar space . . .

Or they could be machines. Just as Hart predicted that civilizations would send robots out to colonize the galaxy, those von Neumann machines may be all that's left of a civilization of once-organic beings. We are well on the way to creating "intelligent" machines ourselves; we have already done so, to some extent. In less than three thousand years since our first scientific questioning we have built a machine that can beat a human at the complex game of chess. In millions of years, will we not have machines that reproduce as well as think? (We ourselves are better at reproducing than at thinking.) Such machines may be the last Earthlings, spreading out and colonizing our neighorhood of the galaxy.

Eventually, however, we may not need machines to survive us. One conjecture often tossed around the SETI table is that advanced beings may have learned the secret of immortality. They may have developed a cure for every disease, including old age, or perhaps they have even found a way to arrest the aging process. As Frank Drake and Dava Sobel explained:

> Most of us view the slow destruction of our bodily organs and our mental processes as an inevitable law of nature. But this is not true. There is nothing in the chemistry of life to require deterioration and death. The system of passing genetic information through the DNA molecule is an extremely robust one, with enormous protections against degeneration. We age and die because we have been programmed to do so, just as salmon suddenly grow old and expire within days of laying their eggs. Death is a way for one generation to make room for the next. But death can be outsmarted.

Of course, death could still come as the result of a physical trauma: being shot or blown up, or crashing an airplane or spacecraft. The consequences to SETI of such semi-immortal beings are worth mentioning. The longer the lifetime of an intelligent species, the more cautious they will be, for they will have more to lose by dying in a dangerous pursuit. Any individual human, for example, would be far less likely to go skydiving, or even ride a motorcycle, if he could expect otherwise to live two hundred healthy years. Make it a thousand, and he might even think twice about driving to the corner for a quart of milk. Extrapolating to forever presents a species that may have forsaken all dangerous activities. It raises questions about the desirability of immortality. But perhaps they have created a virtual-reality system that allows them the full pleasure of all imaginable activities in the comfort and safety of their own home. Such a civilization would certainly not be willing to undertake the dangers of space travel. Rather, if they were interested, they would communicate via electromagnetic radiation. It's not only the quickest and cheapest way to go, it's the safest. This provides a ready answer for people like Michael Hart, who say that if there were any advanced civilizations they would have been here already. Maybe the advanced ones are simply too smart to risk their necks.

One might argue that such a civilization would never announce their presence and incur the risk of attack from other worlds. But, as Frank Drake says, perhaps they see communication as their best bet. Instead of waiting for other civilizations to find them, they broadcast their secrets to immortality. Such knowledge would be impossible for anyone to resist, and then they too would be immortal. And an immortal race would never take the risk of interstellar war.

Of course, any attempt to imagine the shape or minds of alien beings is probably futile. As Sagan pointed out, we can't even hope to accurately predict the appearance of our own race millions of years from now. Imagine an ancient Egyptian or Mayan suddenly transported to present-day New York City or, better yet, to the cabin of an orbiting space station. They would hardly be able to believe that they were still among their own species. The humans of five thousand years from now will be just as mysterious to us. And the humans of 10 million years from now, if they still exist? Magicians, superheroes, gods.

Considering that, how can we even try to predict the nature of alien races, begun with a different biology and metamorphosed through the tangles and turns of a separate evolution, propelled to a point thousands or millions or even billions of years more advanced than we? As Sagan wrote, "extrapolations from existing technology do not suffice. We might restrict ourselves only to what is physically possible, even though it may be technically far beyond our present imaginings. But for million-year timescales, even this procedure is hopelessly modest."

We remain primitive beasts, vainly trying to tune our dim thoughts to the spectrum of intergalactic enlightenment.

4.

What, then, do we do when we contact such superbeings? There are those who say we should do nothing: to transmit a naive "Howdy, neighbor! Here we are!" would be tantamount to sending an invitation to our own annihilation. Why broadcast our presence to those who probably possess the ability to destroy us as we would an ant colony? These pessimists forget, however, that we have been broadcasting our presence regularly with commercial radio and television. If galactic predators are lurking out there, just waiting to find new insects to squash, then we're already doomed.

There's probably no reason to liquidate all your stock holdings just yet, though. For one thing, we've already seen the prohibitive obstacles to interstellar travel. And if there were a civilization that had mastered such transportation, they would probably have found us long ago, and visited us in a way that even the federal government couldn't hide. More germane, though, is the probability that any civilization that survived its own nuclear age without annihilating itself either was peaceful from the beginning by nature, or else learned of necessity to be peaceful.

The question of how to respond when the fateful moment of contact finally comes is something the searchers have pondered as much as anything. As far back as 1961, soon after the Order of the Dolphin conference, Frank Drake began thinking about what sort of message humans might dispatch to their fellow beings. Of all the knowledge we have accumulated, what would be most informative, most representative of ourselves? And how could we make ourselves understood?

Thinking in terms of simple pictures—pictures could convey far more information than words and are accessible to any creature with eyes, whereas language would have to be translated, and the simpler the message the less it might be garbled over the vast distances of space—Drake finally came up with a message that he drew on a piece of paper. The picture consisted of a grid of 551 squares, each one filled in or blank, somewhat resembling a crossword puzzle. He then typed out a series of 551 ones and zeros, corresponding to the dark and light squares. He sent the code to each of the Dolphin members, asking if they could decipher it.

Only one person realized it was a picture. Barney Oliver thought to count the total number of bits and then figured out that this number—551—was interesting in that it was divisible only by one, itself, and two prime numbers, 29 and 19. This property suggested that the ones and zeros formed a 29 by 19 grid. By turning the ones and zeros back into black and white squares, he recreated the original picture. He sent back a similarly coded reply: a martini complete with olive.

Unfortunately, although Oliver had passed the first hurdle, he had been unable to make sense of the picture he had received. The rest of the group, and a second audience, including several Nobel Prize winners, hadn't even gotten that far. And the aliens had been so careful to make their message simple! It was a schematic diagram of an imaginary solar system—a sun and nine planets, but not ours—plus diagrams of the oxygen and carbon atoms, to show the basis of life on their planet. A crude picture of a representative of the species showed that he walks on two feet and has one head and two arms. A diagonal line from his head led to a binary representation of the number five billion on the same line as the fourth planet, to show that there are that many of his kind on that world. The number two thousand on the line of the third planet showed that that many of his kind had colonized it, and the number five by the second planet represented an exploratory party that had been sent there. Elsewhere are the numbers one through five, to set an example of counting. Under the alien is a symbol four bits long, intended to name the being as "Four Bits," so he could be referred to in future transmissions without drawing the picture again. The creature's height was given as 11111, or thirty-one in base ten. This meant thirty-one times the transmission wavelength of 10 cm, of course—about ten feet tall.

This example made it evident to Drake and his colleagues just how difficult it might be to decipher an alien message once we receive it. If a message conceived by a fellow human is that obscure, we can hardly imagine how mysterious a signal from an alien might be.

Since Drake's beginning stab at interstellar communication, a few, rudimentary, galactic telegrams and e-mails have in fact found their way from Earth into space. Remember, in the early days SETI was called CETI—communication, not search. In March 1972, the *Pioneer 10* spacecraft was launched on its mission to Jupiter. The ship would orbit the giant planet and then be flung out into space with enough velocity to escape the solar system and hurtle through the universe for eternity. A couple of years earlier, a planetarium lecturer named Richard Hoagland (who would later lead a group insisting that a particular mountain on Mars that vaguely resembles a human face was placed there by aliens as a message to us) and a journalist, Eric Burgess, had suggested to Carl Sagan that a greeting from Earth be placed on *Pioneer* in case it is ever intercepted by intelligent beings. Sagan jumped at the idea and enlisted Drake to help devise the message.

They quickly decided on a gold-anodized aluminum plaque, which would last for billions of years in a vacuum. It also allowed for real drawings instead of grid blocks, and Sagan's wife Linda produced a drawing of a man and woman standing outside the *Pioneer* spacecraft. Below was a drawing of our solar system, with the *Pioneer* clearly depicted shooting off from the third planet. Drake couldn't help himself, though. To show the location of the Earth and the relative time of the craft's launch, he designed a "pulsar map": lines representing specific pulsars emanate from a central point, with the length of each line designating the distance from our sun. Along each line was written the time period of that pulsar's blinking. The numbers were in base two, of course, and the units of time and distance were the universal "atomic" units based on the hydrogen atom—7 x 10^{-10} seconds and 21 centimeters.*

God knows what aliens will think of our plaque, should they ever find it wandering through the galactic abyss, and if they'll have any chance of locating us via the pulsar map. Here on Earth, however, the message caused a bit of a sensation. Apparently some doubted the morality of exposing alien life to full frontal human nudity. Letters to the editors of major newspapers denounced NASA for sending pornography into space. Their authors never stopped to think that naked humans would appear no more erotic to extraterrestrials than naked penguins do to us. *The Chicago Sun-Times* airbrushed offending body parts out of their published version of the picture, and television networks panicked about whether or not to show the plaque on air. The final word was had by a cartoonist who drew a group of aliens perusing the plaque and one saying, "I see the Earthlings look just like us, except they don't wear any clothes."

Special interest groups had their own complaints. Feminists complained that the woman appeared subservient to the man. Although the couple was intended to be amalgams of various races, ethnic groups each saw their own traits in the couple and, oddly enough, complained. British editorials criticized NASA for allowing two Americans to design the message instead of forming an international consortium for the purpose.

In 1977, however, Drake and Sagan teamed up for another CETI message, this time on the occasion of the launching of the *Voyager* spacecrafts which, like *Pioneer 10*, would sail out to

---

*The time is derived by taking the reciprocal of the frequency: $1/1420$ MHz. The distance is given by the equation that links wavelength (in cm) to the frequency:

$$wavelength = \frac{speed\ of\ light}{frequency} = \frac{3 \times 10^{10}\ cm/sec}{1420 \times 10^{6}/sec} = 21\ cm$$

Jupiter and the outer planets, and beyond them to infinity. What about a sort of "record," they thought, like an LP, but one that contained not only music but photographs and movies, enough to give its audience a real picture of what life on Earth was like. Of course, in 1977 laser discs, which could have done exactly what they wanted, didn't exist. But technology had advanced enough to allow such information to be recorded on a regular phonograph record. And so they recorded photographs of various plants and animals, scenes of deserts and forests, shopping malls and superhighways, and of course, human portraits. NASA wanted no part of naked people this time, but they did at least allow silhouette drawings of a nude man and pregnant woman, showing a fetus inside the womb. Accompanying the photos were spoken greetings and sounds including a kiss, a heartbeat, the humpback whale, and a *Saturn V* rocket at liftoff.

The copper record album was housed in an aluminum jacket, covered with an ultrathin layer of pure uranium, a radioactive element. By noting how much of it had decayed and formed other elements, the discoverers could calculate how old the apparatus was. (The choice of uranium gives a clear indication of what sort of voyage the designers had in mind, since it decays appreciably only over periods of many millions of years.) Included also was a record player with detailed instructions.

The contraption would seem comically primitive only a few years later. Even at the time, Drake was the first to admit how unlikely it was that aliens would actually be able to play the record. The discovery of such a machine, though, obviously of intelligent origin, would be monumental. As Drake wrote, "Had we on Earth received a compact disc from some alien civilization at the time of the Voyager launch, I might not have known what to make of it, yet it would have been the answer to my fondest dreams."

A different medium was used in 1974, when a radio message

was beamed from the newly renovated Arecibo dish. It contained the chemical formulas for the components of DNA, a diagram of the DNA double-helix structure, and a picture of a human figure with an indicated height of fourteen. If the recipients could figure out that this meant fourteen times the transmission wavelength of 12.6 cm, they would know the human was five-foot-nine—exactly Frank Drake's height.

For a target, Drake selected M-13: the Great Cluster in the Hercules constellation, which would be directly overhead at the time of the telescope's inauguration ceremony. With 250 guests assembled outside the great dish, a loudspeaker blared the sound of the signal as it was beamed into space. On its particular frequency, in the direction it was pointed, the signal shone brighter than the sun. "By the time the ceremony and the luncheon were over," wrote Drake, "when everyone had piled back on buses to leave the site, the message had reached the vicinity of Pluto's orbit. Already it was leaving the Solar System, not after a flight of years, as with a spacecraft, but after a journey only a few hours long." The assemblage's excitement about the new telescope and its powerful message was sobered only a bit by the realization that if indeed someone in M-13 were to receive the communication, they would do so 25,000 years from now. If they returned the message immediately, their reply would not reach Earth until the year 51,974 A.D., when the words Arecibo and United States will be long forgotten, except perhaps by ancient historians. That's the nature of SETI. The hope is really not for a dialogue, but for separate one-way communications of information. Even an indecipherable message from outer space would convey the information that we are not alone, a discovery of unimaginable consequence to the human race.

5.

Among the international objections to Drake and Sagan's interstellar communications was one that seemed worth considering. Some editorials in the British press had criticized the two Americans for having the hubris to speak for all of Earth. Indeed, it soon became clear to everyone in the SETI community that the responsibility of preparing such a communication, or even of handling the earthshaking news of the discovery of alien civilizations, should be borne on international shoulders.

John Billingham, Jill Tarter, and Michael Michaud (at the time Scientific and Technical Advisor at the American Embassy in Paris) took the lead in this consideration. They headed the SETI Committee of the International Academy of Astronautics in drafting a "Declaration of Principles Concerning Activities Following the Detection of Extraterrestrial Intelligence." The document was ratified by the IAA in 1989 and has since been adopted by every major international space agency. Written in the formal, ornate language of international treatises, it begins:

> We, the institutions and individuals participating in the search for extraterrestrial intelligence,
>
> Recognizing that the search for extraterrestrial intelligence is an integral part of space exploration and is being undertaken for peaceful purposes and for the common interest of all mankind,
>
> Inspired by the profound significance for mankind of detecting evidence of extraterrestrial intelligence. . . .

The main gist after this was simple: verify that you've got the real thing, and then make it public knowledge without delay. The first part is far from trivial. We've already seen how easy it is to let enthusiasm get the best of you in this field; false alarms are prevalent. The discoverers will need to check their FUDD (follow-up detection device) and every other means to make sure the source is nonterrestrial and unnatural. In the movie of Sagan's novel, *Contact*, Ellie Arroway, the director of Project Argus, detects alien signals while listening to her telescope's input through headphones, lying on the hood of her '58 Thunderbird in the desert night—Hollywood's version of Jill Tarter analyzing incoming data at a computer. Hearing pulses of unquestionably intelligent origin, she jumps into her car and tears through the desert, screaming orders to her staff via walkie-talkie like an Army colonel under heavy artillery fire. (Never mind that such two-way radios are taboo around any radio astronomy site—interference from them and other sources is a major problem.) Sprinting into the control room, she checks out the FUDD and, when it confirms the signal as extraterrestrial, she kisses the monitor and sighs, "Thank you, Elmer."

When the signal really comes, the scene might be just slightly less histrionic (for one thing, the entire staff will probably be far more wary of a possible false alarm—some military satellite, for example, that the government doesn't want to reveal). But Arroway's next move—to call SETI astronomers in Australia and get them on the trail—is right on target, according to the Declaration of Principles: "The discoverer should promptly inform all other observers . . . so that those other parties may seek to confirm the discovery by independent observations . . . so that a network can be established to enable continuous monitoring of the signal or phenomenon."

After the international SETI community has confirmed that the signal is from intelligent extraterrestrials, they will inform the Secretary General of the United Nations and then all other relevant international organizations. No mention is made of informing individual national governments, though of course they would have their own reactions to the discovery. In *Contact*, the U. S. National Security Advisor is incensed at

Arroway for having shared her discovery with other nations, as if the United States had any right to deal with alien civilizations on its own, or as if such an alien species could even differentiate Americans from Chinese.

In 1995, the IAA's SETI Committee covered the next stage, drafting a "SETI Reply Protocol," to deal with the question of whether or not and how to respond to an alien signal. Its proposal followed directly from the earlier Declaration: "The United Nations General Assembly should consider making the decision on whether or not to send a message to extraterrestrial intelligence, and on what the content of that message should be. . . . If a decision is made to send a message to extraterrestrial intelligence, it should be sent on behalf of all Humankind."

How to respond would depend, of course, on the nature of the signal we received. One major factor would be the distance of our correspondents. If the signal came from a star only twenty-six light-years away, say, as in *Contact* (an extremely unlikely possibility), then there would be the possibility of an actual dialogue—a message and response—within the span of one human lifetime. If, as is more likely, our friends were hundreds or thousands or tens of thousands of light-years away, then for practical purposes the signal and our response would be monologues intended to edify their recipients. The full contents of the *Encyclopaedia Brittanica* has already been digitized and could be transmitted in only three hours over a 10 MHz band. Of course, we could also pose questions about the mysteries of the universe to our correspondents, who would probably be thousands or millions of years more advanced than we. Imagine our descendants several millenia in the future awaiting the replies to the questions of the Earthlings of 2000. (Even the humans of the year 5000 will be primitive compared to beings a million years more advanced.)

The consequences of discovering intelligent alien life would go far beyond the mere aquisition of knowledge, beyond the sphere of science. The awareness of another species far more evolved than our own would wreak havoc with the Earth's organized religions, bringing to a head the arguments Thomas Aquinas faced centuries ago, forcing theologians to form a new conception of the universe in which we do not play the part of the chosen people. Not all religious leaders would be dismayed. The Reverend Theodore Hesburgh of Notre Dame, for example, has said, "It is precisely because I believe theologically that there is a being called God, and that He is infinite in intelligence, freedom, and power, that I cannot take it upon myself to limit what He might have done." Yet this view is a minority one, and in the event of extraterrestrial contact all religions would have to adapt a bit to encompass it, if they were to survive at all.

The secular implications are no less transforming. As we have already mentioned, the existence of advanced life would offer proof that civilizations needn't necessarily destroy themselves once they develop the capability to do so. "I can't think of a message more important than that," said Isaac Asimov, "because we need hope." A number of scientists have brought up this implication and examined our current situation in a new light, as though we are in a nuclear trial stage: if we can get through a century or so with nuclear technology, then we're in the pink. We will have proved that we belong with the other advanced civilizations that have passed the test. In 1998, while the outlook is sunnier than it was fifteen years ago, the jury appears still to be out.

A SETI discovery might well alter our fate in that regard. A common enemy has often brought historically antagonistic nations together. Even if, as most scientists expect, extraterrestrials were benevolent, the mere recognition of ourselves as one intelligent species among millions should have a unifying effect. The drastic differences between aliens and us would make terrestrial racial, political, and idealogical differences seem trivial.

The failure to find life might have a similar effect. "Suppose," said Sagan, "we make a long sophisticated and unsuccessful search. That would then calibrate something of the rarity and preciousness of life on Earth, and would again serve to unify the human species."

If we do find someone out there, though, as most SETI scientists feel we must if we try hard enough, there is potentially a vast wealth of information that we might receive from our fellow creatures. Just as we might send the *Encyclopaedia Brittanica*, so Drake expects an "Encyclopaedia Galactica" to come streaming into our radio dishes once we find the right source: "Another, even more stirring Renaissance will be fueled by the wealth of alien scientific, technical, and sociological information that awaits us."

It won't come easy. Translating the Encyclopaedia Galactica into English could prove an intellectual endeavor unequaled in history. Our only hope is that our cosmic teachers take our primitive intelligence into account. "I think that if these civilizations are advanced enough to send a beacon that we can find," says Paul Horowitz, "then they'll have found a way to make themselves intelligible. Also, this won't be the first contact for them. They'll have had millions of years of practice. So I'm not worried about that part."

If the broadcasters consider such low forms of life as us to be negligible, however, and have dumbed down their message only for beings one million years behind them, not ten million, then we may have quite a wait before we can figure it all out. In *Contact*, a World Message Consortium spends several years decoding the Message, which turns out to be the blueprints for building an interstellar spacecraft. In reality, the translation of the Message might be a project that dominates science for centuries, or millennia. It wouldn't take that long, though, for the initial discovery to have an effect. The mere realization that we are not alone would shake the foundations of human thought.

"To discover that we share the galaxy or the universe with other sentient beings would upend our view of life and change our perspective of what it means to be human," wrote Frank Drake. "I see it as an awakening, just like the one that followed Copernicus's revolutionary idea that the Earth was not the center of the universe."

Most scientists who have thought much about the subject believe that this renaissance will occur—maybe tomorrow, maybe in a hundred years, maybe a bit longer. "It will happen," says Paul Horowitz. "No doubt about it. It's not on a millenial scale, either. It's on a generational scale, beginning now."

10 | The Colonel's Lady and Judy O'Grady

Airplanes are interesting toys, but of no military value.
GENERAL FOCH, 1911

This is the biggest fool thing we have ever done. The bomb will never go off, and I speak as an expert on explosives.
ADMIRAL LEAHY to PRESIDENT TRUMAN, 1945

Rock 'n' Roll Will be Gone by June.
Headline in V*ariety,* 1955

No matter what happens, the U.S. Navy will not be caught napping.
FRANK KNOX, Secretary of the Navy, December 1941

1.

As these quotations show, nobody can predict the future. But every newspaper carries the weather forecast, and most of us look at it even though we know it's like reading the daily horoscope. Which some of us also do. So in that spirit, here goes.

The situation regarding Mars is that we have a series of experimental results which can best be explained in toto as the result of biological processes, but not one of the results is definitive for the presence of life; there is no "smoking gun." We already know that there are nonbiological possibilities to explain each of the Martian meteorite results, and it is possible that new data will come forward to make any or all of those explanations more likely than they are today. Still, it is going to be difficult to prove that these explanations are the only ones possible.

The argument that the coexistence of magnetite and pyrrhotite in ALH 8001 indicates the presence of microbial life is based on the knowledge that they are not often found together without life, but we already know that *sometimes* they are. Perhaps future experiments will find more examples of nonbiological affinity; this will still not prove the absence of life in ALH 8001. Perhaps future experiments will find structures resembling the putative Martian bacteria in a terrestrial environment where life does not exist; this will still not prove that the 84001 structure is not a bacterium. An experiment proving that the Martian carbonates formed at high temperatures rather than precipitating from liquid water would disprove the life hypothesis, but such a definite conclusion is difficult to come by. A recent mineralogical study "suggests" that the particular concatenation of minerals found in 84001 does in fact indicate formation at high temperatures, as does also the presence of magnetite whiskers, but the evidence is argumentative rather than compulsory. If the sulfur isotope work were improved to the point where it proved conclusively that the carbonate rims do not contain biologically fractionated isotopes, one could still argue that Martian microbes work differently than do those on Earth.

The situation is similar to one in which your wife wakes you in the middle of the night and says she heard a noise downstairs. You grab a baseball bat and tiptoe downstairs and stand still and listen. You hear nothing. You begin to sneak around the house, and you find nothing. But can you be *sure* there is no one there? In the silent, black, empty rooms each little creak of the heating system makes your heart jump. And finally, when you've been through the whole house and found nothing and you climb back up the stairs, can you convince your wife that there is nothing alive down there?

It's very difficult to prove beyond any doubt the absence of life.

On the other hand, can we reasonably expect definitive proof of the presence of life on Mars from experiments done on meteorites that are, after all, only a random sample of that planet? At the 1997 Houston meeting several scientists expressed the view that evidence of biological membrane structures within the fossilized Martian bacteria of 84001 would probably do the trick, and just half a year later the *Salt Lake City Tribune* carried this headline:

LITTLE GREEN SLIME?

The article told of a talk given that day by David McKay to the Geological Society of America, which was meeting in that city. McKay reported finding in 84001 just such a filmy material, resembling nothing so much as the slime secreted by microbes to attach themselves to each other and to their rock hosts. (The "green" aspect is poetic license, referring to the "little green men" ubiquitous in early science fiction.)

So is this slime a "smoking gun" for the existence of life on Mars? McKay shook his head when asked. "There's a lot of work still to be done before we claim that," he said. For example, in order for Martian slime to last a billion years it would have to be mineralized, and McKay thinks that it "appears to be" mineralized, but nobody is sure yet.

Most scientists don't expect a final verdict to come in from the meteoritic work. The contest will be decided only when we get back to Mars and either find or fail to find signs of life there with a truly exhaustive series of experiments, in which the first line of attack will be to define and find the particular rocks that are most likely to have harbored life.

The *Viking* mission in 1976 was a quick and dirty try, and we know now that such tactics are not likely to be successful. As Carl Sagan pointed out at the time, if a spacecraft from Mars landed in the Sahara Desert and scooped up a handful of sand for testing, it could have decided there was no life on Earth. To properly investigate the question of life on a strange planet, you need a systematic approach. *Viking* was fooled because the surface chemistry turned out to be queerer than we had supposed, and we should have supposed that it might be. But at least we have learned our lesson, and the program going into operation today is starting at the beginning and moving right along: before you search for life you have to know the particular environment you're searching. You have to know what kind of life might conceivably exist, and where to look for it. In short, you have to know the geology, chemistry, and physics of the planet and its atmosphere.

In November of 1996 NASA inaugurated a ten-year Mars Surveyor Program which will launch two spacecraft to the planet every twenty-six months, when Earth and Mars are favorably aligned. It weighs 2,337 pounds, half that of the *Mars Observer.* Compared to NASA's previous mission to Mars, which was lost in space after its 1993 launch due to a ruptured fuel line (at a cost of nearly a billion dollars), the first *Mars Global Surveyor Orbiter* costs less than 20 percent as much, exemplifing the new tendency toward smaller and cheaper and smarter. It reached Mars in September 1997 with not enough fuel left for braking, but made use of atmospheric drag to slow itself down and slip into orbit. It is now scanning the surface of Mars, looking for minerals precipitated from water, and taking atmospheric measurements. It should be able to operate for two years.

Mars Surveyor Pathfinder was launched one month after *Orbiter*, but a more direct trajectory got it to Mars earlier. On

July 4, 1997, it floated down via rockets, parachutes, and airbags to a floodplain named Ares Vallis. It bounced forty meters high, then fell back and bounced another fifteen or twenty times across the surface like a rubber ball. When it finally came to a stop, it folded back its sides, lowered a ramp, and deployed Rover: a six-wheeled twenty-five-pound gadget looking like a child's radio-controlled car.* Measuring two feet by one and a half, less than a foot high, it's powered by solar energy during the day and batteries at night, ennabling it to roam over the surface at a top speed of two feet per minute (less than half a mile per hour). Its main scientific equipment consists of a camera, enabling it to look around and find interesting rocks, and a devilishly clever complete chemical laboratory called the alpha proton x-ray spectrometer. This is a single instrument that bombards with alpha particles (from a radioactive curium source) the rocks *Sojourner* finds and sticks its nose up against. The alpha particles induce three types of reactions. Some of them scatter from the nuclei of the rock's atoms, and this scattering—which *Sojourner* then measures—depends on the nuclear charge of the atoms and is therefore indicative of what atoms are there. This mode of operation can indentify atoms from the lightest (except hydrogen and helium) up to about sodium. Some of the alpha particles knock protons out of the rock nuclei, and *Sojourner* analyzes these also, identifying the elements around sodium. Finally, the alpha particles ionize the heavier elements, and *Sojourner* measures the x-rays that are then emitted, providing identification of the heavier elements. Taken together, the three modes of operation provide a complete chemical analysis of the Martian rocks.

Rover has nothing that can detect life, and it's "a remarkably dumb machine. It can't roll and chew gum at the same time," according to project scientist Matt Golombek. (Its brain is a

*Its full name is *Mars Surveyor Pathfinder Rover Sojourner,* a big name for a little critter. The final tag was given to honor Sojourner Truth, a nineteenth-century runaway slave turned abolitionist and feminist preacher.

computer much smaller and with less capacity than the PCs in our homes; it was picked for its ruggedness and low cost.) It was intended primarily as a technology experiment to test rover possibilities for future work. But it's done a lot more than that. As a public relations probe it's been a fantastic success: on one day, July 8, 1997, its web site received 47 *million* hits, making it the greatest single event in Internet history. And scientifically it's been a success beyond expectations, radioing back hard data that show conclusively that Mars really was wet and warm a long time ago. As Matt Golombek told the Geological Society of America, "Mars was a lot more like Earth than it is now. For those of you thinking about little green buggers up there, that's what you want."

It landed in a region called Ares Vallis, in what appears to be the mouth of an ancient water outflow channel. It's found lots of smooth rounded stone spherules and conglomerate rocks, which look exactly like what you'd expect in an environment where rocks are exposed to running water. Combining the rover results with photographs made from the orbiter, the science team is concluding that there was both a lake there for "a long period of time" and a catastrophic flood involving as much water as there is today in all the Great Lakes combined. Of course, it all disappeared sometime between one and three billion years ago; but for a while conditions there might have been conducive to life.

In December 1998 *Mars Surveyor 98* will be launched. As in 1996, it will consist of two craft: an orbiter and a lander. They will study Martian weather, climate, and the water and CO_2 budget, searching for evidence of climate changes in the past. The orbiter will monitor daily weather, look at surface changes due to wind, measure temperature profiles of the atmosphere, and monitor its water vapor and dust content. It will also serve as a data relay satellite for the lander, which will hit the south pole and look for subsurface water via penetrators. These will detach just before the craft enters the atmosphere and plunge

into the ground at 446 mph. Their forebodies should penetrate one to six feet, while the afterbodies remain above
ground and deploy radio antennae.

Also slated for 1998 is Japan's effort, an orbiter named
Planet-B. In 2001, the United States' *Surveyor 3 Orbiter and
Lander* will take off, carrying a spectrometer into orbit to look
for small-scale mineralogical details of ancient and fossilized
hydrothermal springs, which are hypothesized to be the sine
qua non for the origin of life. But finding such a spring won't
be enough. We will have to put a rover right down there on
that precise spot, and that's hard to do. The best we could do
in 1997 for a rover landing target was an area of 100 by 200 km,
giving an imprecision of nearly ten thousand square miles. We
need to improve our steering and landing capabilities, to
shrink the proposed landing area tightly. Instead of dropping a
lander to bounce randomly across the surface we need to work
out a technique of precision landings.

And we have to improve the rover's capabilities, to make it
able to travel further and faster so that it can wander around
over a large area to find the proper samples. NASA's current
plans call for a $500 million mission in 2008 to bring back a
load of rocks weighing not more than half a kilogram, "about
the size of a baked potato," which isn't much. We can't just
stuff everything we see into a knapsack and bring it all back for
analysis here on Earth; we'll have to know which particular
rocks are most likely to harbor evidence of ancient Martian
life, so precise onsite chemical analyses for organics will be
necessary.

All the above is related to the search for ancient life, which
will be present today only as fossils. Even more exciting would
be evidence of life today on Mars, but since there is no longer
any surface water there can be no living creatures on the surface. To find them (if they exist) we'll have to dig deep inside
the planet; current estimates suggest there may be water at a
depth of about four kilometers. But we have no robotic equip-

ment in our armory that could dig down to such a depth, nor is any planned. Our best automated systems could dig down only a few meters, thousands of times less than is needed. To operate the necessary drills would require a human crew up there on Mars.

The earliest we can possibly send people there would be in 2015, and by then we should know the precise details of the planet, inside and out, so they know exactly where to drill. Then perhaps they will be able to dig out the rocks teeming with subterranean alien life and bring them home.

Or maybe not. Mars doesn't seem eager to give up his secrets.

2.

When the diverse group of scientists gathered in the fall of 1996 at the San Juan Capistrano Conference to discuss the possibilities of life on Europa, enthusiasm was high but expectancy was mixed. There were no experimental results in hand or expected soon that could possibly show the presence of life on that moon, since any postulated life lay hidden beneath an obscuring thickness of ice. Instead we focused on the data that might indicate the presence of an ocean beneath the ice, on other data that showed the existence of life on Earth in what might be a similar environment (hydrothermal springs at the bottom of our oceans), and on theoretical arguments that life can and should and perhaps *must* exist wherever the conditions are permissible.

So first is the question of whether there is a liquid ocean under the ice. Thermal calculations indicate that early in Europa's history, when there was more radioactivity, the mantle was certainly hot enough to melt the overlying ice and produce an ocean. But Europa has been cooling off since then, and the same calculations indicate that if the surface ice layer ever cools to a thickness of greater than ten kilometers it will trigger new mechanisms of heat loss and will then quickly freeze all the way through. It would then stay frozen for all time. So there probably was an ocean at some time in the past; the question is whether it still exists. Steve Squyres presented a summary of thinking prior to the conference, and it went like this:

Date	Existence of an Ocean?	Reference
1979	Probably	Cassen et al.
1980	Probably not	Cassen et al.
1983	Probably	Squyres et al.
1987	Probably not	Ross and Shubert
1989	Probably	Ojakangas & Stevenson.

Not much progress there. At the meeting the consensus ran—or crawled, with lots of hesitation—toward the presence of an ocean. The surface features looked to a lot of people like what you might get if ice is floating on water and convection currents in the water are stretching and pulling the ice apart, leaving circular fractures and chunks that fit together like a jig-saw puzzle. But could we really be sure?

Jim Head, of Brown University, speaking from the audience, said, "We don't have a clue. We don't know what's going on up there. What are the fundamental processes? I look up at the pictures and try to understand them, and I don't know what's happening." And David Stephenson, of Cal Tech, chimed in with, "I don't have a clue what causes these circular features," although he did later admit that the evidence is "fairly convincing" for an ocean there today.

The latest round of *Galileo* pictures has made the evidence more convincing. Michael H. Carr, of the U.S. Geological Survey, sums it up: "This is a very convincing set of pictures with respect to the presence of a liquid ocean." So where do we go from here?

First, we continue with earthly oceanography, learning as much as we can about deep-earth organisms which can exist without the benefit of sunlight. Aside from the hot springs at the bottom of the oceans, there is here on Earth one interesting analog to the Europan system. Lake Vostok in Antarctica, 140 miles long by 30 miles wide, has a permanently frozen ice cover that makes it resemble the postulated Europan ocean. The Russians have already begun drilling deep into the two-mile-thick ice, but have not yet penetrated it to the water below. They've halted the drilling a few hundred feet above the water, fearing they might contaminate it with surface organisms and thus never know what life might be indigenous down there. From here on in we have to go very slowly, calculating the effect of each step before we take it.

In the meantime, a veritable gallimaufry of experiments is being proposed to NASA even as you read this. Perhaps the most exciting is *Ice Clipper,* put forward by Henry Harris, a quondam Las Vegas musician turned JPL engineer, and his colleagues in the physics department at Washington University. With a possible launching date early in the new century, *Clipper* would zoom out for six years before reaching Europa. As it sailed in toward its destination it would bomb it, releasing a twenty-pound metal ball that would fall down, break through the icy surface, and splash up a rain of water and whatever creatures live in that hidden ocean. *Clipper* would sail through the debris and extend its arms like flypaper, catching the bugs and holding them tight. It would also sniff the cloud of rising vapor and analyze what chemical compounds it finds there. It would radio those data back to Earth, and then it would swing around Europa and head home, arriving twelve years after launch and bringing our first samples of Europan life trapped within.

Yes, well, maybe. As exciting as the project is, it does sound a bit like a throwback to *Viking:* a quick and dirty experiment to find life. There are too many things that could go wrong, because we don't yet know enough about Europa. We're not sure just how thin the ice crust is, so how can we calculate the size of the bomb necessary to break through? And on a voyage like that, every extra ounce is begrudged. If the ice isn't broken, or if the water and/or slush doesn't blow up high enough, or if it simply returns without any living creatures in its arms, the $250 million it costs will have been wasted. We think it's more likely that NASA will choose to spend the money on a more systematic—if less dramatic—series of experiments.

Chris Chyba wants to put a radar system on an orbiting spacecraft to detect the presence of the liquid ocean everyone strongly suspects is there, but whose existence has still not yet been proven. Radar at various wavelengths will penetrate solid ice and reflect differently from any liquid water underneath, so such a system could give strong evidence, but the data are subject to interpretation; the system would not give an unambiguous and definitive answer. Chyba points out that no single experiment will: "You must have several different methods because each has problems. It could get very ugly," he warns, "but at least this method works on Earth."

Chuck Yoder wants to put a laser altimeter on the orbiter to measure tidal flexing, which would give variations of up to 30 meters if there is an ocean, but only about 1 meter if there is nothing but solid ice. Other experiments would drop a pod with an internal heat source to the surface, in order to melt through it and actually sample the water that is (hopefully) below.

In other words, the choice is between sending a craft up there to look for life directly—as we did with *Viking*—or going more slowly with a planned series of experiments to investigate the Europan environment and finally to plan an experiment based not on Earth analogies but on what life might look like up there. If we go with *Clipper* and it works, everyone will be echoing the Pope: "Wow!" But if it doesn't work, we have very little results for our 250 million bucks. The old story of the tortoise and the hare still holds, and the smart money—if you're a betting man—is on slow and steady.

3.

Beyond the solar system, the presence of life depends on the presence of planets. (Yes, yes, it's possible to have life without planets, but only in the sense that *anything* is possible; let's try to keep as firm a grip as possible on what we perhaps mistakenly perceive as reality.)

Until recently we thought that we needed not only the mere existence of planets, but that they had to be Earthlike and carefully placed in what was called "the habitable zone." They had to be not too close to their star and not too far away, perfectly positioned so that the heat from their star would not be too much or too little: too little would freeze the water and too much would boil it.

We know now that we were once again making the mistake of geomorphism, thinking only of liquid water on the surface because we thought that sunlight was necessary for life. Now that we know it's not, the habitable zone has expanded nearly without limit, and so has the possibility of life. But we still need planets, and planets with particular characteristics.

The exoplanets that we've seen so far are not likely habitats for life. Some of them have highly eccentric orbits, which means that they sweep in close to their sun and then swerve far away, and this in turn means that their climate is going to vary wildly from extreme heat to extreme cold. This is actually not important if we are talking only about microbial life rather than a technologically advanced civilization, because life might form in the thermally shielded interior of such planets, much as we hope it did in the putative ocean of Europa. But we won't be able to determine if such subterranean life exists on those planets without going there, or at least sending a probe. And we won't be in a position to even think seriously about that during the lifetime of anyone reading this book. The distances are just too great.

There is a stronger argument against these planets as a haven for life, and that is simply their size. The gravity of such tremendous masses renders difficult the chemical and physical processes we think of as necessary for life. But again we're arguing on the basis of the only life forms we know: those that evolved on our own rather singular planet. On the other hand—here we go again, Harry—there's nothing else a scientist can do. Without the constraints imposed by experiment, science would devolve into philosophy, cosmopoetry, or even religion. So until we actually find life forms elsewhere, we are doomed to think like people.

There is one other argument against these planets as havens for life: they may not be planets at all. David C. Black, the director of the Lunar and Planetary Institute in Houston, thinks they may all be brown dwarfs. The distinction is not meaningless. A brown dwarf is a failed star, a mass of hydrogen and helium that contracted gravitationally without sufficient mass to heat its interior to the point where a thermonuclear reaction could begin. We have known for decades that many stars are binary systems, two stars circling each other. To find a system in which one of the stars didn't quite make it to fusion would not be of tremendous significance. To switch on its thermonuclear furnace and become a full-blown, shining star, a mass of about eighty Jupiters is needed. But the star can gravitationally condense into a brown dwarf with only about ten Jupiter masses; or at least that's the present thinking. So a brown dwarf is, at the present time, impossible to differentiate from most of the "planets" that have so far been discovered. And this is not a matter merely of terminology; it goes to the heart of the problem.

We've already pointed out that the massive planets discovered so far are not likely places for life. Their importance is that they exist at all. If planets of any type exist around other stars beside our sun, it means that the process we envisage of planet formation by normal means—as part of the process of formation of the star itself—is valid. And if it is, then Earth-type planets should also be forming around other stars. But if the massive objects that have been found are actually brown dwarfs, all bets are off. The only meaning attached to them would be that binary star systems can form with one partner of low mass, and that would not have any bearing on the possibility of life out there.

The best bet so far for a real, Jovian planet is 47 Ursae Majoris, but even here we can't be sure since it could still turn out to be larger than its estimated mass of ~3.5 J.* So if we can't distinquish between a massive planet and a brown dwarf, what can we do?

Quite a bit, as it turns out. The first leap into the new generation of planet seekers has already been made, by a group led by Michael Shao at NASA's Jet Propulsion Laboratory in Pasadena. They have built an infrared interferometer that has a far greater accuracy than either Doppler analysis or traditional astrometry when searching for planets at habitable distances from their stars. Interferometry is the use of two or more telescopes to mimic the action of one large one. The Very Large Array in New Mexico, for example, is an interferometer with twenty-seven radio telescopes acting like one dish with a fifteen-mile diameter. Shao's infrared interferometer uses two telescopes a hundred meters apart on Mt. Palomar. The previous best interferometry could make out differences in milliarcseconds; this new telescope measures in microarcseconds—the thickness of a human hair as seen from two thousand miles away. The Palomar interferometer has just recently gone into operation, and should be able to detect planets as small as Uranus or Neptune orbiting a sun-like star ten light-years away.

A different approach is to look for planetary systems in the making; that is, to search for the disk out of which planets

---

*The Lalande 11185 system also looks promising, but the data are not only uncertain as to mass but are still awaiting confirmation of the planets' existence at all.

form. The first such system to be seen, in 1983, was Beta Pictoris, one of our nearest stars only 49 light-years away. The disk shows up in photographs nearly edge on, reflecting the central star's light from ice grains and dust, while Doppler measurements indicate that it is rotating around the star. More recently, the Hubble Space Telescope has found a cluster of new stars in the constellation Orion. What it sees in several cases is a black spot with a bright center: a young star at the center with a disk around it that is blocking out the starlight from other stars behind it in the cluster.

If we're right in our ideas of star development, the class of stars known as T Tauri are what the sun was when it began, and so they are prominent targets for protoplanetary disk searches. The technique here has been to search at infrared frequencies instead of using visible light. A star shines about a billion times brighter than its planets with visible light, but the contrast can be only a thousand times in the infrared. In the past few years we have found more intense infrared radiation around such stars than theoretical calculations indicate should be there. Computer models show that mass, size, and temperature indicated by the data all look like planetary disks.

The consensus today is that there are disks around half the young stars, and the size range is 100–1000 AU, which is just right for the beginning stages of a planetary system like our own. We haven't actually seen them evolving into planetary systems; still, that is the most likely thing that's going on.

But even better would be to get away from the Earth, since the atmosphere has bundles of air at different temperatures and densities, which bend light differently and mess up the data. For example, we might use gravitational lenses to see distant planets. We've already discussed Frank Drake's dream of a radio telescope positioned out in the Oort Cloud to detect radio signals magnified and focused by the sun's gravitational field. But the idea of a gravitational lens was first thought of by Einstein in relation to visible light, and in the late 1970s the effect he predicted was actually observed, in the form of "twin" quasars. Astronomers soon realized that what looked like identical quasars six arcseconds apart were in fact double images of one quasar, whose light is being bent around a massive elliptical galaxy by the gravity of its trillion stars. Since then, many more "multiple" quasars have been seen through similar gravitational lenses. These discoveries sustain the hope of someday using such natural telescopes to image extrasolar planets.

This is far in the future, but we already have the technology, if not the public support, to put an interferometer into Earth-orbit. An optical space interferometer that NASA hopes to launch in the first decade of the twenty-first century would have enough precision (without the atmosphere's distorting effects hampering it) to see Earth-sized planets around solar-type stars ten light-years away. An infrared space interferometer would be even more sensitive to planets, which emit most of their radiation in the infrared band (rather than at optical frequencies). And this sort of telescope could also help pick out planets hospitable to life. For atmospheres like our own, ozone (a good indicator of oxygen), water vapor, methane, and carbon dioxide would ring alarms on a good enough infrared telescope. These gases help keep us warm by trapping infrared heat—the greenhouse effect. So an infrared map of an Earth-like planet would show dark spots in the spectrum where each of these compounds traps particular infrared frequencies.

The proposed Space Interferometry Mission, with seven telescopes orbiting along a ten-meter baseline, will be a thousand times more sensitive than anything we have today and will have the capability of finding Earth-sized planets around the nearest stars, or Jovian planets around thousands of stars. It's now in the design stage, and NASA hopes to launch it in 2004, at a cost of half a billion dollars.

Roger Angel and Neville Woolf, of the University of Arizona, want to put a seventy-five-meter long infrared interferometer in orbit. To take full advantage of this instrument, they want to escape the infrared radiation scattered by the zodiacal dust in the inner solar system, so their instrument would have to be put out beyond Jupiter, and would cost two billion dollars.

Even more ambitious is NASA's *Terrestrial Planet Finder*. This is at least ten years away from a possible launch, but if and when it goes it will use four 60-inch infrared scopes along a 240-foot platform to find Earth-sized planets and to analyze their atmospheres for the gases that indicate life. Finally, NASA is beginning to think about a Planet Mapper, a tremendous optical inteferometer that would be able to see and send back pictures of the surfaces of Earthlike planets. But this "is beyond our planning horizons at the present," according to NASA. "Planet Mapper will be a challenge for our grandchildren."

4.

SETI has so far given nothing but null results. Before that search was started, there was speculation that the universe was filled with technologically advanced civilizations talking to each other. The situation was reminiscent of a period in our own development, shortly after we discovered radio technology. Civilizations in London and New York were talking to Australia, while in between them on the islands of Micronesia more primitive civilizations were totally unaware that the air they were breathing was filled with these invisible and undetected radio waves. Before SETI, we envisaged Earth as being in that same situation with respect to the rest of the galaxy.

We know now that this is not true. The space through which Earth travels is not filled with radio conversation between other civilizations. There are three obvious possibilities.

The first is the most likely: that there are indeed others out there, but fewer than we thought at first. Interstellar communication is going on, but the space between the stars is not "filled with radio conversation." The universe is so vast that these technological civilizations are sparsely distributed, and it will take some time and effort to seek them out—particularly if they've chosen a "universal" channel that we haven't thought of, and aren't listening to.

The second possibility is that we are living an even closer analogy to the Micronesians of the early twentieth century: the universe is filled with more advanced civilizations communicating with each other, but just as the Micronesians had no idea of the existence of radio waves, we have no idea of their means of communication. The space through which we travel may be filled with some other source of communication of which we are as yet totally ignorant.

The third possibility is, of course, that there is no one out there, and we are alone.

Actually, a fourth possibility remains. We may be alone in the sense that other civilizations exist scattered throughout the universe, but none of them is as advanced as we; none of them has reached the stage of interstellar searches. Why not? We can imagine that if there are indeed intelligences scattered throughout the universe, one of them must be the most advanced, and why should it not be we?

But this is unlikely, since we exist on a star that is four and a half billion years old and the galaxy is at least four billion years older than that, and the universe another four billion years older than that. If intelligence manifests itself as part of the natural order of things it is extremely unlikely that we on this planet Earth are the earliest form to evolve, and just as unlikely that, being later creations, we overtook all those others who had billions of years head start on us. So if we do not succeed in picking up messages from others, we have to begin thinking that they do not exist.

But it is still possible that, being alone among intelligences, we are still not alone among life forms. On Mars, on Europa, or on other planets and moons around distant stars we may find other forms of life. Perhaps microscopic, perhaps nothing but green slime. Never mind. As the poet says,

> From flying saucers to moldy green slime
> We're all just a row of pins:
> Like the Colonel's lady an' Judy O'Grady
> We're all sisters under the skin.

L'Envoi:

As we approach the end of the twentieth century, we are ending more than two thousand years of speculation about whether life exists on other worlds. By the end of the twenty-first century, we will know.*

---

*And if we're wrong about this, who among you will be around to point it out?

We shall not cease from exploration
And the end of all our exploring
Will be to arrive where we started
And know the place for the first time.

T. S. ELIOT

Chronology

ca. 320 B.C.	Aristotle observes occultation of Mars by moon, therefore Mars is further away.
ca. 300 B.C.	Epicurus: "There are infinite worlds both like and unlike this world of ours. . . . In all worlds there are living creatures. . . "
ca. 50 B.C.	Lucretius: "It is in the highest degree unlikely that this Earth and sky is the only one to have been created. . ."
180 A.D.	First science fiction novel: Lucian's *Vera Historia*.
ca. 1360	Nicole Oresme: "Of course there has never been nor will there be more than one corporeal world."
1510	Copernicus.
1572	Tycho Brahe shows stars are beyond solar system.
1584	Bruno writes *On the Infinite Universe and Worlds*.
1600	Bruno is killed by the Inquisition.
1609	Galileo uses telescope to see planets.
1646	Francesco Fontana discovers markings on Mars.
1659	Christiaan Huygens estimates Martian day.
1666	Giovann Cassini discovers polar caps on Mars.

1675	Antony van Leeuwenhoek discovers microorganisms.
1686	Bernard de Fontenelle: "The Earth swarms with inhabitants. Why then should nature, which is fruitful to an excess here, be so very barren in the rest of the planets?"
1749	Ben Franklin, in *Poor Richard's Almanac*: "It is the opinion of all the modern philosophers and mathematicians, that the planets are habitable worlds."
1755	Immanuel Kant: Most of the planets are certainly inhabited. . . . The intelligence of these inhabitants becomes more perfect the further their planets are from the sun.
1776	Johann Elert Bode: each planet is roughly twice as far from the sun as the previous one, and its inhabitants are twice as spiritual.
1779	Herschel discovers clouds and "a considerable but moderate atmosphere" on Mars. The Martians "probably enjoy a situation in many respects similar to our own."
1806	Organic molecules found in Alais meteorite.
1822	Karl Friedrich Gauss proposes building a giant heliotrope to send light to the moon
1835	*New York Sun* reports life on the moon.
1859	Darwin: *Origin of Species*.
1863	Father Secchi sees *canali* on Mars.
1877	Schiaparelli sees the Mars *canali*, which are translated into English as *canals*.
1886	Someone else finally sees them: J. Perrotin at Nice, France. But many astronomers still do not.
1888	Bright flashes extending over the Martian terminator; ascribed to signals to Earth.
	Science suggests the maria are vegetation.
1892	Camille Flammarion: *La Planète Mars et Ses Conditions d'Habitabilité*: "The present inhabitation of Mars by a race superior to ours is very probable."
	William Pickering sees dark, round "lakes" at the intersection of canals.
1894	Lowell's first theories about Mars presented to Boston Scientific Society. His observatory is founded in Flagstaff, Arizona.
1895	Lowell's first book, *Mars*.

"The Almighty" in Hebrew is seen on Mars, reported by *New York Herald.*

1897   *On Two Planets,* by Kurt Lasswitz, first novel about Martians invading Earth.

1898   *War of the Worlds,* by H. G. Wells.

1900   French Academy of Sciences announces Prix Pierre Guzman: 100,000 francs to whoever first establishes communication with another world.

1901   Tesla detects extraterrestrial radio signals.

1905   Lowell photographs the canals. But they are so faint they cannot be seen on reproductions.

1907   *Wall Street Journal*: "The most extraordinary event of the year . . . is . . . the proof afforded by astronomical obsevations . . . that conscious, intelligent human life exists upon the planet Mars."

Alfred Russel Wallace publishes *Is Mars Habitable?* and says emphatically no.

1911   *Princess of Mars,* by Edgar Rice Burroughs.

1916   Lowell dies.

1919   Robert H. Goddard publishes paper suggesting liquid-fueled rockets for space exploration. *New York Times* says it would never work in a vacuum.

1921   Marconi receives mysterious signals that include the Morse code for V.

1924   United States government asks for radio silence to listen for Martian signals.

1926   Goddard's rocket reaches height of 41 feet over Aunt Effie's Massachusetts farm, reaching 60 mph and traveling 184 feet.

1929   *The Origin of Life,* by J. B. S. Haldane.

1936   *The Origin of Life on Earth,* by Alexander Ivanovich Oparin.

1947   Ken Arnold spots the first flying saucers.

Mysterious object crashes on ranch near Roswell, New Mexico. Air Force says it's a weather balloon.

1956   Sinton finds C-H bonds spectroscopically in the dark areas of Mars; later shown to be hydrogen deuterium oxide in Earth's atmosphere.

1957   *Sputnik.*

1959   Cocconi-Morrison paper in *Nature* begins serious search for extraterrestrial civilizations

1960   Soviet Union attempts to launch two rockets to Mars; both fail.

            Project Ozma begins.

1961   Yuri Gagarin orbits Earth.

            National Academy of Sciences convenes meeting on "intelligent extraterrestrial life."

1962   Three more Russian Mars rockets fail.

1963   NASA sends a rocket to Venus.

1964   First *Mariner* mission to Mars, *Mariner 3*, fails.

            Russia's *Zond* 2 and *Zond 3* fail.

1965   *Mariner 4* gets to within 7,400 miles of Mars and sends back twenty-two close-up photos showing a dead, cratered surface with no canals.

            Russians report receiving ET signal (it is a quasar).

1968   *Viking* project is approved.

1969   *Apollo 11* lands on moon. Two more Soviet missions to Mars fail. *Mariner 6* and *Mariner 7* arrive. Methane and ammonia, signs of life, are found; the data are later retracted.

1971   *Mariner 8* fails and drops into Atlantic; *Cosmos 419* and an unnamed launch fail; *Mars* 2 reaches Mars orbit but finds nothing but a dust storm. *Mariner 9* goes into orbit to wait out the dust storm. Photographs the moons, then finds canyons, lava flows, volcanos, and dry river beds. *Mars 3* orbits okay; lander lands but goes silent after two minutes without returning any photographs.

            First international meeting on ET communication, in Soviet Union.

1973   Pariisky thinks he sees ET signals; turns out to be satellite.

1974   *Mars 4* reaches Mars but fails to orbit. *Mars 5* orbits and returns sixty pictures. *Mars 6* lands but stops signaling seconds earlier. *Mars 7* misses Mars by 808 miles.

1976   *Viking* lands and returns first pictures from surface of Mars.

1977   The "Wow!" signal at Ohio State.

1979	*Voyager* returns photos of Europa: smooth but fractured surface that suggests ice floes in Earth's Arctic. No craters. Speculation: icy surface over liquid water. SERENDIP I at Berkeley.
1987	Infrared observations show disk around Beta Pictoris.
1981	Congress forbids NASA research on SETI.
1982	International Astronomical Union establishes Commission 51 for Exobiology.
1983	SETI put back in NASA's budget. Project Sentinel begins at Harvard, leads to META.
1986	SERENDIP II.
1988	Russia's *Phobos 1* mission to Martian moon fails when wrong command turns photobatteries away from sun. *Phobos 2* orbits Mars but fails to land instruments on Phobos when onboard computer makes the same mistake. NASA conference: Levin says his *Viking* experiment discovered life; Horowitz says no.
1989	President Bush announces Space Exploration Initiative, with plans for manned flight to Mars. Price tag turns out to be $450 billion. Too much, never funded. NASA begins to think of smaller, simpler, cheaper missions.
1991	META II in Argentina. Wolszczan finds exoplanets around pulsar.
1992	SERENDIP III begins at Arecibo. NASA-SETI begins.
1993	Congress cancels NASA-SETI.
1994	NASA-SETI renamed Phoenix, kept alive by private contributions.
1995	BETA begins at Harvard. Jovian planet found around 51 Pegasi.
1996	*Galileo* encounters Europa. Mars *Global Surveyor* launched. *Mars-96* launched and lost; crashes in South America. Mars *Pathfinder* launched, carrying tiny *Sojourner* rover vehicle.
1997	*Pathfinder* arrives safely on Mars. SERENDIP IV begins at Arecibo.

1998   Astrobiology Institute founded at NASA's Ames Re-
            search Center, for research into the origin and develop-
            ment of life everywhere in the universe.

2000   ???

Notes

Chapter 1: Lowell of Mars

1   *"Foolish, philosophically absurd, and formally heretical"* Pannekoep, p. 233.

1   "Wow!" *Science* (January 24, 1997).

4   *"the home of the bean and the cod"* John Collins Bossidy.

4   *Young Percival was born* Hoyt, p. 16.

6   *A subject without an object* For example, Tom Stoppard, *Jumpers* (Grove Press, New York, 1972).

6   *Epicurus* Epicurus, "Letter to Herodotus," trans. by C. Bailey, *The Stoic and Epicurean Philosophers,* W. J. Oates, ed. (New York, 1957). Quoted in Crowe, p. 3.

7   *Lucretius* Lucretius, *The Nature of the Universe,* trans. by R. E. Latham (Baltimore, 1951). Quoted in Crowe, p. 4.

7   *The Pythagoreans taught* Crowe, p. 5; attributed to Pseudo-Plutarch's "Placita," as quoted in *Annals of Science* 1, 1936, pp. 385–430.

8   *Nicole Oresme* Nicole Oresme, *Le Livre du Ciel et du*

Monde, trans. by D. Menut (Madison, 1968). Quoted in Crowe, p. 7.

8   *Philipp Melanchthon*   Stephen J. Dick, *Plurality of Worlds* (Cambridge, 1982). Quoted in Crowe, p. 12.

9   *Bruno*   Frances Yates, *Dictionary of Scientific Biography* (Scribner's, New York, 1972); Zubrin, p. 20.

12   *Tycho Brahe*   Pannekoek; Grant.

14   *But this apparent change in stellar positions was never observed by Brahe*   Brahe's error was that the stars are so incredibly far away that the measurements of the day couldn't reveal their parallax. For example, if the diagram of page 13 were drawn to scale, the star would be not three inches above the Earth but more than three hundred thousand inches: the diagram would have to be drawn on a piece of paper five miles high, and then one could see that the change in angle would be immeasurably small. Instruments capable of accurately measuring the stellar parallax were not built until the nineteenth century; since then such measurements have become routine. [Pannekoep, p. 342.]

14   *Kepler was a small, frail man*   Quote and following description from *Dictionary of Scientific Biography.*

16   *after considerable fiddling and head-scratching*   Kepler's conception of planetary ellipses was based on observations of Mars, which has the largest ellipticity of any of the then-known planets.

16   *Lippershey*   Rovin, p. 28.

17   *He saw four moons revolving around Jupiter*   If Jupiter revolved around the Earth, then so too did the moons. But this would involve an extra motion and so necessitates a philosophic change and a loss of simplicity. See our discussion of *Occam's Razor,* Chapter 4.

18   *Venus showed phases like the moon*   Explained in detail in Fisher, *Birth of the Earth,* p. 34.

18   *the first marking he saw*   Hoyt, p. 2; Wilford, p. 14.

18   *Christiaan Huygens*   Burgess, p. 14.

18   *Bernard de Fontenelle and Ben Franklin and Johann Daniel Titius*   Wilford, p. 16.

18   *Solarians*   Crowe, p. 73.

18   *He never did solve that problem*   Crowe, p. 313.

21   *the great probability*   Crowe, p. 67.

21   *growing substances on the moon*   Crowe, p. 63.

21   *He also looked at the sun*   Burgess, p. 14; Hoyt, p. 3.

21   *Plans were advocated*   Wilford (p. 16) writes "It was said" that this proposition was due to Gauss, but he gives no reference and we can find no confirmation of this. However, see Gauss's heliotrope suggestion in the following paragraph.

22   *Joseph Johann von Littrow and a French inventor*   The Martin Marietta Company, *The Viking Mission to Mars* I (Denver, 1975), p. 11.

22   *Karl Friedrich Gauss*   Crowe, p. 207.

22   *In August of that year*   Crowe, p. 210; Wilford, p. 45.

24   *"I can no longer hold back my chemical urea"*   Quoted in Robin Keen, *Dictionary of Scientific Biography,* s.v. "Wohler, Friedrich."

24   *"Organic chemistry drives me mad"*   Fisher, *Third Experiment,* p. 54.

25   *Wöhler analyzed a meteorite*   *Philosophical Magazine* 18 (1859), p. 213. Quoted in Crowe, p. 402.

27   *Kant had argued*   Horowitz, p. 54.

28   *By 1882 Camille Flammarion*   Wilford, p. 23; Sheehan, p. 170.

28   *J. Perrotin*   Hoyt, p. 7; Wilford, p. 24.

28   *Edward E. Barnard*   Wilford, p. 28.

29   *The flashes were clearly seen*   Hoyt, p. 9.

29   *America's most distinguished scientific journal*   Ley and von Braun, p. 53.

29   *In a book which received a wide readership*   Hoyt, p. 7.

29   *At the next close approach to Earth*   Hoyt, p. 10.

31   *When Mars came into opposition in April of 1894*   Hoyt, p. 57.

Chapter 2: The Gardens of Mars

33   *"That beings constituted physically as we are"*   Lowell.

33   *He told the assembled group*   Crowe, p. 508.

34   *"Misleading and unfortunate"*   Crowe, ibid.

35   *"The [professional] astronomer cannot afford"*   S. New-comb, *Science* (May 21, 1897). Quoted in Crowe, p. 512.

35   *"misleading and unfortunate . . . half-truths"*   E. S. Holden, "The Lowell Observatory in Arizona," *Publications of the Astronomical Society of the Pacific* 6 (June 1894), p. 160.

35   *"his humor, his ready wit"*   Edward Everett Hale, (Boston Commonwealth, 1895). Quoted in Crowe, p. 511.

35   *"an enduring enthusiasm, a proper regard for facts"*   Garrett Serviss, *Harper's Weekly* (September 19, 1896). Quoted in Crowe, p. 512.

36   *Ten years later Maunder would carry out*   J. E. Evans and E. W. Maunder, "Experiments as to the Actuality of the 'Canals' Observed on Mars," *Roy. Ast. Soc. Monthly Notices* (June 1902), p. 488.

37   *A psychiatrist, Charles K. Hofling*   Wilford, p. 24.

39   *In 1898 Samuel Phelps Leland*   Wilford, p. 30; Crowe, p. 517.

39   *Had it not been for the . . . foreknowledge*   Crowe, p. 520.

40   *the words "The Almighty" were seen on Mars*   Hoyt, p. 5.

40   *"It is a burning shame" Popular Astronomy* 3 (1895), p. 47.

40   *Native Americans were laid low by the smallpox and measles*   Robert Utley and Wilcomb Washburn, *American Heritage History of the Indian Wars* (New York, 1977), p. 42.

41   *The pictures were tiny*   Photos reproduced in *Popular Astronomy* 32 (1924), pp. 564ff, show no trace of any canals. The author, E. C. Slipher, adds this comment: "Much of what the original photographs show is lost in the processes necessary to reproducing them here . . ."

41   *"The most extraordinary event of the year"*   *Wall Street Journal* (December 28, 1907).

41   *Alfred Russell Wallace*   Wilford, p. 32.

42   *"The regular and geometric lines"*   Crowe, p. 515.

42   *"Is there life on Mars?"*   Wilford, p. 28.

44   *The French Academy of Sciences*   The prize was first announced unofficially in 1892 by Flammarion; in 1899

the French Academy accepted responsibility and the official announcement came in 1900. [Crowe, p. 395.] Frank H. Winter, "The Strange Case of Madame Guzman and the Mars Mystique," *Griffith Observer* (February 1984).

44   *William Pickering, who had by now left*   Crowe, p. 399; see also Pickering's article in *Popular Astronomy* 32 (1924), p. 580, in which he estimates that the Martians got tired of sending us signals thousands of years ago. In that same year, 1924, he also claimed to have seen moving dark shadows on the moon, which he attributed to swarms of insects. [*Popular Astronomy* 32, pp. 393–404.] It was his last lunar paper, since Harvard decided to dismantle the Mandeville station in Jamaica, where he worked; perhaps because that was where he worked. He retired as an unpromoted assistant professor and died in 1938. Aside from his obsession with life on other planets, he made several important discoveries in planetary science. His main influence on astronomy was his early recognition that observatories should be built in as high and unpolluted regions as possible, rather than where they would be more convenient to university personnel.

44   *Radio had entered the picture*   Colliers (February 9, 1901); *Current Opinion* (1919), p. 170.

45   *Marconi himself said he had received*   New York Times (January 20, 1919).

46   *"Perhaps the Martians tried before"*   Wilford, p. 54.

47   *The ignorance, hysteria, and paranoia*   Rovin, p. 213.

48   *The Roswell apparition was a Project Mogul balloon*   See *UFO Crash at Roswell,* by Charles A. Ziegler and Charles B. Moore, (Smithsonian Press, 1997) for a detailed account.

49   *Other incidents were reported*   Discovery, February 1997.

51   *In 1919 an obscure professor*   Rovin, p. 108.

51   *in defense of Kaifeng-fu*   Grolier's Encyclopaedia.

53   *the slaves arrived from the east*   In 1942, 3,638,056 men and women were brought to Germany as slave labor [Speer, p. 24]. If they did not work hard enough, they

were confined "in a steel box that was so small one could barely stand in it . . . for as long as forty-eight hours, without giving them any food. They were not allowed out to relieve themselves . . . On top of the closet there are a few sieve-like holes, through which cold water was poured on the unfortunate victim in the ice-cold winter." [Schmidt, p. 154; International Military Tribunal records (Nuremberg, 1945), pp. 594–609.]

54   *In 1954 Fred Whipple*   Colliers (April 30, 1954).

54   *Sputnik*   In October 1957 the Soviet Union surprised the world by putting the first artificial satellite (Russian: Sputnik) into orbit. This was followed three months later by the first American satellite, and the space race was on.

55   *William M. Sinton, using the two hundred-inch telescope*   William M. Sinton, *Astrophysical Journal* 126 (1957), pp. 231–39. Also in *Science* 130 (1959), pp. 1234–37.

55   *"The limited evidence we have"*   Wilford, p. 55.

57   *In the initial planning stages*   Wilford, p. 95, Cooper, p. 96.

58   *"most astronomers would probably agree"*   Washburn, p. 66.

58   *"When we were first permitted into the secure planetology science room"*   Washburn, p. 66.

59   *"difficult to believe that free water"*   Wilford, p. 59.

60   *Other regions photographed*   Wilford, p. 60.

61   *One cannot restrain the speculation*   New York Times (Aug. 7, 1969).

61   *"The prospects for life on Mars seemed so dim by 1970"*   Horowitz, p. 98.

62   *As more and clearer photographs came in*   Burgess, p. 24; Rovin, p. 124; Wilford, p. 64.

CHAPTER 3: THE VIKINGS OF MARS

65   *If one stuffs a dirty shirt*   Sullivan, p. 71.

65   *verified to unprecedented levels of accuracy*   Physics Today (March 1997), p. 19.

65   *It is easy to do experiments*   Hoyt, p. 21.

67   *Dr. Felix Pouchet*   Vallery-Radot, p. 98.

68   *The conflict between the two sets of experiments*   Vallery-Radot, p. 108.

69   *Pasteur was possibly wrong*   Dubos, pp. 167, 175.

70   *spontaneous generation must have occurred*   In 1871 Hermann von Helmholtz introduced the idea of panspermia: "We may enquire whether life has ever originated at all or not . . . whether its germs have not been transported from one world to another, and have developed themselves wherever they found a favorable soil." [Hoyt, p. 32.] But this does not solve the problem of "whether life has ever originated at all or not," it merely transposes it to another world. For somewhere, at some time, life must have originated from nonliving stuff, since it could not have been present at the instant of the Big Bang.

71   *"the idea of a God is unnecessary"*   Vallery-Radot, p. 109.

71   *theoretical ideas put forward by a Russian biologist*   The ideas of Oparin and Haldane were published in a series of articles during these years. Oparin's first article appeared in 1924; his magnum opus—"The Origin of Life on Earth"—was published in 1936. Haldane's seminal article, "The Origin of Life," appeared in the *Rationalist Annual* of 1929 and was reprinted in a collection of essays, *The Inequality of Man,* in 1932.

75   *Since it lived in a Garden of Eden*   Horowitz, p. 52.

76   *the original presence of life on Earth*   Mojsis et al, *Nature* (Nov. 7, 1997). (Radioactive dating provides a timescale for the early history of the Earth. Briefly, a radioactive atom of element X decays into an atom of a different element Y at a distinct rate which can be accurately measured and which is not changed by any geologic conditions. Various tricks allow us to estimate the initial presence of Y (the amount of Y that was not formed by the decay of X), and so measuring the relative amounts of X and Y in a rock, together with the rate of change, give us directly the age of formation of the rock. Various X and Y pairs prove useful in different kinds of rocks and for

different geologic histories; some examples are uranium decaying into lead, potassium into argon, or rubidium into strontium. This enables us to estimate the age of the oldest rocks with identifiable fossils at about 3.8 billion years.)

78   *Mars answered by replying*   Zubrin, p. 33.

80   *These two experiments were "terrestrial in orientation"*   Horowitz, p. 130.

82   *Sagan suggested putting a camera on board*   Shklovskii and Sagan, p. 293.

84   *The biology experimenters wanted to land*   details from Burgess, pp. 34ff; Horowitz, pp. 123ff.

87   *"You can almost imagine camels"*   Wilford, p. 90.

88   *No one wanted to break the spell*   *The New Solar System* (1981); Wilford, p. 92.

88   *It had been designed to work*   Horowitz, p. 123.

89   *Just two days later Oyama's instruments*   Cooper, p. 133.

91   *"If it is a biological response"*   Rovin, p. 179; Fisher, *Third Experiment*, p. 146.

91   *"The data is conceivably of biological origin"*   Fisher, *Third Experiment*, pp. 137, 143.

93   *Dr. Horowitz had argued from the beginning*   Horowitz, p. 140.

Chapter 4: The First Martians

96   *Charles B. Lipman was a Russian immigrant*   Sullivan, p. 129.

97   *In 1932 he reported the results*   Charles B. Lipman, "Are There Living Bacteria in Stony Meteorites?" *Am. Mus. Nat. Hist. Novitates* 588 (Dec. 31, 1932).

98   *A few years later, in 1935*   Roy, quoted in J. M. Hayes, "Organic Constituents of Meteorites," *Geochimica Cosmochimica Acta* 31 (1967), pp. 1395–1440.

98   *"You say you have found"*   Warren G. Meinschein, quoted in David Bergamini, *Life* (May 5, 1961), pp. 57–62.

101   *Sisler never published*   Shklovskii and Sagan. Carl Sagan points out that one of the microbes Sisler found showed

a preference for using oxygen. Since Earth is the only place in the solar system with abundant free oxygen (that is, oxygen not tied up in more complex molecules) it is extremely unlikely that any microbe could evolve with a biochemical makeup which knew how to utilize oxygen. Therefore the microbe is probably a terrestrial contaminant.

101   *In 1963, with Frank Fitch*   Fitch and Anders, "Organized Elements: Possible Identification in Orgueil Meteorite," *Science* (June 7, 1963), p. 1097.

103   *In 1964 Anders presented to a scientific meeting*   E. Anders et al., "Contaminated Meteorite," *Science* 146 (1964), pp. 1157–61.

104   *This is the way the world ends*   T. S. Eliot, *The Hollow Men: Collected Poems* (Faber & Faber, London, 1963).

104   *None of the five common [compounds] present in biological organisms*   James G. Lawless, Clair E. Folsome, and Keith A. Kvenvolden, "Organic Matter in Meteorites," *Scientific American* (June 1972).

104   *In a review written in 1973*   E. Anders, *Science* 182 (1973), pp. 781–90.

104   *In a review the previous year*   Lawless et al., ibid.

104   *In 1975 Nagy himself wrote a comprehensive book*   Bartholomew Nagy, *Carbonaceous Meteorites.*

106   *What the two grad students noticed*   The glass, called maskelynite, is known to be formed when a particular silicate mineral (plagioclase, formed during crystallization from a volcanic melt) is shocked at pressures of greater than 300,000 bars; that is, 300,000 times normal atmospheric pressure, or four and a half million pounds per square inch. The maskelynite and plagioclase is intergrown in these meteorites, which was interpreted to mean that a shock had converted the plagioclase to maskelynite after the rocks had solidified from the molten state.

106   *a body that small can't retain its internal heat*   Planets are heated on the inside by radioactive elements, mainly potassium, thorium, and uranium, and by residual heat

left over from their formation. The smaller the body the less radioactivity and the higher the surface-to-volume ratio, which means both that there is less heat generated and that heat escapes more rapidly to space. Earth is still hot inside, but the moon has long since cooled off.

107   *In 1980 McSween and Stolper wrote*   *Scientific American* (June 1980), pp. 54–62. Although some other workers were playing around tentatively with the idea, McSween and Stolper were the first to state the proposition in concrete terms.

107   *In private conversation, they later said*   Bill Cassidy, Houston meeting, March 1997; McSween, personal communication, 1997.

108   *a light came on over my head*   Cassidy, ibid.

109   *Bogard and Johnson weren't all that certain*   D.D. Bogard and P. Johnson, "Martian Gases in an Antarctic meteorite?" *Science* 221 (1983), pp. 651–54. Further work has tied down the Martian connection positively. See Wiens and Pepin, *Geochimica Cosmochimia Acta* 52 (1988), p. 295; Clayton and Mayeda, *Geochimica Cosmochimia Acta* 60, 1999, 1996; and NASA, <http://www.jsc.nasa.gov/pao/flash/marslife/>.

110   *Viking found no life on Mars*   Horowitz, p. 145.

112   *Six years later David Mittlefehldt*   He analyzed a particular mineral in the meteorite—chromite—and found that it contained iron in the trivalent +3 state rather than the usual divalent +2. (Romanek, personal communication, 1997).

112   *He realized that something was wrong*   Mittlefelhdt was analyzing another diogenite and found that it contained the unusual mineral FeS_2, which is not found in diogenites. This time he carried out further tests which showed that this meteorite was mislabelled: it was actually a specimen of ALH84001. Suddenly he realized that 84001 had both trivalent iron and FeS_2 (with its iron in the divalent state), and the only meteorites with those characteristics were the SNCs. [D. Mittlefehldt, "The Source of ALH84001," *Planetary Report* (January/February 1997), pp. 8–11.]

113   *if this newer work was right*   When the JSC group got these results they wanted confirmation, so Gibson called a group headed by C. T. Pillinger in England, who already had pieces of the meteorite slated for study. When told that they hadn't run their samples yet, Gibson suggested they "move this rock to the top of your list," without telling them what results he had already obtained. A few days later Monica Grady called him:

"What did you get?"
"What did *you* get?"

Finally they each disclosed their results and found they had the same data. They decided on a joint publication in *Nature* 372 (December 15, 1994), pp. 655–57. [Gibson, personal communication, 1997].

114   *Meanwhile, Chris Romanek began thinking about a meeting*   this and other details from Romanek, personal communication, 1997.

114   *Dr. Robert Folk, of the University of Texas*   "SEM Imaging of Bacteria and Nannobacteria in Carbonate Sediments and Rocks," *Journal of Sedimentary Petrology* 63 (1993), pp. 990–99.

115   *He took the photographs to Everett Gibson*   Gibson, ibid.

118   *a double laser beam mass spectrometer*   In this apparatus the sample is untouched by human hands and therefore is as free of contamination as possible (although this still doesn't remove the possibility of contamination before it gets to the laboratory). In a vacuum chamber the sample is zapped with a burst from an infrared laser which heats a tiny portion of its surface so quickly that it induces a mini-explosion, vaporizing a puff of material. A second laser, operated at a carefully selected energy, zaps this cloud of vaporized stuff, selectively ionizing some of the molecules. These molecules, now having lost an electron, are positively charged and so when an electric field is applied they are accelerated down the vacuum chamber and into a time-of-flight spectrometer where their masses are identified by the simple process of timing them: the lighter molecules move faster and reach

the detector sooner. The combination of selective ionization and mass determination gives the precise molecular identification.

118   *the 84001 data show that organics do exist*   In 1998 data were presented showing that most of the organics in 84001 are terrestrial, but some at least are definitely Martian. (The PAHs were not tested by this technique.) [A. J. T. Jull et al, *Science* 279 (1998), p. 366.]

118   *The second line of importance is that PAH's are usually*   If the PAHs are to be used as evidence for past life in 84001, they must be shown to be indigenous to the meteorite rather than being terrestrial contamination, and must be of the particular types that are formed by bacteria. The contamination problem is divided into two parts according to time and location: before the sample reached the laboratory, and after. The after part is clearly settled by control experiments. Zare's lab ran other meteorite samples intermittently with the 84001 samples, and found the PAHs only in 84001. And in 84001 they found them only in the carbonates, not in the silicate minerals that form most of the meteorite's mass. So their experimental procedure did not introduce them. The before part was a source of early suspicion since 84001 had lain on the Antarctic ice for thousands of years and could have sucked up all kinds of contamination. But if the PAHs were terrestrial, introduced into the meteorite as it lay on the ice or as it was handled before it got into Zare's lab, the abundances should have been greatest at the surface and less deeper into the rock. On the other hand, if the PAHs were Martian they should have been vaporized from the surface as the meteorite burned its way through the Earth's atmosphere; the abundances should be greater in the interior. And that is exactly what the measurements showed, an increasing abundance as interior pieces of the rock were measured. Furthermore, the PAHs found in 84001 were distinctly different from the most common ones found in our atmosphere; terrestrial sulfur-bearing PAHs, for

example, are abundant on Earth, but not found at all in 84001. So these compounds do seem to be Martian, not terrestrial, but it must be added that not everyone is yet convinced by these arguments.

120   *"One evening we were looking at phases"* "Life on Mars," Discovery Channel.

122   *"the only explanation consistent with the observations"* J. W. Valley et al., "Low-Temperature Carbonate Concretions," *Science* 275 (1997), p. 1633.

124   *McKay says that one day his bright thirteen-year old* Gibson, personal communication.

125   *"Shape is almost useless as an identifiable characteristic"* R. L. Folk, "SEM Imaging of Bacteria and Nannobacteria in Carbonate Sediments and Rocks," *Journal of Sedimentary Petrology* 63 (1993), pp. 990–99.

126   *The real problem is not the length*   See discussion in Letters section of *Science* (June 20, 1997).

127   *"when bacteria become stressed, they condense"*   Folk, ibid.

127   *a surprisingly strong magnetic field*   R. A. Kerr, "Martian Magnetic Whisper Detected," *Science* (Sept. 26, 1997).

CHAPTER 5: THE OCEANS OF EUROPA

134   *In a 1992 paper titled "The Deep, Hot Biosphere"*   Proc. Nat. Acad. Sci. 89 (1992), pp. 6045–49.

137   *In the mid-1870s HMS Challenger sailed the seven seas* Broad, p. 36–37.

138   *Things that do not love the sun*   Eisley, p. 38.

139   *Edson S. Bastin, a University of Chicago geologist*   J. K. Fredrickson and T. C. Onstott, "Microbes Deep Inside the Earth," *Scientific American* (October 1996).

141   *I had been working on methods of measuring the age*   Two other groups, one at Hawaii and the other in Oregon, discovered this independently. See J. G. Funkhouser, D. E. Fisher, and E. Bonatti, "Excess Argon in Deep-sea Rocks," *Earth Plan. Sci. Lett.* 5 (1968), pp. 95–100; G.B. Dalrymple and J.G. Moore, "Argon-40: Excess in Submarine Pillow Basalts from Kilauea Volcano, Hawaii,"

Science 161 (1968), pp. 1132–35; and C.S. Noble and J.J. Naughton, "Deep Ocean Basalts: Inert Gas Content and Uncertainties in Age Dating," *Science* 162 (1968), pp. 265–68.

141   *Claude Zobell of the Scripps Institution in California* Fredrickson and Onstott, ibid.

145   *Cornelius Drebble, court engineer to James I   Grolier Encyclopaedia.*

145   *Sergeant Ezra Lee volunteered*   Quotations are from Sgt. Lee's memoirs, as quoted in Marshall Cavendish, *The First Book of Firsts* (London, 1973).

146   *America's first commercially profitable steamboat*   A Scottish steamboat, the *Charlotte Dundas,* had been hauling coal along the Forth and Clyde canal for several years before Fulton's Clermont steamed up the Hudson. [Cavendish.]

146   *In the Civil War the Confederate Hunley*   Kloeppel; Ragan.

146   *both England and France had built truly maneuverable submarines*   Submarine history from *Britannica Encyclopaedia, Grolier Encyclopedia, The First Book of Firsts,* and Kaharl.

147   *In 1963 the nuclear submarine USS Thresher sank*   Broad, pp. 50–78.

147   *in a symposium held in Washington, DC on February 29, 1956*   Kaharl, pp. 9ff.

147   *"a blind and cumbersome prototype"*   Picard and Dietz, flap text.

148   *the Reynolds Metals Company had designed such a craft*   Kaharl, p. 16.

148   *Designed to carry three men*   Broad, p. 97; J. M. Edmond and K. Von Damm, "Hot Springs on the Ocean Floor," *Scientific American* (April 1983), pp. 78–93.

149   *"by rubber bands hooked onto plastic milk-carton carriers"* Broad.

149   *a suspected but not yet observed phenomenon known as submarine hydrothermal systems*   The phenomenon had

been predicted by J. W. Elder of the University of Manchester. [Edmond and Von Damm, ibid.]

150   *"We all started jumping up and down"*   Edmonds, quoted in Kaharl, p. 175.

151   *hyperthermophiles*   The most hyper of all the thermophiles found so far is a bacterium living in a black smoker in the Atlantic, thriving at a temperature of 113o Centigrade. [*Science* (Feb. 14, 1997), p. 933.]

152   *That was the highest temperature ever recorded*   Broad, pp. 109–110.

152   *temperatures reaching up to an incredible 350° Centigrade*   Bacteria do not live at these temperatures, but in the regions close by at temperatures ranging up to perhaps 150° Centigrade, which for theoretical reasons is currently supposed to be an upper limit for life processes.

152   *evidence of ancient life forms have been discovered on Earth*   The evidence consists of skewed carbon isotopic abundances in sediments from Greenland. [*Science* (Aug. 22, 1997); *Nature* (Nov. 11, 1997).]

153   *life formed deep in the ocean floor*   J.B. Corliss, J.A. Baross, and S.E. Hoffman, "An Hypothesis Concerning the Relationship Between Submarine Hot Springs and the Origin of Life on Earth," *Oceanologica Acta* supplement (1981), pp. 59–69.

156   *That same year Ray Reynolds and his coworkers*   R. T. Reynolds et al., "On the Habitability of Europa" *Icarus* 56 (1983); pp. 246–54.

161   *A new theory was needed to explain the fuzzy edges*   R. A. Kerr, "Galileo Turns Geology Upside Down on Jupiter's Icy Moons," *Science* 274 (1996), p. 341.

162   *When those photos came in*   R. A. Kerr, "An Icy World Looks Livelier," *Science* (Jan. 24, 1997), p. 478.

163   *"It's becoming more and more clear"*   Ronald Kotulak, "New Life?" *Chicago Tribune* (April 10, 1997).

Chapter 6: The Princess of Oz
169   *The known is finite*   Wilford, p. 5.
171   *I think it's part of the nature*   Wilford, p. 52.

171-72   *"I must have spent at least forty-five minutes"*   Drake, p. 35. This and other Drake quotes and Drake material, unless otherwise noted, are from *Is Anyone Out There?* by Drake and the science writer Dava Sobel. While we recognize the great contribution of Sobel to the book, it is presented in Drake's voice and mainly reflects his ideas. For simplicity's sake, therefore, we will refer in the text only to Drake.

172-73   *Cocconi and Morrison background*   Swift, pp. 19-24, 35, 49-51.

175   *a "hydrogen line" at 1420 MHz*   The most abundant and the simplest atom in the universe is hydrogen, consisting of a single proton orbited by an electron. When heated, as in a star, the electron absorbs the heat energy and jumps from its ground level to any one of many excited states. These states are not stable, however, and the electron quickly cascades back down to the ground state, releasing the energy it absorbed in bursts of electromagnetic radiation. Each burst, or quantum, carries away the energy difference between the level it came from and the one it dropped down to:

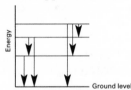

The frequency of the radiation emitted is directly governed by the energy, in the basic equation of quantum mechanics: Energy equals Frequency times a constant. (The constant is named Planck's constant, after Max Planck, the first pioneer of quantum theory, who derived the equation in 1901.) Since the energy exists in discrete amounts, so too do the frequencies; instead of a continuous band of radiation, the hydrogen atom emits only particular frequencies. This is true of all the elements in the universe; each type of atom emits particular frequencies; the pattern becomes, in effect, a fingerprint identifying the element.

Wouldn't it be reasonable, Cocconi and Morrison suggested, that a civilization choosing a broadcast channel would choose one of the hydrogen frequencies (since any civilization capable of detecting such signals would surely know about the hydrogen spectrum). But which frequency in particular? How could we pick just one of the many available?

The problem was solved by Iosef Shklovskii's prediction that the ground state of hydrogen should be split in two, and that the atom can shift from the higher ground state to the lower, so that hydrogen atoms in interstellar space emit radiation at the frequency given by this transition. The difference in energies for this split ground state is so small that in the figure, if drawn to scale, it would be less than the width of the line. Therefore its frequency is quite low; it turns out to be 1420 MHz, which lies right in the "window of opportunity" between the cosmic radio noise and that of the planets. Finally, the transition is so infrequent that the natural background of the universe would not overwhelm any artificial signal.

175   *tune in to join the party*   Not everyone agrees on this choice of wavelength. Stuart A. Kingsley, at Ohio State University, thinks we may find success at optical wavelengths. See <http://www.coseti.org>.

176   *"could see no reason"*   Drake, p. 5.

176   *he was inspired by a series of guest lectures*   Drake, p. 11.

184-192   *Order of Dolphin conference and Drake equation discussion*   Drake, pp. 52-64; Sagan and Shklovskii, pp. 409-18; White, p. 196.

185   *Oliver had flown*   Swift, p. 93.

188   *(at least until a species . . .)*   Sagan and Shklovskii, p. 412.

188   *the fact that two intelligent species*   Swift, p. 29.

188   *(Other workers, including Sagan . . .)*   Sagan and Shklovskii, ibid.

192-96   *Kardashev/Sholomitskii blunder*   Drake, pp. 102-4; Sagan and Shklovskii, pp. 394-96.

196   *black holes at their centers that are swallowing up*   Black holes are incredibly dense astronomical objects sur-

rounded by such a strong gravitational field that nothing—not even light—can escape from them. If massive enough, their gravity is thought to be capable of swallowing up nearby stars. As these stars disappear into the black hole, some of their mass is converted into the radiation that we observe as a quasar.

197-99   *Discovery of pulsars*   Drake and many other sources. Explanations for the award range from prejudice against women, to prejudice against graduate students, to prejudice against irritating iconoclasts, to a feeling that the person responsible for putting together and maintaining a world-class lab deserves the credit for anything discovered there, to the feeling that while Jo Bell made the initial discovery it was Hewish who followed it up perfectly. The Nobel committees don't talk about their selection process, so speculation is all there is.

198   *Arecibo, formerly best known*   Charles Willeford, *Cockfighter* (Crown, New York, 1972).

200-01   *Cyclops*   Swift, pp. 251-54, 320; J. S. Hey, *The Evolution of Radio Astronomy* (Science History Publications, New York, 1973), p. 67.

205-07   *The "Wow!" signal*   Dixon, "The OSU Program," in Bova and Preiss, pp. 199-204; Swift, pp. 241–45; Drake, pp. 142-43.

207   *Space could be full of [such] signals*   Drake, pp. 43-44.

Chapter 7: Guaranteed to Fail

213   *"Well," Sagan reported*   McDonough, pp. 149-51.

214   *Jill Tarter, a Berkeley graduate student*   Drake, pp. 152ff.

216   *"He did the Sagan and Shklovskii book"*   Swift, p. 381.

216   *His idea was to look for very narrow-band signals*   Horowitz designed a receiver to look at a band centered on the 1420 MHz frequency, with a resolution better than that of any previous equipment. It also ran under the assumption that the sender had our solar system in mind and had thus shifted its frequency to allow for the motion of its own star relative to ours, producing a heliocentric signal. If the sender simply broadcast at 1420 MHz, that frequency would be Doppler-shifted by the

time it reached Earth, depending on the velocity of the sending planet relative to Earth. This makes our search that much more complicated and difficult, for those relative motions are extremely complex. Planets are moving around their stars, those stars are moving through their galaxies, and galaxies are hurtling away from each other at unthinkable speeds. The resulting Doppler shifts would require a fantasy computer to unravel. By assuming a targeted beacon, however, Horowitz didn't have to worry about this almost infinite range of shifts. (The shift due to the motion of the Earth around the sun might still be left if the alien civilization didn't know the Earth's orbit, but this is a relatively small correction.)

216   *Suitcase SETI*   Harvard SETI, <http://mc.harvard.-edu/seti/setihist.html>.

220-23   *NASA SETI*   *Science* (July 20, 1990), p. 249.

221   *$200 billion savings and loans bailout*   *USA Today* (July 17, 1990).

221   *$160 billion albatross*   *Newsday* (July 26, 1990).

221-22   Congressional battle   William Triplett, "SETI Takes the Hill," *Air & Space* (Oct/Nov 92), pp. 80-86.

223-24   1992 struggles   Richard A. Kerr, "SETI Faces Uncertainty on Earth and in the Stars," *Science* 258 (Oct. 2, 1992), p. 27.

225   *Transformation to Phoenix*   SETI Institute, <http://www.seti-inst.edu>.

225-26   *First, he went down to Argentina*   Drake, p. 220.

226   *BETA (billion-channel extraterrestrial assay)*   To be fair, BETA would not really analyze a billion channels simultaneously. The name was symbolic of the system's great advances over META. BETA would actually look at 80 million channels, but it would do so with three separate antennas—two on the 26-meter dish at Harvard and a third on a nearby terrestrial antenna—to eliminate any chance of being fooled by satellites or terrestrial sources. That made 240 million channels, and since it operated in eight 40MHz steps covering the entire 320MHz-wide "water hole" (1,400MHz to 1,720MHz), the Harvard group (for eponomymous reasons at least) could say

they had about two billion channels. Each of the eight hops took 2 seconds, so the entire water hole would be covered every 16 seconds.

226   *"SERENDIP is probably the most sensitive"*   Horowitz interview (June 4, 1997).

Chapter 8: Strange New Worlds

230-3   Theories of star formation   Fisher, *Birth of Earth*, pp. 137-39.

232-34   Discovery *of Neptune*   Croswell, pp. 40-45; see also M. Grosser, *Discovery of Neptune* (Harvard University Press, Cambridge, 1962) and A.F. O'D. Alexander, *The Planet Uranus* (Elsevier, New York, 1965).

234-35   Discovery of Pluto   Croswell, pp. 48-55; see also Hoyt, and C.W. Tombaugh and P. Moore, *Out of the Darkness: The Planet Pluto* (Stackpole Books, Harrisburg, PA, 1980).

235-37   Wobbles, astrometry, and Bessel   Croswell, p. 87; Halpern, p. 90.

237-38   *Barnard's Star*   Halpern, pp. 94-106; Croswell, pp. 90-99.

238-42   Entire pulsar-planet discussion   Croswell, pp. 136-58; Halpern, pp. 133-20; *Sky & Telescope* (May 1992).

242-51   Doppler-shift method and discovery of Pegasian planets Croswell, pp. 180-208; Butler, *The Planetary Report* (July/August 1997); Boss, *Physics Today* (September 1996); Powell, *Scientific American* (May 1997); MJF visit to DPS meeting, Cambridge, MA, July 31, 1997; David Gray, <phobos.astro.uwo.ca/~dfgray>; Geoff Marcy, <cannon.sfsu.edu/~gmarcy/planetsearch/nrp.html>; J. Glanz, *Science* (Jan. 9, 1998), p. 170.

Chapter 9: Where Are They?

254   *When he won the 1938*   T. Wasson, ed., Nobel Prize Winners (H. W. Wilson Company, New York, 1987).

255   *"They are among us"*   McDonough, p. 191; Drake, p. 130; White, p. 28.

254-56   *Hart, Tipler's arguments*   Drake, pp. 202-4; Halpern, p. 254.

257   *In order for a raindrop to form*   White, p. 31.

259   *"cast[ing] [of] nets"*   Heidmann, pp. 132, 142.

260-61   *Estonia meeting*   Drake, p. 158.

261   *"It doesn't seem that someone"*   White, p. 190.

262   *"it was as if the Niña"*   SETI Institute, "Project Phoenix," <http://www.seti-inst.edu>.

262   *"all the people like Columbus"*   Swift, p. 44.

263   *a radio telescope in orbit . . . Space Cyclops*   Drake, pp. 229-30.

264-65   *Eddington/Dyson experiments*   Hoffmann, p. 129; Encyclopaedia Britannica.

266   *"They won't be too much different from us"*   Swift, pp. 83-84.

266-67   *"[Alien civilizations] are guaranteed not to look like us"*   White, p. 188.

267   *"There seems no reason for extraterrestrial organisms to have . . ."*   Sagan & Shklovskii, pp. 350-52.

267-68   *"floaters"*   Sagan, *Cosmos,* p. 30.

268   *Nature always has more imagination*   Swift, pp. 311, 324.

268   *Among other things . . . such as comets*   The Planetary Report (March/April 1996), p. 6.

268   *Or they could be machines*   McDonough, p. 116.

269   *"Most of us view the slow destruction"*   Drake, pp. 160-61.

269-70   *Immortality*   Drake, p. 161.

270   *As Sagan pointed out . . . "hopelessly modest"*   Sagan & Shklovskii, p. 361.

271-76   Communication with aliens   Drake, pp. 175-85; McDonough, pp. 120-21.

279   *The full contents of the* Encyclopedia Britannica   Heidmann, pp. 201-3.

280   *"It is precisely because I believe"*   Hesburgh, in introduction to Goldsmith, p. vii.

280   *"I can't think of a message more important than that"*   White, pp. 180-81.

281   *"I think that if these civilizations"*   Horowitz interview (June 4, 1997).

282   *"To discover that we share the galaxy"*   Drake, p. 160.

282   *"It will happen"*   Horowitz, ibid.

Chapter 10: The Colonel's Lady

284   *magnetite and phrhhotite*   A recent study by Posfai et al. (*Science* 280 (1998), pp. 880-83) found no pyrrhotite in terrestrial magnetic bacteria, concluding that the magnetic iron sulfide mineral *greigite* is "the most abundant sulfide in [terrestrial] magnetotatic bacteria and could be the best, although not an unambiguous indicator of past biogenic activity." This mineral was also indicated, though not positively identified, in ALH 84001 by the original JSC team (Gibson, personal communication).

284   *A recent mineralogical study "suggests"*   E. R. D. Scott, A. Yamagauchi and A. Krot, "Petrological Evidence for Shock Melting of Carbonates in the Martian Meteorite ALH84001," *Nature* 387 (1997), pp. 377–79. For a discussion of the most recent arguments for and against the evidence for fossil life in 84001, see the pro and con articles by Gene McDonald and David McKay et al. in the May/June 1998 issue of the *Planetary Report* (available from the Planetary Society, 65 N. Catalina Ave., Pasadena, CA 91006-2301).

289   *a $500 million mission in 2008*   J. Kaiser, *Science* (Feb. 27, 1998), p. 1309. The material returned from Mars will be placed in immediate biologic quarantine, to prevent both the spread of possible Martian microorganisms into the Earth environment and the contamination of the Martian material by terrestrial life forms. Ed Anders, arguing before the first lunar return mission against the possibility of lunar microbes, offered to eat some of the returned material. No one is offering to eat pieces of Mars.

289   *At the 1997 meeting*   The 1997 Houston meeting was dominated by papers relating (both pro and con) to the claim of life in 84001. The 1998 meeting was much more subdued in this respect. Some feel that this reflects a loss of interest due to lack of substantive proof, while others suggest that it's a normal part of the scientifc process: there's just not too much more that can be done until we bring back samples from Mars.

291   *Existence of an ocean?*   Cassen et al., *Geophysical Research Letters* 6 (1979), pp. 73–74; Cassen et al., *Geophysical Research Letters* 7 (1980), pp. 987–88; S.W. Squyres et al., *Nature* 301 (1983), pp. 225–26; M. N. Ross and G. Schubert, *Nature* 325 (1987), pp. 133–34; G. W. Ojakangas and D. J. Stevenson, *Icarus* 88 (1989), pp. 193–204.

298   *So an infrared map of an Earth-like planet*   Croswell, pp. 218–28.

298   *someday using such natural telescopes to image extrasolar planets*   Halpern, pp. 189–93.

301   *From flying saucers to moldy green slime*   After Rudyard Kipling's *The Ladies*.

302   *We shall not cease from exploration*   T.S. Eliot, "Little Gidding," *Four Quartets,* (Harcourt Brace Jovanovich, New York, 1943). It seems that half the people at NASA and absolutely everyone involved in the search for life "out there" knows this poem and will quote it at the drop of a proposal. Eliot's views of life are little to be admired, but in these lines he seems to have got it right.

Glossary

AAS   American Astronomical Society

achondrite   A class of meteorites without chondrules. (See *chondrite*)

AGS   American Geological Society

AGU   American Geophysical Society

ALH   Prefix for meteorites found in the Allan Hills section of Antarctica (followed by number indicating year and order of cataloguing)

asteroids   A group of rocks in orbit between Mars and Jupiter

bacilli   Elongated bacteria

biogenic molecules   Molecules produced by living creatures

carbonaceous chondrite   A class of chondrites containing abundant organic molecules

carbonates   Chemical compounds based on the $CO_3^=$ ion; usually calcium or magnesium carbonate (Although containing carbon, these are not considered organic molecules.)

CETI   The attempt at communication with extraterrestrial intelligence

chondrites   A class of stone meteorites containing small rounded pebbles (chondrules) that are thought to have originated in the primitive solar nebula

cocci   Spherical bacteria

DOE   Department of Energy

exobiology   The study of life on other worlds, also known as astrobiology

extrasolar planet   A planet that is not part of our own solar system

gyr   Gigayear: a billion years

HRMS   High Resolution Microwave Survey

hydrocarbons   Molecular chains or rings composed primarily of hydrogen and carbon atoms; the basic building blocks of life

IAU   International Astonomical Union

JPL   California Institute of Technology's Jet Propulsion Laboratory, operated for NASA

JSC   Johnson Space Center (Houston, Texas)

Life   ???

light-year   The distance light travels in one year: 186,000 miles/sec x 60 sec/min x 60 min/hr x 24 hrs/day x 365 days/yr = 5.9 x 10^{12} miles, or about six trillion miles.

LPSC   The Lunar and Planetary Science Conference, held every spring in Houston

meteorites   Rocks or chunks of iron that fall to Earth from space. Thought to be mostly asteroids, but may also be cometary nuclei or pieces of the moon or Mars.

micron   One millionth of a meter

nanometer   One billionth of a meter

nm   Abbreviation for nanometer

NRAO   National Radio Astronomy Observatory

NSF   National Science Foundation

organic molecules   Molecules built around carbon atoms, generally hydrocarbons. Originally thought to be purely biogenic, but can be produced abiogenically or in the laboratory.

PAHs   Polycyclic aromatic hydrocarbons

SEM   Scanning electron microscope

SETI   The search for extraterrestrial intelligence

SNC   Pronounced snick. The Shergotty-Nakhlite-Chassigny class of meteorites, now referred to as Martian meteorites

TEM   Transmission electron microscope

Bibliography

BOOKS

Bova, Ben and Byron Preiss, Eds. First Contact: *The Search for Extraterrestrial Intelligence.* Penguin Books, New York, 1990.

Broad, William J. *The Universe Below.* Simon & Schuster, New York, 1996.

Burgess, Eric. *Return to the Red Planet.* Columbia University Press, New York, 1990.

Carr, Michael H. *Water on Mars.* Oxford University Press, New York, 1996.

Chandler, David L. *Life on Mars.* Dutton, New York, 1979.

Clark, Ronald W. *Einstein.* World Publishing, New York, 1971.

Cooper, Henry S. F. Jr. *The Search for Life on Mars.* Holt Rinehart Winston, New York, 1980.

Croswell, Ken. *Planet Quest.* The Free Press, New York, 1997.

Crowe, Michael J. *The Extraterrestrial Life Debate 1750–1900.* Cambridge University Press, Cambridge, 1986.

Davoust, Emmanuel. *The Cosmic Water Hole.* MIT Press, Cambridge, 1991.

Drake, Frank and Dava Sobel. *Is Anyone Out There?* Delacorte Press, New York, 1992.

Dubos, Rene. *Louis Pasteur: Free Lance of Science.* Little, Brown, Boston, 1950.

Edelson, Edward. *Who Goes There? The Search for Intelligent Life in the Universe.* McGraw-Hill, 1979.

Eisely, Loren. *The Immense Journey.* Vintage Books, New York, 1959.

Fisher, David E. *Birth of the Earth.* Columbia University Press, New York, 1987.

Fisher, David E. *The Third Experiment.* Atheneum, New York, 1985.

Goldsmith, Donald, ed. *The Quest for Extraterrestrial Life: A Book of Readings.* University Science Books, Mill Valley, CA, 1980.

Grant, Robert. *History of Physical Astronomy.* Johnson Reprint, London, 1966.

Halpern, Paul. *The Quest for Alien Planets.* Plenum Press, New York, 1997.

Heidmann, Jean. *Extraterrestrial Intelligence.* Cambridge University Press, Cambridge, 1995.

Hoffman, Banesh. *Creator and Rebel.* Viking, New York, 1972.

Horowitz, Norman. *To Utopia and Back: The Search for Life in the Solar System.* Freeman, New York, 1986.

Hoyt, William Graves. *Lowell and Mars.* University of Arizona Press, Tucson, 1976.

Kaharl, Victoria A. *Water Baby: The Story of Alvin.* Oxford University Press, Oxford, 1990.

Kloeppel, James. *Danger Beneath the Waves: A History of the Confederate Submarine.* Sandlapper Publishing Co., Orangeburg, SC, 1992.

Ley, Willy and Wernher von Braun. *The Exploration of Mars.* Viking, New York, 1956.

Lowell, Percival. *Mars.* Macmillan, New York, 1895.

———. *Mars and Its Canals.* Macmillan, New York, 1911.

McDonough, Thomas R. *The Search for Extraterrestrial Intelligence.* John Wiley & Sons, Inc., New York, 1987.

Morrison, Philip, John Billingham, and John Wolfe. *The Search for Extraterrestrial Intelligence.* NASA, Washington, DC, 1977.

Nagy, Bartholomew. *Carbonaceous Meteorites.* Elsevier, New York, 1975.

Pannekoek, A. A *History of Astronomy.* Allen & Unwin, London, 1961.

Piccard, Jacques and Robert S. Dietz. *Seven Miles Down.* Putnam, New York, 1961.

Ragan, Mark K. The Hunley: *Submarines, Sacrifice, and Success in the Civil War.* Narwhal Press, Miami, FL, 1995.

Regis, Edward Jr. *Extraterrestrials: Science and Alien Intelligence.* Cambridge University Press, Cambridge, 1985.

Rood, Robert T. and James S. Trefil. *Are We Alone?* Charles Scribner's Sons, New York, 1981.

Rovin, Jeff. *Mars!* Corwin, Los Angeles, 1978.

Sagan, Carl, ed. *Communication with Extraterrestrial Intelligence: CETI.* MIT Press, Cambridge, 1973.

————. *Cosmos.* Random House, New York, 1980.

Schmidt, Matthias. *Albert Speer.* St. Martin's Press, New York, 1984.

Sheehan, William. *Planets and Perception.* University of Arizona Press, Tucson, 1988.

Shklovskii, I. S. and Carl Sagan. *Intelligent Life in the Universe.* Holden-Day, San Francisco, 1966.

Speer, Albert. *Infiltration.* Macmillan, New York, 1981.

Sullivan, Walter. *We Are Not Alone.* McGraw-Hill, New York, 1966.

Swift, David W. *SETI Pioneers: Scientists Talk about Their Search for Extraterrestrial Intelligence.* University of Arizona Press, Tucson, 1990.

Vallery-Radot, Rene. *The Life of Pasteur.* Garden City Publishing Co., Garden City, New York, n.d.

Washburn, Mark. *Mars at Last.* Putnam, New York, 1977.

Wilford, John Noble. *Mars Beckons.* Knopf, New York, 1990.

White, Frank. *The SETI Factor.* Walker and Co., New York, 1990.

Zubrin, Robert. *The Case for Mars.* Free Press, New York, 1996.

WEB SITES
Extra Solar Planets
Observatoire de Paris. <http://www.obspm.fr/planets>.

Marcy, Geoff. <www.physics.sfsu.edu/~williams/planetsearch/planetsearch.html>.

Life on Mars and Martian Meteorites
Federation of American Scientists. <http://www.fas.org/mars/index.-html>.
Planetary Society. <http://planetary.org>.
NASA. <http://www.jsc.nasa.gov/pao/flash/marslife/>.

SETI
Berkeley. <http://sag-www.ssl.berkeley.edu/serendip/>.
Horowitz, Paul. Harvard University. <http://mc.harvard.edu/seti/>.
Ohio State. <http://everest.eng.ohio-state.edu/~klein/ro/>.
The Optical Search. <http://www.coseti.org>.
Planetary Society. <http://seti.planetary.org>.
SETI at Home.<http://www.bigscience.com/setiathome.html#new>.
The SETI Instiute. <http://www.seti.org>.
The SETI League. <http://www.setileague.org/>.

Miscellaneous
NASA. <http://www.nasa.gov/NASA_homepage.html>.
The Astronomy Net. <http://www.astronomy.net/>.

Index

PLAGUE YEAR

PLAGUE YEAR

STEPHANIE S. TOLAN

MORROW JUNIOR BOOKS · NEW YORK

9 1 0 3 7 7

Printed in the United States of America.
1 2 3 4 5 6 7 8 9 10

Library of Congress Cataloging-in-Publication Data
Tolan, Stephanie S.
Plague year / Stephanie S. Tolan.
p.   cm.
Summary: Sixteen-year-old David becomes intimately involved when a
scandal is discovered about the strange new boy in his high school
and everyone else turns against him.
ISBN 0-688-08801-5
[1. Prejudices—Fiction.   2. High schools—Fiction.
3. Schools—Fiction.]   I. Title.
PZ7.T5735Pl   1990
[Fic]—dc20   89-13605   CIP   AC

Fic
Tolan

ACKNOWLEDGMENTS

Thanks to Carl Heiner, principal of Waterford/Halfmoon High School, Waterford, New York, and Janet Lynch of Albany County Social Services for their assistance.

For Frances Carpenter and Barriss Mills,
teachers who made a difference

1 I may have been the first person in Ridgewood to see Bran Slocum. It was the first Saturday in October, about seven o'clock in the morning. The sun was turning the edges of the hills outside of town orange, and some wisps of mist were still rising from the damp ground. I was up and outside at that ungodly hour, pounding the sidewalk and sweating in the chilly air, mainly because of Molly Pepper.

I'd been dating Kristin Matthis since about the middle of the summer, and when Kristin made the cheerleading squad, Molly, who takes better care of my social life than I do, pointed out that it wasn't reasonable for a cheerleader to date Watson the Wimp. If I wanted to keep Kristin, I had to become a jock. The problem was that I'm tall and skinny ("free of excess body fat," Molly said) and have these really long arms and legs. I'm not crazy about heavy body contact, so football was out. It was the wrong season for basketball even if I could shoot baskets, which I can't.

But I can run. I'm not fast, but I have terrific endurance. Once I get started, I can run practically forever. So Molly insisted I try out for the cross-country track team, and I did, just to keep her off my back. To my surprise, I made it. There have to be at least five runners on a cross-country team if the school wants to compete officially and get ranked. Only four of the guys who tried out were fast, and none of them could go as far as I could, so Coach Morelli took me on in the hope that he could find some way to increase my speed. That's why I was out that Saturday morning. Coach had given me a stopwatch to time my runs. I was supposed to try to take a few seconds off my time every day, running the same course.

You've heard of that movie, *The Loneliness of the Long-Distance Runner*? I've never seen it, but the title's right on. Long-distance running isn't the kind of thing you do in groups. I mean, who else in their right mind would leave a nice, warm bed and an X-rated dream about Miss October to put on some holey sweats and run through the empty weekend-morning neighborhoods of Ridgewood, New York?

I'd been running half an hour, and I hadn't seen a single person except an old lady who was standing on her porch in her bathrobe, waiting for her pug dog to finish his business in her rhododendrons. That's why I particularly noticed when a cab drove past me, going real slow, and pulled up in front of a tired-looking, aluminum-sided two-story house down almost where Larch Street ends at the back of Ridge Lawn Cemetery.

My route takes me through the cemetery because the

gravel road is better for running than sidewalks, if you watch out for the ruts and dips, and because it looks terrific in there in the fall, when the leaves are changing. That's something you can't say about the neighborhood in general. Or most of Ridgewood for that matter.

I kept on running toward the cab as its back door opened and somebody heaved a duffle bag out onto the sidewalk and then climbed out after it. He must not have heard me coming. He reached into the cab and grabbed a backpack, swung it over his shoulder, and then backed right into me, as I started to swerve to avoid the duffle.

In the moment it took to get my balance and avoid falling, I got a good look at him. He seemed about my age, shorter than me by a couple of inches, but built better, with good shoulders and a tough-looking body. His hair was an ordinary sort of brown, but so long he'd tied it into a ponytail that hung a good six inches down the back of his jeans jacket. He had two rolled red and white bandannas tied around his neck. And he had an earring in one ear. You couldn't miss it. It was a big gold hoop.

He grunted when we hit, and turned toward me, his face startled—scared. For just a minute, he was like an animal in the middle of the road when your headlights hit it, eyes wide and terrified. Then he blinked, and it was as if he'd slipped a mask on. His face went blank.

"Sorry," I muttered. He frowned at me. At least I think he was frowning at me. It was hard to tell. His right eye was off somehow, angled sort of over my shoulder, while the other seemed to bore into my forehead. His mouth was a straight, tight line. "Sorry," I said again, though it hadn't really been my fault.

Still frowning, he settled his backpack more firmly onto his shoulder and I turned away and started running again. I glanced back over my shoulder once, and saw a balding guy with a paunch come out of the house and start down toward the kid, who was paying the cab driver, and then I ran on.

By the time I was through the service gate and onto the cemetery's gravel road, I'd just about forgotten the whole thing, watching the coils of mist that were rising from among the headstones like anemic ghosts.

I'd been running through the cemetery every day, and I don't think I'd ever thought about the dead people. But that morning, I had a couple of second thoughts. When I came out the main gate and checked the stopwatch, I saw that I'd made it through in record time. Not great time for cross-country maybe, but great for me.

I found out Slocum's name and that he'd come to Ridgewood to stay when I got to school Monday morning. He was in my homeroom, and he got there late, because of having to go to the office first and get enrolled. Maybe it would have helped if he'd been there right at the bell like everybody else, so he wouldn't have had to make that solitary entrance, but with the way he looked, maybe not.

When the classroom door opened, Mrs. Campbell had just finished taking the roll and was reading off the morning announcements for the junior class. People were shifting around and whispering, the way they always do during homeroom, and a couple of the conscientious types were finishing up homework. But the minute Slocum came through the door, everybody stopped whatever they were doing and it got real quiet. People just stared.

Maybe back in the sixties there were guys at Ridgewood who wore their hair in ponytails. But not now. Some guys wear their hair pretty long, but they mousse it and let it hang sort of full and wavy over their shoulders. Most everybody else wears their hair like mine, short enough that even in a high wind, it doesn't move a whole lot. So the hair was one thing.

Then there were the scarves—and that earring. And the eye. He stood there in the doorway, his left eye looking at Mrs. Campbell, and the right seeming to stare out the window. His mouth was that hard, straight line and he had no expression on his face at all, as if he'd been carved out of stone.

The silence didn't last. As he walked over to Mrs. Campbell's desk to hand her his late pass and his enrollment card, Matt Singleton, who sits right in front of me, whispered that the guy ought to be wearing a dress to go with his hair and his earring. People who hadn't heard asked others, and pretty soon Matt's observation had made its way up every aisle.

Slocum turned and looked at the class when Mrs. Campbell pointed to an empty desk at the back of the second row, and everybody got a clear look at the eye that turned slightly out and up and seemed to move on its own.

"Oh, n-no!" Matt said, out loud this time, "P-P-Princess Charming has the evil eye." There were a few giggles.

"Matthew!" Mrs. Campbell said, pushing herself to a standing position behind her desk. "That's enough out of you. This is . . ." she glanced at the card in her hand. "This is Bran Slocum, and I want you all to make him welcome."

There was silence again for about two seconds. Bran Slocum stood looking out over the faces turned toward him, then sort of ducked his chin and started toward his seat, his backpack hanging off one shoulder. "Bran muffin," somebody behind me muttered, and laughter swept the room.

"Bran flake," someone else said.

"Class! Settle yourselves immediately!" Mrs. Campbell banged her roll book on the desk, but the laughter didn't stop.

"Raisin bran" was next, and then Matt let out a huge guffaw. "*Fruitful B-B-Bran!*" he shouted, and doubled over his desk. That did it. Anybody who hadn't laughed before was laughing now, me included.

"Class!" Mrs. Campbell said again, but she didn't have a chance of getting order now, and she knew it.

"Fruitful Bran," someone else repeated, and the laughter overwhelmed Mrs. Campbell's voice completely. She just gave up, and stood at the front of the room, reading the rest of the announcements into the din. I twisted around in my seat to see how Slocum was taking it, thinking how glad I was not to be in his position. I remembered that look I'd seen when I'd bumped into him. That moment of fear. But he just sat there, his mask in place, perfectly quiet except for one finger, tapping on the chipped formica of his desk top.

The all-school announcements were made over the loudspeaker, and then the bell rang to end homeroom. Mrs. Campbell turned her back and began writing the assignment for her first-period class on the board as everybody grabbed their stuff and shoved out into the hall, talking and giggling. Everybody except Slocum, who stayed at his desk, probably waiting till we were all gone. I had a feeling, as I left, that for as long as he was a student at Ridgewood High, he was going to be called Fruitful.

At lunch, when Kristin had left our table to go to a cheerleaders' meeting, Molly picked up her tray from the table where she usually sits and came to join me. I was surprised, because even though Molly and I are good friends, hardly anybody at school even knows about it. That's her doing, mostly.

I've known her all my life, or at least as long as I can remember. When we were little our parents spent a lot of time with each other, so we were together a lot. In grade school we were pretty much best friends.

It changed in junior high. That was when kids started paying a lot of attention to how people looked. And Molly wasn't pretty or cute. She was short and squat and had a plain face. When the other girls were getting figures, Molly wasn't. When the other girls started wearing makeup and paying lots of attention to clothes and hair, Molly didn't. She wore her dark hair straight and parted in the middle, and always dressed in jeans and oversized T-shirts or sweat shirts. Nick Bruno called her Goblin Girl one day, and the name stuck.

She started avoiding me in school. At first my feelings were hurt, but she went on calling me on the phone a lot, and we still did things together on weekends sometimes. She said we could stay friends, but there wasn't any point in both of us being unpopular. "If people get the idea that we're going together or something, you won't have a chance," she told me. And it didn't take me long to discover that things were easier for me if people didn't connect me with "Goblin Girl."

By the time we started high school, we both wanted to keep things the way they were. It worked out pretty well for me. I got to keep Molly as a friend, which I wanted to do. But I could date other girls without having to worry about hurting her feelings. "You do the usual high school stuff for both of us," Molly said during our sophomore year. "If anything really interesting happens you can tell me all about it. That way, I won't miss anything, but I won't have to figure out how to change myself into somebody I'm not just to fit in." And she didn't change, either. She was a little taller, though not very much, and built more like a girl, but she still didn't wear makeup or worry about clothes. And she really didn't seem to mind that she didn't have much of a social life. She contented herself with overseeing mine.

"Have you seen him?" she asked that day, when she'd plunked her plastic tray down next to the remains of my peanut-butter-and-jelly-and-apple lunch.

"Who?"

"Slocum, of course. You think there's any other guy around here I'd be asking about?"

"Well—there's your crush on Dr. Towson."

"Oh, yeah. It's my fondest dream to marry an uptight, by-the-book, middle-aged high school principal. Have you seen Slocum?"

"In math, like you."

"I meant during lunch. Where is he?"

"How should I know? Maybe he brown-bags it like us poor folk. He could be eating outside by the fountain. Why?"

"Just wondered. 'Fruitful.' That is so disgusting." She flipped her hair behind her ears and took a bite of creamed corn. "I suppose you think hassling Slocum's just good, clean fun."

I gathered up my empty sandwich bag, apple core and juice box and stuffed them into the crumpled brown bag. "No, but a little teasing isn't going to kill him. He'll survive it, just like the rest of us. You survived 'Goblin Girl.' "

"Sure, but not everybody's as tough as me."

"What makes you think he'd even appreciate your concern? He's not exactly Mr. Personality." Just then I saw Bran Slocum come through the cafeteria doors. "Speak of the devil."

He was greeted by whistles and catcalls. "Hey, Fruitful!" somebody yelled. I recognized the voice—Nick Bruno, junior class loudmouth, middle linebacker on the football team and all-round jerk. "We don't have a faggot table yet, Fruitful. Why don't you start one?"

Slocum, a lunch bag in one hand and his backpack

over his shoulder, didn't even turn his head in Nick's direction. He just walked, with a kind of disinterested slouch, toward an empty table in the corner.

Molly pushed her hair behind her ears again, put her milk carton into the empty segment on her tray, and stood up. "Let's go sit with him," she said.

"No, thanks. He's your project. Anyway, I told Kristin I'd see her after her meeting."

"Okay, okay. Don't let me interfere with young love. See you later." Molly picked up her tray and headed for Slocum's table.

"Looky there," Nick called out as Molly said something to Slocum and put her tray down across the table from him. "The Goblin Girl's a fag lover!" More catcalls and whistles from the guys at Nick's table.

I headed for the door, tossing my bag into the trash can as I went. The last thing I heard before the doors swung shut behind me was Molly's voice with an edge to it that cut through all the cafeteria's usual background noise. "Lay off, scumbag."

I grinned. I didn't know about Slocum, but Molly could handle Nick Bruno.

2 A couple of times that week I wondered what makes the Nick Brunos of the world do what they do. It's as if there's a trigger that sets them off. If you've got it, you're in trouble. If you don't, you're okay. Bran Slocum had it. And it wasn't only Nick Bruno that got set off. His buddies, Matt Singleton, Jerry Ritoni and Gordon Krosky, took their cue from him and went after Slocum any way they could. Besides that, other guys who wanted to get in good with Nick were busy imitating them.

By Tuesday, someone had painted "FAGGOT" in pink letters on Slocum's locker. The next day the word was spangled with gold glitter. He was bumped and jostled in the hall and sometimes crowded into the lockers. Once he was nearly pushed down the steps when Jerry Ritoni faked a fall behind him. And always the name Fruitful followed him, called out down the halls, whispered in class, chalked onto blackboards in rooms where he had classes.

I didn't pay any of it much attention, because I had other things on my mind. Running was one. The more I ran, the better I liked it. I was getting faster. Coach Morelli told me that week that if I kept at it, I just might have the makings of a marathoner. It's amazing what being good at something does for the way you feel about it. We had practice after school every day, but I was the only one on the team who got up and ran every morning, too. It wasn't just for the team. Running made me feel good.

Kristin was on my mind a lot, too. Molly had said I needed to be on a team if I wanted to keep on dating a cheerleader, and that's why I'd started with cross-country in the first place. But during football season, football was the sport all the cheering was for. The football team got to see the cheerleaders all the time— the squad practiced right by the field. The cross-country team hardly ever even ran at school. Except for an occasional speed trial on the track, we ran a course Coach had laid out through a state park south of town. So I didn't get to see much of Kristin after school.

Besides that, she was a sophomore and we didn't have any classes together. If it hadn't been for lunch, we would hardly have seen each other at all. And there's just so much romance you can get into a conversation in the cafeteria. We usually went to a movie or a party Saturday night, but that was starting to get old. Boring. I didn't feel the way I had when we started seeing each other in the summertime, and I didn't think she did either. Something needed changing.

There was a dance after the pep rally and football

game on Friday night, and I was hoping that would at least break the routine. I was planning to take her in the old pickup truck that is Dad's and my sole transportation, and if the weather was nice, we were going to go off by ourselves instead of to a party afterwards. I thought we could take some food and blankets and have a sort of picnic in the back. But on Wednesday Dad announced that there was a craft fair in Vermont on Saturday, and he was taking his stuff over in the truck Friday night to get set up.

It wasn't the first time my dad's peculiar profession had made problems for me. In some ways it does that every day of my life, and if I let myself think about it too much, it would cause real trouble between us. So I try not to. He's a wood carver, and what he mostly carves are totem poles, sort of contemporary ones, and smaller sculptures he calls "house and garden totems." Sometimes he does them on commission, but mostly he does them on his own. There isn't a booming market for totems, so they fill up our garage, which is also his workshop, and our backyard, which looks like some kind of petrified zoo. Bears, mountain lions, owls, eagles, possums and raccoons stand alone or on each other's heads and stare out at the world beyond our back fence.

There's a sign in front of our house that says, James Watson, Totems and Wood Sculpture, which is supposed to bring customers. But the bottom line is there isn't much bottom line. So he works part-time, seven till two, for a printer he's known since they were both hippies together twenty years ago. This gives him afternoons and evenings to carve, but it also means that we live a very frugal life. I have to work at the grocery store weekends if I want to have any cash at all. Even taking Kristin to the movies every so often was always a strain on my budget.

My mom used to try to get him to take some other kind of job—some forty-hour-a-week, full-time job that would pay real money, so things wouldn't always be so tough. She said he could still carve evenings and weekends. But carving isn't just something my dad likes to do. It's his mission in life, a philosophical, ideological thing I don't quite understand. Mom didn't either. She left him to marry a lawyer when I was ten. They let me choose, and I stayed with Dad because I couldn't stand the lawyer. I don't regret my choice, really, but it hasn't been easy.

Mom left her totem, a fox, behind. It still stands in the front hall with Dad's dove and my wolverine. When she left he was working on a pole that had all three of them together, but he never finished it. Dad says it wasn't just the carving and the money that drove her away. The trouble with their marriage was that foxes and doves don't make very good companions. That's an example of the way he thinks about totems.

I tried very hard to persuade Dad to let me use the truck Friday, pointing out that I needed it sometimes, too. I suggested he could leave for Vermont really early Saturday morning. He said I could have a car of my own the minute I had the money for it, but meantime, the truck was his, and his friend the printer was coming over to help him load on Friday afternoon. So I was scrambling to find wheels for Kristin and me, and that's another reason why Bran Slocum wasn't on my mind a lot that week.

He was, however, on Molly's mind. She'd taken him on as one of her rescue projects. Every day at lunch he'd sit at the same table, always alone, and every day Molly would go over and sit with him, ignoring the taunts of Nick and his gang. I'd glance over at them from time to time, and Molly was always talking as she ate. She has about a million different interests, and she must have told Slocum about every single one of them that week, because she didn't seem to be getting much of anything back from him. Theirs weren't actually conversations. He just sat there, eating, occasionally glancing at her, in that disconcerting way that didn't quite connect.

The only thing really unusual about that first week of Bran Slocum's life at Ridgewood High was that by Thursday, with everyone gearing up for Friday's big pep rally and game and dance, the harassment still hadn't faded out.

As Slocum came out of his last-period class Thursday, Gordon Krosky tripped him, and Jerry Ritoni and Matt Singleton grabbed his backpack as he went down. They kept it away from him, tossing it back and forth between them while he just stood and watched, until they had to go to football practice. Then they took it with them and dropped it into a toilet in the locker room. Slocum, having said nothing the whole time, followed them, picked it up, slung it dripping over his shoulder and walked out. I was sitting there, putting on my running shoes, and as I watched him go, his back straight, his head up, that ponytail moving slightly with each step, I thought it couldn't go on much longer. Not even Nick and his goons could keep it up with so little reaction from their victim. Or so I thought.

Friday morning I still hadn't solved the problem of wheels. I'd asked Artie Weston, who was also on the track team, if I could borrow his beat-up Volkswagen, but he'd asked Cheryl Heroux to the dance, so he was using it himself. A couple of guys offered to let us double, but the whole point was for Kristin and me to get off by ourselves for a change, and maybe get a little spark back into the fading fire, so I turned them down. It looked as if to be alone together, Kristin and I were going to have to walk. I was sure Kristin was going to loathe and despise that idea.

Maybe that was why my morning run went so badly. I was wiped out a couple of blocks after I started. Every step seemed to send the pavement up through my legs, the jolt reaching clear to my brain. I couldn't get my breathing timed right, and I was slower than ever. Besides, it was colder than it should have been so early in October. My hands, my nose and my ears felt as if they were going to fall off, even while the rest of me was sweating.

When I came out of the cemetery and saw Molly, a stocking cap pulled down over her ears, walking Muttsy, the three-legged stray she'd adopted, I slowed to a walk and joined her.

"Don't break training for me," she said. "Muttsy and I are having a great time all by ourselves."

"Is this your subtle way of telling me to get lost?"

"Nope. I just don't want to be blamed if you get in trouble with Coach Morelli."

"No problem. He's crazy about me." Muttsy sniffed a tree trunk and squatted, balancing precariously. "Is the invalid all better?"

"Dad says she's fine. She just needs to get used to getting around without the leg." Molly's father is a veterinarian, and her mother is his nurse, so maybe it comes with her genes to be a pushover for a sad, furry face.

I blew on my hands and started running in place. "I've got to get inside and get warm. You want to come over for breakfast? Dad's probably gone to work by now."

"What've you got?"

"Good stuff. Juice and dry cereal. High-fiber, low-salt, low-sugar stuff all full of nuts and berries and twigs."

"What if I want waffles? Or Sugar Pops?" Molly punched my arm. "What happened to bacon and eggs and pancakes, anyway? Your dad used to make the most enormous breakfasts—"

I punched her back. "He still does. I didn't say Dad eats this stuff, I said we *have* it. I eat it. And it's what I'm offering you. It's good for you. Capital G good. Also easy."

A few minutes later, we were on our way up the gravel driveway that leads through a scraggle of hedge to the bungalow with the peeling yellow paint that is Dad's and my house. Molly brought Muttsy inside.

It was obvious that Dad had been up, fixed his breakfast and gone. The egg carton was sitting open on the stove, greasy paper towels were next to the frying pan, dirty dishes and the orange juice pitcher and the jumbled wreck of the newspaper littered the table. Our small black-and-white television was on to the "Today" show. Willard Scott, wearing some sort of a turban, was pointing to the northeast on his map and talking about a strong warming trend.

"Good," I said, getting out the cereal and a carton of milk, "it's too early to be so cold." Molly, who knew her way around our kitchen as well as I did, set out bowls and spoons. "Sorry about the mess," I said.

"Don't worry about it," she answered, clearing Dad's dishes to the sink. She dropped a bacon scrap onto the floor, and Muttsy pounced on it with the agility of a panther. "I guess I can stop worrying about you," she said to her.

"That car wrecked her leg, not her head. She still recognizes bacon when she smells it."

"Only some things are worth the effort, I guess. You should see how hard it is for her to get up when she has to go out and it's raining!"

While we ate the kitchen was quiet except for the gurgling of our sick old coffee pot and the low drone of the television. "How's the boyfriend?" I asked Molly, when I'd finished my cereal and drunk the rest of the milk out of my bowl.

"What boyfriend?"

"Don't play games with me, Molly Pepper. I've seen you at lunch with him every single day."

"Boyfriend is hardly the operative word," Molly said, frowning. "He's not the most communicative person I've ever met."

"I'm sure you more than make up for that. So, what's he like?"

She shook her head and finished her orange juice. "It's funny, David, but Bran's different. Really different."

"Yeah. That's his whole problem. He could at least cut his hair."

"I don't mean the way he looks. I mean the way he *is*. You know what Bruno and his gang have been doing. He doesn't react. He doesn't get mad, he doesn't get upset. It's as if he's above it all. Or beyond it."

"Nobody's beyond that kind of thing," I said.

Muttsy whined, and Molly put her bowl, with a little milk in the bottom, on the floor for her. "I've been trying all week to figure out what it is about Bran, and I can't. He's just—still. That's a good word for it. *Still.*"

"He hasn't had a chance to get a word in edgewise. Every time I've seen the two of you together, you've been the one doing the talking."

"He talks. Sometimes. As long as you don't ask direct questions. Anyway, it isn't about talking or being quiet. That's not what I mean. It's something about his whole self—who he is. He's just *still.*" She sat for a moment, staring off into space. Molly does this from time to time, and you just have to wait her out. "Remember what I'm telling you," she said, after a bit. "There's something special about Bran Slocum. You'll see."

"Right." It sounded to me as if Molly was falling for the guy. I should have been happy for her. I had Kristin, now Molly would have Bran. I should have been happy, but for some reason, I wasn't. I wondered if I might be jealous, but dismissed the thought. It was probably just that I was used to knowing Molly only had me. That wasn't really jealousy, was it? "I've got to take a shower," I said, getting up. "If you wait, I'll walk you over to your place."

She checked her watch. "Okay, but we don't have much time. Tell you what. You walk me home, and I'll drive us to school. Then you can keep the keys and take my car for your date with Kristin."

I just looked at her.

"Well, don't you want it?"

"Of course I want it. You think I'm nuts?"

"Slow, maybe, never nuts. Hurry up!"

I patted her on the head and hurried off to the bathroom. That, I thought, feeling better, is a real friend. When I'd showered and dressed, I came back to the kitchen to find that Molly had cleaned it up. She was standing now, with the dish towel in her hand, staring at the television with an expression of supreme disgust. "What's the matter?" I asked.

She waved the cloth at the screen. "That creep."

On the screen a bunch of cops were hurrying someone in handcuffs through a crowd of jeering people.

Whoever it was had a suit jacket draped over his head. "Who?"

"That serial killer. The one who killed all the kids in New Jersey. Buried them in his yard. His trial starts today." Molly shuddered. "He didn't just kill them, you know. He tortured them first. His youngest victim was only eleven."

"I thought he only killed runaways. Do kids run away as young as that?"

"I guess so. Nobody even knows how many there were. Kids disappear every year and are never found." Her face was still creased with disgust. "I wonder how many steps it takes to get from Nick Bruno to that."

"Don't get melodramatic," I said. "Bruno's no psycho. He's just your normal, average bully. There's a big difference between hassling people and killing them."

She turned off the television and hung the dish towel on the rack. "I don't want to think about it. I wish they wouldn't put the story on the news. It probably gives weirdos all over the country ideas. Come on, Muttsy." She leaned down and picked Muttsy up, talking to her while the dog licked her cheek. "At least the guy who hurt you did it by accident."

"And then had the good sense to take her to your father."

Molly set Muttsy back on the floor and clipped on the leash. "If Dad hadn't been able to save her, he'd have put her to sleep," she said. "He can't stand to see an animal suffer. What kind of person gets his kicks torturing kids?"

"Like you said. A psycho."

"Psycho pervert creep."

"Let's get going."

"Yeah. And forget about it. I don't want to know there are guys like that in the world."

"There aren't very many," I assured her, as I pulled the door shut behind us. "The world's not like that."

3 By the time we'd parked in the school lot, back next to the drivers' ed cars where Molly always parks when she drives, the sun was out and the day had warmed up considerably. It was the kind of fall day that makes the covers of nature calendars—all reds and golds and deep blue. We walked up through the lot with the juniors and seniors who drive and a couple of faculty members hurrying to the side door to keep Dr. Towson from knowing they were late. The buses were pulling up, one after another in a great caravan, dropping their passengers at the side of the building.

As we turned the corner and started up the walkway toward the flagstone courtyard in front of the school, we saw a crowd of kids standing around the old memorial fountain. They were watching something, but we couldn't tell what. The kids coming up with us and the ones who'd gotten there just before crowded up, asking what was happening and shoving to see. Everybody else was quiet. There was a tension in that quiet that I didn't like. It reminded me of a crowd I'd seen once, gathered around a car accident.

Molly went ahead of me, squeezing past kids, using her elbows when she had to, and I followed. "Quit shoving," somebody said, and the kids in front of Molly squeezed closer together, keeping her from getting any farther.

"Can you see anything?" she asked me, after trying in vain to push on ahead.

Over the people in front of me I could see, all right. There were four people in the open space by the fountain. Gordon Krosky, Matt Singleton and Nick Bruno were facing the fountain with their backs to the crowd. Bran Slocum was backed against the crumbling stone basin, his face unreadable, his bad eye angled so far off to the side that you could see mostly white. His backpack was on the flagstones a few feet away.

"Well?" Molly asked, gouging me in the side with her elbow. "What's going on?"

"Nothing at the moment," I said, hoping nobody would let her through. I had a pretty good idea what she'd do once she found out what was happening and who the target was.

"That hair's greasy," Nick said. "Don't you think so, Matt?"

"That's Bruno, isn't it?" Molly asked.

"G-g-greasy!" Matt repeated.

"We don't like greasy hair, do we, Gordo?"

"Nah." Krosky, a huge guy with a small head and

brain to match, glanced over at Nick as if to be sure
he'd given the right answer. Nick nodded and Krosky
smiled. "Nah, Nick. We hate it!"

Molly was jumping up and down, trying to see.

"It's a shame that such long, gorgeous hair should
be allowed to get so"—Nick shuddered broadly and
made a sound in the back of his throat as if he were
about to be sick—"disgustingly greasy. Let's do him a
favor and wash it!"

With that, Nick lunged toward Slocum and grabbed
him by the shoulders. Slocum tried to pull away, but
Matt caught him from one side and Gordon from the
other. They didn't have an easy time holding him, even
though there were three of them, and they were all
bigger than he was. Slocum flung himself from side to
side, cracking Gordon into the fountain with a sound
that echoed through the courtyard.

Molly tried once more to push her way forward, and
the kids in front pushed back again. Giving up, she
turned and began shoving her way back the way we'd
come. Kids moved readily out of her way, and then
pushed in after her. I stayed where I was, not the least
interested in trying to follow her. I felt sorry for Slocum,
but I had no intention of having a confrontation with
three ticked-off football players.

Nick ripped the rubber band off Slocum's ponytail
with a jerk that looked as if it would tear half the hair
out, and then they managed to get Slocum, still struggling to get away from them, turned toward the fountain.

"Hold him," Nick said to the other two, as he shoved Slocum's head into the filthy, debris-choked water that was left in the basin from the September rains. Slocum jerked and fought, but Nick pushed with both hands, jamming his face against the bottom, then scooped handfuls of dirty water over Slocum's hair.

There was a commotion in the crowd over to my left, and I saw Molly's short, stocky figure barreling through, using her backpack like a battering ram. She burst into the open and went straight for Nick, bashing him on the back with her pack.

"Let him up, you creeps!" she hollered.

Nick pushed her away, but she came right back, aiming her pack now at Gordon, who let go of Slocum to defend himself. Bran, with only Matt holding him now, jerked loose, muddy water and cigarette butts running down his tangled hair and over his face, and knocked Nick sideways. Nick tripped over his own feet and went down, cursing. Jerry Ritoni came out of the crowd to help Nick up.

Just then the first bell rang, jangling across the courtyard, and the big double doors swung open as Dr. Towson unlocked them. He does this with great ceremony every morning, personally welcoming his students to another day in the hallowed halls of Ridgewood High. Towson, a tall, broad figure in a pin-striped blue suit with a red bow tie, propped the doors open and took up his position in the middle, where he could nod and speak to everyone who went in.

Immediately, the crowd broke up into couples and knots of kids, talking, laughing, jostling each other, heading around the fountain and up the wide front steps as if nothing unusual had been happening.

"Cowards!" Molly yelled, at no one and everyone.

Nick, Matt and Gordon brushed themselves off, patting each other on the back and grinning. As they skirted the fountain, Jerry Ritoni aimed a kick at Slocum's backpack and sent it across the flagstones into the sparse grass.

Slocum just stood, wiping his face with the sleeve of his jeans jacket, while Molly went to retrieve his backpack. Above his head, on the cracked obelisk at the center of the fountain, the brass plaque proclaimed the glory of the "boys from Ridgewood High who gave their lives in the Great War."

Kristin, who'd been nearer to the front of the crowd, saw me and came over to where I was standing and letting people move around me. She slipped her arm through mine. "Those guys don't let up."

Molly and Bran were walking up toward the stairs, next to each other, but not touching and not talking. Other kids were giving them a wide berth.

I looked down at Kristin, and she grinned, her green eyes crinkling at me. The sun struck highlights of pure gold in her long, blond hair, and I grinned back. "Molly Pepper's not afraid of anybody," she said. "She'd take on Godzilla. I can't figure out what she sees in that guy, though. Talk about weird!" She tugged at my arm. "Come on, Davey, I can't be late for homeroom again or Fergie'll kill me."

We went around the fountain, heading up toward Dr. Towson, who was nodding and smiling, greeting people by name every so often, and ostentatiously checking his watch every few seconds. "I've got wheels for tonight," I said.

"I knew you'd think of something." Kristin did a little double skip, still hanging onto my arm. "Let's make it a really terrific night." She reached up and kissed me on the cheek, her pale, soft hair brushing my face. She smelled fresh and sweet—like a field of wildflowers, and I remembered how I felt the first time we went out. It shouldn't be so hard to get back to the way we were then, I thought. Some time alone ought to do it. By the time we got inside, Molly Pepper and Bran Slocum were the farthest thing from my mind.

Bran was late to homeroom, but he came, a few minutes after the bell, his hair wet but combed and slicked back into a ponytail again, his face clean. The only evidence of what had happened was his jacket, wet half the way from neck to elbows, and a purplish swelling on his upper lip. He handed Mrs. Campbell a late slip and went to his seat, his head up, shoulders straight. I thought he paused for just a second before he sat down, looking at Matt Singleton, but I couldn't be sure.

I had to hand it to him. I'd have gone home. It isn't that I'd have been running away exactly. I'd have gone home to take a shower and change. And then I just wouldn't have come back. I glanced over at him, thinking about the word Molly had used. *Still.* She was right. While Mrs. Campbell finished taking the roll, he just sat there, not moving, quiet, contained and expressionless. He might have glanced at Matt occasionally, but with that eye, there was no way to know for certain.

I didn't see Molly till math class, and then she lit into me. "Why didn't you come up there and help me this morning, you clod, you wimp?"

I started to tell her I'd been hemmed in by people who wouldn't move, but she didn't give me a chance.

"You let me go after those guys all by myself. You and every other person here. Doesn't anybody understand? It would only take a couple of guys *not* letting them get away with stuff like that, and they'd have to stop. God, what cowards people are!"

I just shook my head. "Molly, get real. Bruno is Bruno. Nobody's going to go up against him when he's by himself, let alone when he's surrounded by his goons. Nobody even wants to."

Molly slammed her books onto her tablet-arm chair. "Yeah, nobody including you."

Ms. Caitlin started class then, so Molly didn't have a chance to say anything else.

I did a crude sketch of a lion in my math notebook. It was smiling, and out of its smile drooped the leg of an antelope. At least that's what I meant it to be. In a balloon over its head I wrote, "It's a jungle out there."

4 That afternoon I was glad Molly was taking English history instead of Contemporary Issues with me. We were in the middle of a unit on justice in America, and when we got to the room seventh period the bulletin boards were covered with newspaper and magazine clippings about the story that had upset her so much at my house. She'd said she wished they wouldn't put it on the news. From the look of those clippings, she was probably the only person in America who didn't want to know every gruesome detail. Papers don't print what people don't want to read. And plenty had been printed about this case.

Mr. Byrd was standing behind his desk when we came in, and he flapped one long, thin hand at the boards. "Take some time to look at these before we get started. I want everybody to be thoroughly familiar with this story before we begin our discussion."

"Who isn't familiar with it?" Zachary Lewis asked.

"You'd have to have been in a cave for about a year to miss this story."

Mr. Byrd shoved his hands into his jeans pockets and leaned back against the blackboard. "Just look. And read. There are a lot of viewpoints here. Different perspectives, different tones. The *New York Times*, the *National Enquirer*, *People* magazine, our own *Ridgewood Courier*, the *Reader's Digest*. A little bit of everything. And if you don't get enough from the print media, I've got some videotapes we can look at to see what television has to add."

Jennifer Logan groaned. "They must have shown men bringing bodies out of that guy's yard in those plastic bags about a million times. We've seen enough of that!"

Zach Lewis laughed. "Yeah, and cops digging with masks over their noses. I'll bet that was gross duty. You suppose they got paid extra?" Zach could find something funny in almost anything.

Cheryl Heroux squealed. "Do we have to talk about this, Mr. Byrd? It's too disgusting. What's it got to do with justice in America, anyway?"

Mr. Byrd sighed loudly. "Even you, Miss Heroux, ought to be able to figure that out. Joseph Collier's trial begins today. Four counts of murder."

"I heard he killed ten or fifteen kids," Nick Bruno said. "How come only four counts? Why don't they get him for all of them?"

"Eight bodies were found in his yard, but some of them had been there a very long time. Two haven't even been identified. Apparently, the police think they have enough real evidence to convict him on only four, and one of those was found somewhere else. As for how many more he may have killed, we'll probably never know, unless Collier decides to confess. Serial killers are more likely to get away with murder than anyone else, if they dispose of the bodies carefully."

"I guess that doesn't mean in their own backyards," Zach said.

"Don't be too sure. The investigation that led the police to Collier in the first place was about the disappearance of a boy who was not one of the ones in the yard. His body was found in an abandoned gas station. If it hadn't been for that one, the bodies in the yard might never have been found. Now, take ten minutes, and let's get on with our discussion."

For the next ten minutes, we looked at pictures, scanned news articles, read captions. By the time Mr. Byrd waved us to our seats, I was about ready to agree with Molly. I didn't want to know any more.

The tabloid papers were the worst. Somehow a photographer had managed to get close-up shots of the body from the gas station and they'd printed them half a page high in color. The only decent thing they'd done was to show only arms and legs, so you couldn't tell much about the kid that body had been. Arms and legs with cigarette burns and bruises—and a hand missing two fingers. When Cheryl Heroux saw that one, she had to excuse herself and run out. When she got back, she was sort of sickly pale, and I was pretty sure she'd thrown up. Not even Nick teased her when she came

back and slipped into her seat. We all felt kind of sick,
I guess.

What made it worse was that all of his victims were
kids. Lots of them younger than us, some of them
our age. I wondered what the guy had said to them
to get them to go with him so he could do that to
them. In every one of the pictures, Collier was wearing a suit and a tie. He was just this ordinary-looking
middle-aged man with a bald spot and glasses. Somebody you wouldn't look at twice. Not my idea of a
psycho killer.

Even so, wouldn't they have been able to tell, just
looking at him, that he was somebody to run from, not
somebody to get into a car with? Or go into a strange
house with? Surely they would have seen something
in his eyes, heard something in his voice. All the kids,
even the youngest, were too old to fall for being offered
candy. That's what everybody used to warn us about—
taking candy from strangers. This guy couldn't have
been handing out gumdrops.

"Okay," Mr. Byrd said, when we were all settled in
our seats. "Let's talk justice. Joseph Collier's trial begins
today, after three weeks of jury selection. And he's being
tried in New Jersey, where the murders took place. Do
you think he can get a fair trial there?"

"Wh-wh-who cares?" Matt Singleton said.

Mr. Byrd scratched his beard and glared at Matt. "I
imagine Joseph Collier cares. What if he didn't do it?"

At that everyone was talking at once. "Of course he
did it." "How'd the bodies get into his yard if he didn't
do it?" "The police got him, didn't they?"

Byrd just leaned on the board until the noise subsided. "The first thing you learned about the justice
system in America is that a defendant is innocent until
proven guilty." Scott Handleman groaned. "Mr. Handleman, you have an objection to that?"

Scott's face reddened. "Well—no. I mean, not usually. It's better than the way they do it in Russia or
someplace—"

I expected Byrd to pounce, to ask him exactly how
they do it in Russia. He liked nothing better than to
catch somebody saying something when they didn't
really know anything about it. He's good at embar-
rassing kids, which is why most of the time I just listen
in his class instead of talking. But he let it go and waited
for Scott to go on.

"It's okay to assume somebody's innocent if he's
being tried for shoplifting or something. Robbery,
maybe. I don't know—even murder, I guess. But this
is different. This is—" Scott faded out.

"Too grotesque a crime for regular justice?" Byrd
looked around the room. "Anybody want to disagree
with Scott?"

Nobody did. I was tempted to say that there couldn't
be different kinds of justice for different kinds of criminals, but I kept remembering those cigarette burns.
He'd tied the kids down and made those burns, one
after another. And he'd chopped off that boy's fingers—
my stomach flip-flopped just thinking about it. What if
by presuming this guy innocent and following all the
legal technicalities, they somehow ended up having to
let him go, so he could do it again? Maybe there was

such a thing as a crime that was just too horrible for the normal rules.

"I don't think I understand," Jennifer Logan said. "The cops dug up all those bodies from Collier's own backyard. One of them, the papers said, couldn't have been there more than a month. So how could he be innocent?"

"Somebody else could have buried them there," Mr. Byrd suggested.

I imagined someone sneaking into a stranger's yard and burying a body there while the guy slept. Not once, but eight times, over months and years. Impossible.

"And I suppose the guy never noticed that every so often he came home and found all his grass dug up and his tomato plants gone," Zach said.

"All right, then, what if he buried the bodies, but didn't kill the kids?" Mr. Byrd asked.

This time he was greeted with silence. I hadn't thought of that, and apparently nobody else had either. "Why would he do that?" Jennifer asked.

Mr. Byrd shrugged. "I don't know. Maybe to cover up for the person who really did it? Let's say he had a crazy relative who every so often went off the track and killed somebody. And let's say he didn't want to see this person he cared about go to prison or get the death penalty."

"That would still be a crime," Zach pointed out.

"But not murder. Different crime, different penalty."

"D-d-do you think that's what h-h-happened?" Matt asked.

Mr. Byrd shook his head. "I'm just trying to get you

to think about some possibilities. None of us knows what happened, so we can't just assume automatically that things are the way they appear. That's the basis of our whole system. We attempt to look for truth, not appearances." He glanced around at the bulletin boards. "Do you think the man can get a fair trial in New Jersey, or anyplace else for that matter, after all this publicity about the case? Most of those articles take it for granted that Collier's guilty."

"None of those k-kids got a fair trial," Matt said. "Wh-wh-why should he?"

"Okay, okay." Mr. Byrd wasn't usually one to give up, but he must have seen that he wasn't getting anywhere with the idea that Collier could be innocent. Nobody was ready to give him that. "Let's try a different angle. What if he did do it, but is crazy?"

"He'd *have* to be crazy to do it," Jennifer said.

"So you want to equate evil with madness? Does that make Hitler crazy? Pol Pot? Al Capone? Terrorists?"

"It don't matter whether this Collier is crazy or not," Nick said. "I don't care if somebody hears voices or sees Elvis in the grocery. He kills kids, he goes to the electric chair. Simple as that."

On that, there was agreement. Mr. Byrd looked around at all our nodding heads and shook his. "If justice were simple, we wouldn't be doing this unit. That's why I brought all these articles in. A case like this tests our system to its limits. Guarantees that seem right in every other situation are called into question. Just let me tell you this. No matter how awful the crime a person is accused of committing, he has the absolute right in this country to a fair trial. That's what this whole unit is about. Even Joseph Collier has a right to a fair trial. Now, you want to talk about the death penalty?"

"Does New Jersey have the death penalty?" Scott asked.

"Yes. And if Collier's convicted, the prosecution will certainly ask for it."

"*If* he's convicted?" Nick said. "Somebody ought to just lynch the guy right now and make sure!"

Just then the bell rang.

"With that enlightened suggestion, we end," Mr. Byrd said, his eyebrows knit together. "And God bless America!"

5 Because of the pep rally that night, after-school practices were canceled. It was one of the few times all semester that when school was out at three-fifteen it was really out and just about everybody went home. I grabbed my books and met Kristin at her locker, Molly's car keys in my pocket. She was gathering her books as other kids opened and shut lockers and pushed past us. As close to her as I was, there was so much noise in the hall I had to practically yell to be heard.

"I'll run you home now and pick you up again about five. Will that give you time to get ready?"

Kristin looked in the mirror on her locker door and ran her hand through her hair. "Look at this! My hair's a disaster." I looked. To me it looked the way it always did—but I knew better than to argue. "I have to wash it, that's all. And press my outfit. It's been squashed in my locker all day. Better make it five-thirty. Where are we going to eat?"

"Peroni's, I thought. Is pizza okay?"

"As long as we don't have to have anchovies. Last time I went there they put anchovies on and I was sick the whole night."

"No anchovies. I promise."

"Anchovies! I should hope not!" Molly's voice boomed from behind me. I turned and found her at my elbow, clutching her books and dodging kids trying to get by on their way out to the buses. "Anchovies are aphrodisiacs. Like oysters. They're nothing but trouble. Sorry to interrupt, but can I have a word with you, Watson?"

"Sure. What's up?"

Kristin went back to collecting her belongings from her locker, and I let Molly pull me out of the traffic into a classroom doorway. "I don't want to dampen the romance of the moment or anything, but would you give me and Bran a ride home?"

I glanced over at Kristin. She was folding her cheerleader outfit and paying no attention to us. She couldn't have heard. "How come?"

"Ritoni was bragging last period about what they're going to do to him after school. Nick and Matt and the rest of those guys are waiting for him at the main gate. If he walks he has to go right past them."

I could imagine how Kristin would feel about being seen with Slocum. But it was Molly's car, after all. "Where is he now?"

"He's talking to Towson about a placement test for Spanish. I told him I'd meet him outside the office."

"I'm ready!" Kristin joined us, her sweater tied by

the sleeves over her shoulders, her pink nylon tote bag on her arm. She smiled brightly at Molly.

The contrast between the two of them was startling. Kristin, all blonde and pink and glowing, clothes color coordinated, hair curling softly around her face, and Molly, in her dark sweat shirt and faded jeans, her face pale, her black hair pushed behind her ears. I felt caught, somehow, between them. Molly nudged me. "Ask."

"Kris," I said, "you know that new guy? Slocum?"

"Fruitful Bran?" she said, as casually as if that were his real name. I winced, but Molly managed not to react. "Sure. What about him?"

"He needs a ride. Nick and some of the guys are waiting for him by the main gate. Planning to beat him up. If we give him a ride, they won't get him."

Kristin looked from me to Molly and then back. "You're kidding, right? You want *him* to ride with *us?*"

"Him and Molly, actually," I said.

She frowned. "Isn't there anybody else who could take him?"

Molly shook her head. "He'll ride in the back and stay down. Nobody'll see him."

"You sure?"

"That's the whole point," I said, knowing that Kristin's main problem was that someone might see him with us. Then I thought of a way to clinch it. "You remember that old movie we saw that you liked so much, where the hero smuggled aristocrats out of France? *The Scarlet Pimpernel,* it was called."

Kristin grinned. "Of course I remember. It was so

romantic! He kept all those people from getting their heads chopped off. I told you, it was just about my favorite movie ever."

"This'll be just like that. We'll take Slocum out right under their noses."

"Okay," Kristin said. "It'll be like a mission of mercy." She giggled. "It's about time somebody got around Nick Bruno, anyway. He thinks the whole world has to do anything he says."

"You'll be heroes," Molly said. "But we'd better get going or we'll be the only ones left on campus."

She was right. Already the halls were nearly empty. Half the buses would be gone by the time we got outside. Tricking Bruno was one thing. Facing him was something else. "So, what's the plan?" I asked.

"You two go get the car and bring it up by the big doors behind the cafeteria. Bran and I will meet you there." She hurried off down the hall.

"Do you know this guy?" Kristin asked, as we walked down the school's wide front steps.

"Nobody really knows him, except Molly a little bit."

"What the guys did to him this morning was bad." She smiled. "It was funny though, in a way."

"I doubt that Slocum thought it was funny."

"I know," Kristin said, "but don't you think he asked for it sort of? I mean, he *is* weird. You suppose he's a faggot like they say?"

I didn't answer. I just took her elbow and angled her past the fountain toward the parking lot.

"Seems to me," she went on, "that the smart thing to do when you're new to a place is to check it out.

You know—you come on the first day in something really plain and ordinary. Jeans and a sweater maybe. And you see how everybody else is dressed. I mean, Davey, why would anybody want to stand out like he does? He'd probably be okay if he got rid of that earring and cut his hair—of course, there's nothing he could do about his eye."

"No."

"But I don't think anybody would beat somebody up just because he had a weird eye."

"Don't be too sure."

"Well, maybe Nick would," Kristin conceded. "But not the others. Not Matt, anyway, or Jerry."

"Are you kidding? The others do whatever Nick tells them to. There's the car." I pointed to Molly's Civic, crouched like a silver hedgehog next to the beige drivers' ed sedans. "The glorified roller skate."

"It's cute!" Kristin said. "Except they're going to be pretty crowded in back."

"Yeah, well, Molly's small."

We got in and I started the engine, its putter-roar making me feel terrifically conspicuous. Several kids looked in our direction. Kristin waved to one of the cheerleaders, who was tangled with a guy in a letter jacket, both of them leaning on a beat-up Firebird. "You sure nobody'll see him?" Kristin asked.

"We'll get him inside quick," I assured her. The sun had been shining down on the car all afternoon, and it was hot inside. I rolled down the window and drove around to the back of the cafeteria. I pulled up as close to the doors as I could, got out and leaned the seat forward. Then I stood back and out of the way.

The double doors opened and Molly came out, pulling Slocum behind her. They hurried to the car, and Molly stepped out of the way and practically pushed him in first. "Scooch over and I'll get in this side, too. And keep your head down!"

When they were both in and I could put the seat back, I got in, too. In the rear view mirror I saw Slocum, slouched down, with his knees jammed against his chin. His face, as always, was unreadable. And he hadn't said a word.

"Kristin," Molly said, when she had settled herself so that her knees jabbed into my back through the seat, "this is Bran Slocum, Bran, this is Kristin Matthis. She's a cheerleader."

Kristin turned around, a smile firmly plastered on her face. "I'm very pleased to meet you." Slocum nodded.

"And this is David Watson," Molly went on, "cross-country runner. He runs like a tortoise—you know, slow but steady."

"Thanks a lot!" I started the car again and revved it for a moment. "We met once," I told Bran. "Or at least I sort of ran into you."

"I remember." His voice was low but strong, easily heard over the engine's noise.

"Wagons ho," I said and moved out.

"Head down!" Molly commanded.

Slocum frowned, but ducked his head.

We drove out of the parking lot and turned down

the street in front of school. Sure enough, most of the defensive football team was standing around by the front gate. They were roughhousing among themselves, scuffing around in the leaves, pushing and shoving, pretending to throw each other out into the street in front of the cars that were passing—slowly—beneath the flashing yellow lights.

I drove past as fast as the car in front of me would allow. Kristin looked the other way, and I put a hand up to my face. Molly, in her subtle way, turned her head away but stuck a hand out my window and gave them all the finger.

There was a clatter against the back window. "They're throwing stones," Molly said, and turned to look back at them, shaking her fist.

"You're lucky they didn't break the window," I said. "That was dumb."

"As long as they didn't see Bran," Molly said. "I can't stand those creeps."

"Larch Street, right?" I asked Slocum, who had sat up, now that we were safely past.

"No. I have to pick up my cousin's kids at day care. On the other side of the cemetery. Birch, just off Broad."

"What do you do—baby-sit?" Kristin asked, incredulously.

"Till she gets off work," Slocum said. "Why?"

"Oh—no reason. You don't seem to be a baby-sitter type, that's all." We drove for a while in what could only be called an uncomfortable silence. Then Kristin turned around. "Can I ask you something?" she said to Bran.

He grunted, and Kristin took that to mean yes. "Where'd you live before you came here?"

There was no answer. Slocum was staring out the window now.

"Bran? Where'd you live before?" Kristin must have thought he hadn't heard her.

Still he looked out the window.

Kristin tried another tack. "Did you go to a big school before?"

"Not very," he said, finally.

"Bigger than Ridgewood or smaller?"

There was a long silence except for the sound of the engine. "Well?" Kristin wasn't getting the message. I reached to turn on the radio, and remembered that Molly's antenna was broken. I began humming.

Finally, Slocum answered. "Bigger."

"I guess things were different there, right? Fashions and like that? I mean, *lots* of guys at your other school probably wore their hair the way you do—and big earrings."

I stopped humming and fought the impulse to laugh. Kristin, I was pretty sure, hadn't been trying to make a joke.

"A few," Slocum said.

"Well, but it was like a whole style, right?" Kristin's voice was all cheerleader enthusiasm. She had discovered an explanation for the way he looked that she could understand. "The way most of the guys at Ridgewood wear their hair short or else long and moussed—depending on who they hang around with."

There was another silence. I kept expecting Molly to

say something, but she didn't. Even over the roar of the engine, I imagined I could hear everybody breathing. I looked in the rear view mirror. Molly had her hand over her mouth and Slocum was studying the floor, avoiding looking at Kristin, who was still turned around in her seat, facing him.

Finally, he answered, his tone intensely serious. "Style," he said, "has always been very important to me."

Molly spluttered and coughed into her hand. Slocum's face was as solemn as ever. I turned and smiled out at the bus that was passing us going the other way. Kristin didn't seem to notice our reactions. She turned and settled back into her seat, satisfied.

Hoping she wouldn't decide to ask him about his eye next, I went back to humming until I turned onto Birch and Slocum pointed over my shoulder. "The center is up there—just past the green house. The white one with the gate on the porch—you can see the sign in the front yard."

"How will we fit them in?" Kristin asked as I pulled up to the curb across the street from the house he'd pointed to.

"No problem," Molly said. "We'll both get out here and walk the kids home. You guys go on."

Kristin got out and opened her door so Slocum and Molly could scramble out. As they started across the street, the front door of the white house opened and two little blond boys, dressed alike in grubby overalls and T-shirts, tumbled over each other onto the toy-strewn porch, yelling "Bran, Bran, Bran!"

"Oh, twins!" Kristin squealed. "Aren't they adorable?"

Slocum broke into a run, loped up the steps, pulled open the gate, and knelt in front of the boys with his arms wide. They flung themselves against him, the first one nearly choking him with both arms around his neck. The other tried to climb over the first and dislodge him. Molly stopped at the bottom of the steps and just watched, as Kristin and I did, our mouths open.

Bran snatched the boys up, one under each arm, and whirled around with them while they whooped with delight. He was grinning almost as broadly as they were. "How's it going, you little beasts?" I heard him say, before a passing car drowned him out.

"Would you look at that," Kristin said.

"I guess there are at least two people who like him even if he is weird," I said. Three, I thought to myself a moment later, catching the expression on Molly's face as I pulled the car away from the curb.

6 From the beginning I suspected that the evening wasn't going to do for Kristin and me what I'd hoped. When I got to her house at five-thirty, she wasn't ready, so we were running late by the time we got to Peroni's. Then she mentioned anchovies so often while we were ordering the pizza that the waitress got confused and there were anchovies all over it. There wasn't time to order another pizza, so I picked them off, but Kristin said it still tasted fishy and wouldn't eat anything but the crusts. By the time we got to the pep rally, she was in a lousy mood, and I was having trouble not reminding her it was her own fault.

Things got a little better at the rally. The warm afternoon had turned into a warm night, with a piece of a moon and high, thin clouds that streamed over and around it. The cheerleaders did a new routine that ended with a pyramid, and it came off just right. Lots of kids were there, the bonfire was huge, the band sounded better than usual and by the time it was over everybody was full of school spirit, cheering and laughing and waving their blue and gold Ridgewood pennants.

"This is what high school's supposed to be like," Zach Lewis said, as we headed for the game. "Maybe we'll even win tonight. All I need now is a date."

I sat in the bleachers with Zach and Scott Handleman, but I mostly watched Kristin instead of the game. She's the kind of cheerleader any team would want. She throws herself into it because she really cares. She yells her heart out, and when she does her jumps, her short skirt showing off those very nearly perfect legs, I find myself yelling for our team to win, if only to make her happy. Watching her, I was looking forward to the rest of the evening again.

Unfortunately our team lost—twenty-one to thirteen—and Kristin cried. That's how serious she gets. She and Jennifer Logan hugged each other and sobbed on each other's shoulders till you'd have thought somebody on the team had died. I was pretty sure the cross-country team could be shut out at the state meet, and they'd barely notice. I don't know what it is about football.

I had high hopes for the dance, but it turned out to be worse than the game. The seniors had hired a band instead of getting a DJ, and it was a disaster. The lead singer couldn't stay in tune with his own guitar, and there was something wrong with the electrical system. The speakers either squealed or blanked out. Zach said if he were going to find a date at the dance, it would have to be one of the band members since they were the only ones who were willing to hang around. Kristin and I tried dancing, but gave it up and went out into the parking lot with everybody else.

The team had never even gone inside. They just went straight from the locker room to the parking lot, where somebody had a keg of beer and lots of empty soda cans stashed in his car trunk. When Towson and the faculty chaperones came out all they saw were kids sitting and leaning on cars, drinking soda. It was pretty loud out there, with all the car radios tuned to a Top 40 station from Syracuse, but as long as the adults didn't take a whiff of those soda cans, they didn't have much to complain about.

I'm not much of a beer drinker. I'll drink one, just to go along with the crowd, but I don't like the taste. Kristin does. She insisted on drinking with the rest of the cheerleaders, who were all clustered around the team while the guys hashed the game over. According to them, we hadn't lost because Hamilton played better, we'd lost because the officials called everything in their favor. It was the same argument I'd heard after every loss all season. Nobody thought to wonder why the officials seemed to like every team in the conference except Ridgewood.

Leaning against one of the cars, watching Kristin giggle with Myra Cunningham and Jerry Ritoni, drinking her third or fourth Pepsi can full of beer, I decided it was time to get out of there. Without the truck, my idea for a cozy little picnic was shot, and a Civic's not the best car in the world for parking. But I had an idea.

I went over to Kristin, who was assuring Jerry that

Ridgewood would cream Hamilton next year, and took her free hand. "Let's go. I think I've got a way to save this whole night."

"What's so bad about the night? There's people, and beer—" she looked up at the sky. "And a moon—sort of."

"This is a lousy place to see it, though. I've got someplace much better in mind. Besides, I thought we wanted to be alone tonight."

She giggled. "Okay. Let's go be alone and look at the moon." She grabbed a bag of pretzels off the hood of a car. "I'm taking these along, though. I'm starving."

In the car, Kristin leaned her head against my shoulder. "I'm sorry about the anchovies," she said, and took a sip of her beer. "Are you sorry about the anchovies?"

I kissed the top of her head. "Very sorry."

As we drove out of the parking lot Kristin ruffled the short hair on the back of my neck, giving me goosebumps. "So come on, David, where are we going?"

"Someplace secret. A very nice place nobody else knows about so we can be sure we'll be alone. Just us and the moon."

"Sounds good to me." She fed me a pretzel, had one herself, and then finished her beer. "I should have filled this up before we left," she said, and tossed the can out the window.

"Kristin!"

"Oh, don't be such a stick. It's not as if Ridgewood was the Garden of Eden. One little can's not going to ruin it."

I noticed a pair of headlights pulling out of the school

parking lot behind us and another pair after that. Apparently we weren't the only ones giving up on the dance.

I turned up Union Street and headed for the abandoned quarry on the hill above town. The quarry is a great place to swim in the summer—deep and cold and forbidden. Parents tell their kids as soon as the kids are old enough to go off on their own that they must never go near the quarry. So, naturally, that's where everybody goes.

There are signs posted around the place that used to say No Swimming. Now the *no*s are spray-painted out and replaced with *good* or *free*. A couple of kids have drowned in the quarry because the sides are steep, and it's hard to climb out if you don't know where the rocks are that you can use as platforms and steps. But even drownings don't keep kids away.

It isn't just for swimming that the quarry's popular, though. The gravel road that goes up the hill and around it widens out on the high side into what's become a kind of parking lot. The view's terrific. Ridgewood spreads out across the valley and up a hill on the other side of the river that runs through downtown. At night from up there you can't see all the crumbling porches or the boarded-up shop windows. You just see the lights twinkling, and the river like a silvery ribbon reflecting them.

"I thought you said we were going someplace secret," Kristin said as we wound our way up through the curves and over the ruts of the gravel road. "The quarry's about as secret as the dance. That's where every-body's going to be."

"Secret I said, and secret I meant. Trust me."

Kristin shrugged, put her head on my shoulder and closed her eyes.

I was taking her to a place Molly and I had found when we were little. It was only a few hundred yards from the road, and even though there were signs other people knew about it, we never saw anybody there. So we called it our secret place. It was a shack perched just back from the rock lip of the quarry, halfway around the rim from the parking area. We figured it had been some kind of office for the company that had abandoned the quarry; it was just big enough to have held a desk or two and some file cabinets. Now its only furniture was a rickety bench. The shack was made of sheet metal that had been shot full of holes by people doing target practice, and the corrugated roof was rusted through in places.

We'd spent a lot of time there when we were kids. The door had been torn off the hinges and thrown over the edge. It had hung up on a ledge about halfway down, stuck against a small tree. Molly and I climbed down there when we were eleven or twelve or so and wedged the door between the wall of the quarry and that tree so that it formed a flat platform for jumping off into the water. It wasn't safe to dive because there were rocks under the water, but those rocks were what made it a great place to swim because you could use them for climbing out again.

Molly and I didn't go there together much anymore, but I still went back sometimes, when I felt like getting off by myself. The view was just as good as the one from the parking lot. The shack was overgrown with morning glory, and the sumac trees around it had gotten so tall that unless you knew it was there, you weren't likely to notice it from anywhere else on the quarry's rim.

There was no place to leave the car next to the path that led there, so now I drove past it and around a curve to where I could pull off the gravel onto grass.

"This is it?" Kristin asked.

"We have to walk there," I explained, and got the flashlight Molly keeps under the driver's seat. "It's not far."

When I shone the light on the overgrown path that led through the tall grass and under the trees, Kristin pulled back. "I'm not going in there. Are you nuts? What if there are snakes?"

"Trust me. There aren't any snakes." At least not at night in October, I thought, remembering the time Molly had found a huge garter snake and tried to scare me with it. Instead, we'd taken it home and she'd kept it in an aquarium until she realized she'd have to feed it live food. She'd brought it right back.

Kristin held onto my arm so tightly she practically cut off the circulation, but she did come with me. It wasn't easy picking my way along the path, slipping on rocks and leaves, with Kristin hanging on me like that, but finally we came out into the clearing and I shone the flashlight on the shack itself.

"Nothing lives in there, does it?" she whispered, her lips against my ear.

"Not so much as a chipmunk."

"We aren't going in there!"

"Not unless you want to. Come on." I led her through the sumacs, the light flashing on the brilliant red of their leaves, around the side of the shack until the view of the water and the lights of Ridgewood opened out in front of us. I clicked off the flashlight. The moon was a silvery glow behind a fast-moving cloud. Kristin caught her breath. "It's beautiful," she said, as a cloud swept past and the moon touched the ripples on the water with silver.

"I told you it was a nice place."

I put my arms around her and was about to kiss her when something moved by the edge of the quarry. A figure stood up, dark against the sky. Kristin screamed and my heart was suddenly pounding in my throat. I clicked the flashlight back on, and shone it on denim jeans and jacket. A pair of hands, white in the light, moved to shield the face behind them. "Get it out of my eyes!" I recognized the voice—Bran Slocum.

My hands were shaking as I lowered the light. "You scared us half to death."

"I could say the same."

I felt Kristin's hand slip into mine. In the darkness, Slocum looked somehow menacing. The eye that angled to the side showed more white than the other and his lip, still swollen from the attack that morning, gave his mouth a kind of sneer.

"How'd you find this place?" I asked.

"Molly told me about it. She said it was a secret. Guess not."

I shone the light on the bullet holes in the shack, then on the litter of faded beer cans and soggy paper scraps around it. "Somebody else knows about it."

"You said it was secret, too," Kristin said. "What'd you come up here for?" she asked Bran. "Is somebody with you?"

Slocum made a sound that was almost a laugh. "Hardly." He looked out over the water. "I just like to get off by myself sometimes. This seemed like a good place."

Kristin tugged at my arm. "Let's go back to the car. You can have it to yourself," she said to him.

"No, you two stay. I was just about to start back anyway. The twins get me up at dawn."

I shook my head. The night was shot, and I didn't have the energy or even the desire to try to save it anymore. All of a sudden I was just tired. "It's a long walk in the dark. You want a lift?" I asked him. Kristin dug her nails into my arm, but I pretended not to notice.

He stood for a second, as if he wasn't sure I'd meant it. I was almost ready to take back the offer when he nodded. "Sure. Thanks."

Kristin muttered something that I didn't catch. I figured she was mad, but I just wanted to get out of there. I put my arm around her and guided her back around the shack. On the narrow path I led the way, holding Kristin's hand, and Bran followed. When we were nearly to the gravel road, I stopped and Kristin bumped into me. "What?" Slocum asked.

I didn't have to answer. When we stopped moving, they could hear what I'd heard. Voices, and the sound of a car engine idling. "So much for secrecy," I said.

"Hey! Th-there's a l-l-light down that way!" a voice called.

"Matt Singleton," Kristin said.

"Come on!" The sound of feet pounding on gravel came toward us. "Nick! Down here!"

Matt and Jerry Ritoni appeared in the light, and from the sounds behind them, the others were on their way.

Jerry squinted. "Watson? We wondered who was in that car, and where they'd gone."

"Is it Goblin Girl and the faggot?" Nick called.

"Nah!" Jerry yelled.

Behind me, I heard Bran turn around and head back down the path toward the quarry.

"D-d-don't just stand there," Matt said. "C-c-come on out."

7 Bran's movements, slow and careful as they were, sounded as loud as gunshots to me, but the guys didn't seem to hear them. Maybe it was a good thing they'd been drinking. He needed to move faster, though. The chance for another go at Bran seemed to be their idea of a great way to end a lousy night. I'd seen what they did to him in the middle of the day on the school grounds. No telling what they'd do to him up here.

Or to me, for that matter. If they found him up here with us, they wouldn't ask any questions about how we happened to be together. Kristin and I stepped out onto the road, and I squeezed her fingers, trying to let her know she shouldn't mention Bran. She squeezed back. I shone my light on the guys clustered there.

"Get it out of our eyes, dork," Nick said, blinding me with a light of his own. "What are you doing up here?"

"Ever hear of the right of privacy?" I asked.

"Where's Goblin Girl?"

"You mean Molly Pepper? I have no idea. Why?"

"We s-s-saw the car and f-figured she was up here someplace," Matt said. "Flipped us off from that c-c-car this afternoon."

"Nick said she'd have that faggot with her," Jerry said.

"If he's a faggot, what would he be doing up here with her?" I asked.

"Maybe they're both some kind of perverts." Nick blinded me again, then swept the light onto Kristin. She closed her eyes, but didn't say anything. "Matthis, why the hell do you go with this wimp?"

"What do you care?" she said. "You've got Jennifer, haven't you? Or did she dump you tonight?"

"Nobody dumps me," Nick flared. "She's in the car." He waved his light at the path behind us. "Where does that go, Watson? Is there some place back there we should know about?"

"Maybe it's a s-s-secret hideaway," Matt said. "A love n-nest."

I moved so that I was blocking the path more completely. "It's just a path to the quarry," I said as casually as I could. The sound of Bran's movements had faded, but he hadn't had time to get entirely away yet. Maybe he could hide out in the shack. As overgrown as it was, there was a chance they wouldn't see it. "It's nothing."

"Let's check it out," Nick said, and shoved me out of the way. The others followed him, kicking their way through the weeds to keep up with his light.

"You're in for a big disappointment," I yelled after them. So much for the Scarlet Pimpernel, I thought, looking down at Kristin. We listened to them for a moment, then Kristin tugged at my hand.

"Let's get out of here," she said.

There was a shout from Nick and the sound of someone running. "It's him!" Nick yelled. "The faggot. Come on!"

"Wait up with the light!" Jerry shouted. "We'll break our necks!"

"Let's go," Kristin insisted, and pulled at me.

"I can't," I said, and even as I heard myself saying it, I didn't know why. "Go find Nick's car and wait with Jennifer."

"David Watson, don't you dare leave me alone up here—" she said, and I pulled loose and headed down the path after the others, trying to tell by the racket they were making whether they'd managed to catch Bran yet. If they had, there'd be nothing I could do for him except get myself beaten up, too.

But they hadn't. Nick was cursing as he went, and the others were sticking together, apparently stumbling over each other trying to keep up with Nick and the light. I turned my own light off so they wouldn't see me, and followed as close as I could.

They had come out into the clearing near the shack, and Nick stopped, flashing his light this way and that. The others fanned out around him. "Give it up, faggot," he shouted. "There's no place to go."

"Hey, Nick!" Krosky said. "Look over here."

The light swung toward the shack, and I shook my

head. If Bran had managed to get inside they'd have him for sure.

"T-t-trapped like a r-rat," Matt said.

Nick laughed. "Gotcha!" he yelled. He motioned for Jerry and Gordon to go around the shack on one side and he and Matt went around the other. "We're gonna have a little fun." They knocked sumac branches out of the way and stamped their feet as if they were trying to scare their quarry into making a run for it. I watched, holding my breath, as they disappeared around the front of the shack.

And then, as I was trying to think of something—anything—I could do, I saw a movement off to my left at the edge of the clearing. In the pale moonlight I could make out a figure moving as quietly as possible farther in under the trees. He hadn't gone into the shack after all.

The guys would know it was empty in another second or two, though, and they'd be back. There was no way he could get away, or safely enough hidden, in time.

The idea came in a flash, and there wasn't even time to think about it. I took off and ran straight across the clearing and in under the trees on the far side. I stumbled a couple of times in the undergrowth and nearly crashed into a tree that was leaning at an angle between two others, but I kept on, my flashlight off so there'd be nothing to follow except the noise I was making.

I couldn't help grinning when I heard Nick screaming curses behind me. It had worked. They thought I was Slocum. And they were coming after me.

I went on for a few more seconds in the darkness, tripping and dodging the looming darker shapes of the trees, and then put on the flashlight. I had enough of a lead not to worry that they'd catch me easily, and I knew I could be no more than a shape and a light to them. I glanced over my shoulder every now and again, and their light followed, jerking and jolting through the darkness. There were occasional curses as somebody tripped or crashed into somebody else, but still they came. I angled left after a while, to be sure I didn't go back toward the quarry, and the ground began to slope downward under my feet. I was heading for the highway.

It wasn't long before I had to slow down because the light was falling farther and farther behind. Finally, I heard Jerry say that he was quitting. Gordon must have stopped with him. "Krosky, you're a wimp!" Nick shouted. I could tell by his voice that he was tiring, too. The football game and the beer were taking their toll. When Matt and Nick finally stopped, too, Nick's last words were, "We'll get you later, faggot!" I slowed to a walk, hoping Bran would be far away by the time they got back to the shack.

Then I thought about Kristin. She'd be furious—even if Jennifer was waiting in Nick's car like he said, even if Jerry's girlfriend, Myra, was there, too.

They'd probably give Kristin a ride home. The only other choice was to leave her in the car, to wait for me. Either way, though, she was going to be ticked. It looked as if I'd chosen Bran over her. That wasn't it, but I didn't know how I could explain it to her. I wasn't sure I could explain it to myself.

I started back, making my way by the fast-fading flashlight. I was in no hurry. Kristin probably wouldn't have told Nick anything, but I hoped he wouldn't do a lot of thinking about where I'd gone.

When I finally saw the Civic, gleaming faintly in the pale moonlight at the edge of the road, Nick's car was gone and so was Kristin. I hoped that by now she was safely tucked into bed. And that when I called in the morning, she'd understand. The trouble was, I was pretty sure she wouldn't. She thought the Scarlet Pimpernel was romantic, but I doubted that she'd feel the same about being abandoned on the quarry road in the dark.

I pictured her at the game, all sparkle and smile and gorgeous body. It was a great image, and I loved it. The trouble was, I was beginning to think I wanted something more.

On the drive back I kept watching for Bran along the road till I realized that if he was out there, he'd probably be dodging headlights, knowing that Bruno was out there somewhere, too. I decided to drive by his house, just to check. I couldn't go home without knowing for sure that he got away.

Except for a yellow light burning next to the front door, the house he lived in was dark when I pulled up to the curb. I considered ringing the bell, but decided against it. No sense stirring everybody up. Maybe the light had been left on for him. If so, he wasn't home yet. I decided to wait, but I was so tired, I fell asleep with my head against the window. I don't know how long I slept, but when I woke up, nothing had changed. The house was still dark except for that light. I figured I'd done enough. Since Dad wasn't home, I wouldn't have to explain being out so late, but I did have to work the next morning.

I was turning the car around when my headlights showed Slocum stepping over the chain across the cemetery road. He ducked behind the brick gatepost, and I turned off the car and got out. "It's me. Watson. Are you okay?"

He came toward me, moving stiffly. "Yeah. Thanks."

"It's okay. I just wanted to be sure they hadn't found you."

He shook his head. "When they went after you, I climbed down into the quarry."

"In the dark?" I thought of the long drop to the ledge where the door was. I wouldn't have gone down there for the first time at night.

"There was some moonlight. Molly told me about that platform, so I knew where it was."

I just looked at him. "That took a lot of nerve."

Bran smiled an odd smile that was gone almost as soon as it appeared. "Not so much. Got to get some sleep." He stuck out his hand and I took it. His grip was firm and solid. "Thanks."

"No problem," I said.

He started away, then turned back. "I don't fight," he said. "But not because I can't. I just wanted you to know that."

There was an awkward pause, and I nodded.

"Well, thanks again."

"Any time," I said, and he turned away again.

I got into the car and watched till he'd let himself in and turned out the porch light. Then I drove away, thinking that I hadn't meant that at all. Not at all.

8 "Why do you keep getting involved with Slocum anyway?" Kristin asked Saturday morning when I called to apologize. "Everybody's going to think you're buddies. Is that what you want? Why don't you just let him handle things on his own?"

"I don't know," I said. And it was true. I didn't. "Maybe because he's outnumbered."

Kristin sighed. "Even if you take his side, he'll still be outnumbered. They'll just beat up both of you. I don't understand the point."

"They didn't beat up either one of us, Kristin. He got away from them, and he wouldn't have if I hadn't gotten them chasing me. Nobody got hurt. It would be nice if we could keep it that way." She didn't say anything, and the silence went on too long. "I guess the guys took you home," I said, finally.

"Thanks to Jennifer. I had to sit on Gordon's lap all the way home. I was *not* thrilled. And my dad was in

a snit about how late I was. If I'd waited for you, I'd
be grounded."

"Lucky you didn't, then."

"Yeah, lucky."

"I'm sorry," I said. "Really."

"Me too."

I was at the grocery store pricing the week's specials
when I realized that I hadn't asked Kristin to go out
with me after work that night. And she hadn't mentioned it either. It would be the first Saturday night we
weren't together in nearly three months.

"Well, if it isn't the Scarlet Pimpernel!" Molly was leaning out the window of her car when I came out of the
store later. "You got plans for the next hour or so?"

"If you call going home to bed plans."

"You can't be tired this early on a gorgeous day like
this. I know perfectly well you didn't get up to run this
morning and you didn't have to be at work till nine
o'clock."

"I had to leave a certain vehicle at a certain person's
house first and walk over here," I said. "Besides, as
you apparently know, I was out a little later than I
expected to be last night."

"I've heard all about it and I'm very impressed. Hop
in."

"So what's the plan?" I asked, when she pulled out
of the parking lot and headed away from town.

"A small social gathering." That's all Molly would
say until she drove through the gates of the state park

and pulled up next to the picnic area by the river. "Come join the picnic."

A plastic tablecloth covered one of the picnic tables and a cooler sat at one end. Judging from the paper plates and cups, the half-empty cider jug, potato chip bags and bread crusts, the picnic had been in progress for a while.

"Molly's back!" The twins came running up the river bank. They were barefoot, their faces smudged with dirt, their overalls rolled up to their knees. "Come see what Bran and us builded in the water!" The first one to reach Molly started pulling on her hand.

"Hold it. I want you to meet a friend first. David, this is"—she squinted at the one who had hold of her hand—"this is Keith. He's the one with the scratch on his cheek. And that's Kipp. Twins, this is David."

"I'm the oldest one," Keith said, still pulling at Molly.

"Only two minutes," Kipp said, scowling. "Bran says two minutes doesn't hardly count."

"Does so, and I'm bigger."

"Does not—"

Bran, also barefoot, came up the bank. "Stop fighting and say hello like civilized human beings," he said to them. "Then get back here. We've got a leak!"

Keith let go of Molly. "Hi!" He tossed the word over his shoulder as he plunged back down toward the river.

"Hello," Kipp said, and followed his brother.

Molly and I joined Bran and walked down to the river's edge, where the twins were bent over their project, scooping mud from the bank. A stone and stick dam stretched from the bank to a log that lay parallel to the bank and a couple of feet out in the shallow water.

"They wanted to dam the whole river," Molly explained, "but Bran told them it was a bigger project than they could handle in one afternoon."

"I also suggested that Ridgewood wouldn't be real happy to have its downtown flooded out." Bran went down to help the boys, who were arguing about whether mud was enough to stop the leak.

"You'd better get it fixed and be done," Molly said. "It's getting late and chilly, and your grandma's going to skin you if you come home all wet and get pneumonia."

Kipp wiped his nose with a muddy hand. "She won't skin us," he said. "She'll skin Bran. Won't she, Bran?"

"She'll skin us all. And then your mother'll skin us again afterwards."

Molly and I left them arguing about whether they could be skinned twice and went up to the table. She handed me an apple. "Guaranteed fresh. We went to an orchard this afternoon and picked them ourselves. Aren't the twins great?"

"Yeah. Great." I bit into the apple and glanced around. There were people at several of the other picnic tables. Families, mostly. But a hundred yards farther down was a bunch of kids about our age. They were too far away for me to see whether I knew any of them.

"Don't get paranoid," Molly said. "Nobody's going to see you with us. Or care."

I should have known Molly would guess what I was thinking. "Everybody's going to think you're buddies," Kristin had said.

I'd known Molly long enough to know what she was up to. In spite of her protestations when I'd called Bran her boyfriend, she liked him. A lot. You couldn't miss it. And she couldn't leave it at that. What she wanted was for me to like him, too.

"The boys are cute," I said, hoping the subject change wasn't too obvious. "Their mother's his cousin?"

"Angela, her name is. Angela Ridley. She's staying with her parents—Bran's aunt and uncle—till she can afford a place of her own. The father's not around. I gather he never was. Anyway, Bran's aunt says Bran's practically saved all their lives."

"How so?"

"The twins are crazy about him. You wouldn't know it to watch them with him, but they were holy terrors till he arrived. Nobody could do anything with them. But when he came they latched onto him and won't let go. They'll do anything he tells them to do. If they didn't go to day care, they'd want him to stay home with them every day."

"The way things have been for Bran at school, that wouldn't be such a bad idea."

"Snack break," Bran's voice called from the river bank. A few moments later he appeared, grunting and puffing his way up toward us, like an overburdened donkey. One twin was perched on his shoulders, the other was hanging onto his back. He carried a pair of shoes and socks in each hand and his sneakers hung by their laces from his teeth. "Get these dam engineers some chips," he said.

"Hey, Molly! We're dam engineers," Keith hollered. "Dam, dam, dam engineers."

"Get us chips!" Kipp said.

When Molly found out I wasn't doing anything with Kristin that night, she suggested we take the twins home and then all go to a movie. I told her I was too tired. Before she had a chance to argue or question my motives, Bran declined too. "It was one very long night last night," he said.

"Oh, all right, if both of you are going to fink out on me—" Molly said.

I looked at Bran over Molly's head and thought I detected the ghost of a smile. "Some other time," he said.

"Right," I agreed, gratefully. "Some other time."

9 Monday was another brilliant fall day, the air crisp but the sun so bright that it was too warm for jackets by lunchtime. Just about everybody who brown-bagged lunch went outside to eat instead of to the cafeteria. Frisbees sailed over the fountain and music blared from a car that had been brought up to the edge of the courtyard.

Zach Lewis and I were leaning on the fountain. He was telling me, between mouthfuls of corn chips, that he'd heard the miniplaza going up across from the grocery store where I worked was going to have a Friendly's and he was going to apply for a job there, when I noticed a guy who looked around thirty or so talking to a group of kids down by the curb. Zach saw that I wasn't listening and stopped talking about ice cream. "What's the matter?"

"That guy over there, talking to those freshmen."

Zach looked over his shoulder. "The one in the tasteful plaid sport coat?"

"Yeah. Who is he?"

Zach shrugged. "Somebody's dad, maybe. How should I know?"

"Too young. Besides, there's something odd about him. Watch him for a minute. He doesn't really look at the people he's talking to. His eyes are moving all the time, like he's watching for somebody. You think he's selling drugs?"

"Right here in the courtyard?"

"Why not? Half the school's outside today." As I said that, the guy moved away from the kids he'd been talking to and headed for one of the groups playing Frisbee. "I think I'll go check him out," I told Zach.

"Are you in the market for what he's selling?" he asked.

"Don't be stupid. I just want to see what's up. Coming with me?"

"No thanks. I stay strictly away from that stuff." He popped another handful of corn chips into his mouth. "Food's all the addiction I can handle."

Whatever he'd said to them, the first group of kids was buzzing excitedly when I reached them. "What did that guy want?" I asked.

A chubby, freckled kid answered eagerly. "You're not gonna believe this. Nobody's gonna believe it."

"What?"

"You know that psycho killer who's in the news all over the place?"

"Collier," another kid prompted.

"Yeah. Joseph Collier. Well, get this! His son is going

to this school. He's right here in Ridgewood! Living here.''

Two or three kids started talking at once, and I overrode them. ''Don't be dumb. Everybody in town would know if that was true. Did that guy tell you that?''

''Sure,'' the freckled kid said. ''And he knows. He's a reporter. From one of those big national newspapers—''

''Like in the grocery store,'' someone else said.

''—he's looking for him for an interview. The reporter doesn't know his name, because he changed it, but he said the kid would be easy to recognize because he has a bad eye and he'd be new in the last couple weeks—''

The sun was still beating down on the courtyard, but I felt a chill go through me as if a cloud had blotted it out. Bran Slocum. It couldn't be.

''—says he's staying with relatives here till the trial's over. His paper wants to get to him before any other reporters do. He says anybody who helps him find the kid'll get his name in the paper.''

''What'd you tell him?'' I asked.

The boy straightened his shoulders importantly. ''I told him he's here all right—that weirdo with the earring. But I didn't know his name. He went to find somebody who does. Didn't even take my name, the jerk.''

The kids were all talking again, as excited as if a movie star had come to town. By now the man was deep in conversation with a group of seniors who'd been smoking next to the fountain. He would have Bran's name in no time. Probably had it already. Could it be true? I started up toward the building, my mind running so fast I could hardly keep up with it.

Serial killers didn't have families, did they? They weren't normal people leading normal lives. They were some kind of monsters. I remembered the articles in Mr. Byrd's room. The neighbors said that Collier was so ordinary. An ordinary guy could have a son. But even if he did, how could it be Bran?

I thought of Bran with one twin on his shoulders and one on his back, his shoes dangling from his teeth. Joseph Collier's son? Impossible.

Impossible or not, the story would be all over school by the end of lunch hour. I needed to find Molly. The way she'd reacted to the spot about Collier on the "Today" show, I couldn't imagine how she'd feel when she heard this. Worse, how she'd feel if it turned out to be true. It would just about kill her. I didn't want to be the one to tell her, but I didn't want her to hear it from anybody else, either.

A Frisbee came out of the sun and nearly clipped me as I started up the steps. I tossed it back, and saw that the reporter had gone to still another group of kids, a notebook out now. If it was true, Molly should know it—right away. I ran up the steps, through the front doors and past Towson's office.

"Track's outside, Watson!" he boomed as I tore past. I slowed to a walk till I was around the corner and then ran again, dodging kids coming out of the cafeteria. I pushed through the swinging doors, nearly knocking down a short kid who was coming through the other way. Molly was at her usual table, with Bran, of course. They both looked up as I burst in. I hesitated, then walked reluctantly over to the table.

"What's up?" Molly asked.

Bran smiled. The smile was there and gone again, leaving his face expressionless, the right eye staring past me. I remembered Collier's face from the front page of *Life* magazine on Mr. Byrd's wall. There was no resemblance. Maybe it was all just a practical joke Bruno had thought up. He'd probably paid that guy to pretend to be a reporter and spread the story. That had to be it.

"Well?" Molly said. "You came in here like the devil himself was after you or something. Now all of a sudden, you're paralyzed?"

"I wanted to talk to you," I said, my voice cracking. "Alone."

"That's rude, Watson. Really rude."

Bran stood up. "I was just going," he said.

Molly grabbed his hand and pulled him back down. "No way." She turned to me, her dark eyes narrowed, and pushed her hair behind her ears. "So, what's this secret you don't want my friend to hear?"

"Nothing. I—It's not my secret." I turned to Bran. "Maybe it's yours. There's a reporter outside, looking for you. Asking questions. He says he knows who you are, and he's telling everybody he talks to."

Bran stood up again, pulling away from the hand that Molly put out to stop him. "Thanks." He grabbed his pack off the table and stood for a moment, looking toward the main doors, then he turned and headed for the kitchen. He ducked past one of the workers coming out with a bucket and sponge, and was gone.

"What was that all about?" Molly asked.

I dropped into a chair across the table from her, my stomach churning. The story was true.

"David!"

I looked at Molly and swallowed. She had to know. I swallowed again. I could hear the words over and over in my brain, but couldn't make myself say them. Finally, I looked down at the table, away from Molly's eyes, and got the words out. "Bran is Collier's son."

She stared at me blankly for a moment. Then it connected. The color drained out of her face. "Joseph Collier?"

"The serial killer."

10

"No wonder he won't talk about himself," Molly said, finally. We'd been sitting there, both of us trying to take in the truth of it as the normal noise and movement of the cafeteria went on around us. "What kind of life do you suppose he's had?"

I just shook my head. I had no way of picturing it. Ordinary, the neighbors had said. What would that be? Molly's father, a vet, was ordinary. Zach's, a pharmacist. Kristin's, a factory foreman. Those were ordinary fathers. A lot of times, over the years, I'd envied that.

Once, in junior high, I'd gone with my father to a craft fair and some of the guys from school had come by the booth. They'd listened while Dad, who still had a full beard then, and wore a buckskin shirt and moccasins, had rambled on to a customer about the mystical importance of knowing one's totem, of tuning in to our spiritual connections with the animal world. And the guys had laughed at him. I'd spent the rest of that day pretending to be interested in a potter at the other end of the tent. Wishing for a father who was ordinary.

In spite of Dad's hippie weirdness, though, our life together wasn't so very different from anybody else's. I wondered what kind of life somebody would have with a father who killed kids, burned them with cigarettes and cut their fingers off and buried the bodies in the backyard. Did they sit down in their kitchens to eat macaroni and cheese and talk about what kind of day they'd had?

"Maybe he never even lived with his father," I said. "His parents could be divorced. Maybe he lives with his mother—"

"Then why isn't he with her now? Why did he come to Ridgewood?" Molly kept pushing at her hair, though it hadn't moved from behind her ears. It was as if her hands were acting on their own. "I thought I was getting to know him, that he was beginning to open up a little. But what did I know? That the Ridleys are his aunt and uncle, Angela's his cousin and he takes care of the twins. That's all. Nothing about him. Nothing at all about him." She rubbed a hand across her face. "I feel like I'm going to be sick."

The bell for sixth period rang, and the cafeteria began emptying. Molly stood up and looked around her at the kids talking, laughing, roughhousing as they carried their trays back and gathered their books. "Every single person in the school's going to know about it by the end of the day."

On my way to my free period after lunch, I had to pass Towson's office. The reporter was there, leaning

over the counter, talking to Mrs. O'Neil, the school
secretary. I didn't see Dr. Towson. I wondered if they'd
tell the guy anything about Bran. Probably that was
against the law.

I couldn't concentrate on the Spanish homework I
was supposed to do that period. Too many questions
were chasing each other around in my brain. There was
no way to answer them, but they wouldn't go away.

If Bran *had* lived with his father during the time his
father had been killing kids, burying their bodies in the
backyard, wouldn't he have known? No matter how
ordinary Collier looked to the neighbors, his son would
know better.

If every so often Dad brought somebody into the
garage and killed him with a chisel between the eyes,
I'd know that, wouldn't I? I'd see it in him somehow,
even if I was never there when he did it. Even if he
cleaned everything up and got rid of every piece of
evidence, I'd know. When he looked at me, I'd know!

And if Bran *knew*, then why hadn't he gone to the
police to keep it from happening again? I tried to think
if it had been Dad, whether I'd have been able to report
what I knew, to send him to jail. Or to die.

I drew a hangman's noose in my notebook, then
scribbled it over. My father wasn't Joseph Collier. He
caught spiders in his coffee cup and put them outside
instead of squashing them. One winter when a mouse
nested in the base of one of his favorite totems he
wouldn't even set a regular trap. He caught it alive and
took it clear up near the quarry so it wouldn't come

back. I couldn't compare myself to Bran. My father to his.

Maybe Bran's father tortured him, too. Beat him up. Made him afraid that if he told he'd be buried right along with the others. I shuddered. Bran had grown up in a house with a homicidal maniac. I didn't know what that had been like. I couldn't imagine it. Finally, I just couldn't imagine it.

The news got around fast, all right. Everybody was talking about it as we got to class seventh period. "Did you hear?" Jennifer Logan asked Mr. Byrd.

"If you mean did I hear the rumor that Joseph Collier's son is here in Ridgewood, yes." His voice was tight.

"It's B-B-Bran Slocum," Matt said.

"I *said* he was a pervert," Nick added.

"Sit!" Mr. Byrd said. "And be quiet." He didn't slouch against the blackboard now, waiting, in his usual relaxed way, till we settled down. He stood very straight and very tense, his face stern. "I said, be quiet!"

By the time the bell rang, everyone was seated and nobody was talking. "First of all," he said, "the source of this 'news' would seem to be a reporter from a tabloid newspaper. Not what you might call an unimpeachable source. So in the first place we don't know that Collier's son is here at all.

"From what I understand, the reporter arrived in Ridgewood acting on a tip, without so much as a name to go on. He got Slocum's name from students—students he promised would get their names in his newspaper for helping him. So in the second place, we don't know that Bran Slocum is the person, *if* there is such a person."

"It isn't as if there were lots of new kids with one bad eye," Scott Handleman said. "Who else could it be?"

"I repeat that we don't know for a fact that Collier's son is even here." He gestured toward the clippings on the board. "It's true that Collier has a son. That's mentioned in two of these articles. One merely mentions the son's existence, the other says that the boy went to live with friends when Collier was arrested. That was months and months ago. We don't know that he's left New Jersey, let alone come to Ridgewood."

I thought of the way Bran had jumped up when I told him a reporter was looking for him. That was enough for me.

"But what if he is here?" Cheryl Heroux asked. "What if it *is* Slocum?"

Mr. Byrd spread his hands. "All right, what if? Keeping in mind that we have no facts to support this claim, what would it mean if it were true?"

"It would mean he should get kicked out," Nick said. "Out of school. Out of town."

There was a murmur of agreement from some of the other kids.

"For what reason?" Mr. Byrd asked.

" 'Cause he's a sick, perverted weirdo."

"Is that a clinical diagnosis, Mr. Bruno?"

"Because w-w-we don't want n-n-no *psycho's* k-kid in our school!"

Mr. Byrd nodded. "Now there's an enlightened reason, Matt. What possible harm could come of Collier's son attending this school?"

"It wouldn't help our image," Zach said. "Can you see the chamber of commerce brochure? 'All the best serial killers send their sons to Ridgewood High.'"

Mr. Byrd snuffed out the laughter almost before it started. "I doubt that this will turn out to be a laughing matter if the story's true."

"What about what you said the other day?" Cheryl asked. "About how maybe Collier was innocent and he was just covering up for a crazy relative. Maybe Bran Slocum's the crazy relative."

"What a clever thought, Miss Heroux. Except that since the first killings took place more than a decade ago, Bran Slocum would have had to commit a murder before his sixth birthday."

Cheryl flipped her hair back and looked out the window as everyone laughed. The tension that had been building in the room seemed to ease a little.

"We haven't time to spend the whole period on this issue," Mr. Byrd said. "But I want every one of you to remember that we don't know the truth yet—"

"Suppose it is true," Jennifer interrupted.

"If it is"—Mr. Byrd went on, his voice fairly bouncing off the walls—"if it is, I want you to remember that Collier himself has not yet been found guilty. His son, who may turn out to be just another victim in this case, is almost certain to be guilty only of having been born."

The subject wasn't brought up again, but I was glad to get out of the room when the period was over. The whole atmosphere had made me uncomfortable. More than uncomfortable. As we left class, a few of the kids were grumbling at Mr. Byrd for taking Bran's side.

After school Kristin stopped me as I passed her locker on my way to track practice. "I guess you'll stay away from him now," she said.

"You mean Bran? Why?"

"Come on, David. I suppose you think he didn't know what his father was doing. Some of the kids are saying he probably even had a hand in it."

"That's stupid, Kristin, and you know it. Collier started killing kids more than ten years ago."

She pulled her cheerleading pompoms out of her locker and slammed it shut. "Which means the kid was raised on murder. Think about it."

"I have been thinking about it, and I don't have any answers. But neither does anybody else. Do you know everything *your* father does?" I asked her. "If he cheats on his taxes, do you help him do it?"

Kristin turned on her heel and stomped down the hall. When she was nearly to the door, she turned back. "My father doesn't cheat on his taxes," she yelled. "And he sure doesn't torture children!"

"That's funny," Zach Lewis said, coming up behind me, "I thought all fathers did both of those things."

"Mr. Byrd's right, Zach. This isn't funny."

"It's all in how you look at it, Watson," he said.

During practice we ran speed sprints on the track. I put every ounce of concentration into running and broke my best time twice. After that Coach gave us eight laps and called it a day, but when the other guys left, I kept on. Somehow, running around and around the track I could keep the questions out of my mind.

The sun was angling down toward the tops of the trees, tinting everything a dark orange, when I saw Molly sitting on a bench by the track. The hood of her sweat shirt was up, her shoulders hunched against the wind that had sprung up. When I finished that lap, I went over to her. "You want to walk a couple laps while I cool down?"

She joined me. "I went over to Bran's house when school got out. He wasn't there. So I went down to the diner where his aunt works to tell her about the reporter. I figured they'd better be ready, because he's bound to show up at their house."

"What'd she say?"

"She thanked me. She couldn't get off work early, so she said she'd call her husband and see if he could go home—so Bran wouldn't have to handle it alone. They knew this might happen, so they'd planned what to do—they're just going to stonewall. Refuse to talk about it."

"That won't be too hard for Bran," I said.

We walked awhile, Molly kicking at the cinders of the track with every step. "I suppose you've heard people saying that Bran probably had something to do with the murders."

I shrugged. "A few."

"It isn't true. Couldn't be. He couldn't even have known about it. You've seen him with Kipp and Keith. He's not somebody who could torture kids." She turned to me, her face tense. "I admit I don't know much about him, David, but I know that for sure. For absolutely sure."

I nodded. She was probably right. No matter how strange he was, the idea that he might hurt kids just didn't fit. We walked on around the turn and back toward the bench. I was beginning to feel the chill of my damp sweats. "Do you have the car?"

"Yeah, you want a ride?"

"Sure."

"Good. I don't feel much like being by myself right now."

Halfway to my house, Molly turned left around the cemetery. "I want to go see if Bran's home yet," she explained.

I shifted in my seat, not knowing how to say what I wanted to say. All of a sudden it was very important to me to go straight home, to smell the smell of fresh wood and hear the smack of the hammer against the chisel in Dad's shop. I wanted to go in and see him bent over whatever piece he was working on, his hair drooping into his eyes. I wanted him to ask me how practice went, and whether I was ready for the Olympics.

But I couldn't explain it to Molly. Couldn't even explain it to myself. In a few minutes the house loomed up ahead. Bran was out front, pushing the twins on a homemade go-cart.

Molly stopped the car and got out. "Hi, twins!" she called. "Cart's looking good." Bran gave the cart a final shove and stood up.

"Lookit how fast we go," Keith yelled, turning to look back at Molly.

"Keith! Watch out!" Bran called—too late. The cart angled off the sidewalk and jolted over the curb, throwing both boys onto the street. Bran hurried over and helped them up, checking them over, dusting them off, his hands sure and gentle. "You have to keep your eyes on the road when you're driving," he told them. "It's the first rule."

"I'll remember," Keith said and went to tug at the front of the cart to get it back onto the sidewalk. "Come help me, Kipp," he yelled.

Kipp frowned, his lower lip jutting out. "You made me hurt my knee. You don't steer good."

"Do too!"

"Do not!"

Bran grabbed Kipp around the waist and slung him against one hip, leaning down with the other hand to help Keith with the cart. "You're both getting good. Let's go get a Band-Aid for that knee. And an apple."

Kipp, hanging nearly upside down, grinned. "A big Band-Aid? Do I get a big one?"

"Biggest we got," Bran said.

"Me too!" Keith yelled, holding up his arm. "I scraped my elbow."

"Two big Band-Aids," Bran said, and slung Keith across his other hip, lugging them both, giggling, up the front steps. "Be back in a minute," he called back to Molly and me. "You want apples?"

"Sure!" Molly said, just as I was about to say no.

Molly looked in at me, and I shrugged and got out of the car. Bran didn't seem to be reacting to the fact that his secret was out. He was just the same as ever.

He came back, the two boys frolicking around him, their pants rolled up. They sported Band-Aids on all four knees.

"Serious injuries," Bran said, handing us each an apple.

Molly nodded, her face grave. "That was a near thing, that crash."

Keith grinned, apple juice running over his dimpled chin. "Good thing we wasn't going faster. We coulda been killed dead."

Bran patted him on the head. "How about putting the cart away and resting your wounds awhile. Go see what Grandpa's watching on TV. But don't you go changing channels without asking."

When the boys had gone in, the three of us sat down on the porch step and bit into the apples.

"Has that reporter found you yet?" Molly asked.

"No." Bran glanced up and down the street and now I could see the tension in his face. "That's why I'm outside. Any sign of a stranger, and I'm out of here. I don't feel like talking."

We sat awhile, the chirring of a cricket filling the spaces between us. Bran finished his apple and threw the core in a high arc across the street, where it fell directly into the sewer. A phone rang in the house. It rang again, and then stopped.

"Maybe you should talk about it," Molly said. "Not to a reporter, but to somebody—"

"I've got to be getting home," I said, and stood up. "I need a shower."

"Will you be in school tomorrow?" Molly asked, as they stood up.

"Yep."

I thought about the kids in Mr. Byrd's class. He shouldn't come to school. Not now. The phone rang again.

"We'll see you there, I guess." Molly was still holding her apple core.

Bran took it from her and threw it. It landed short, bounced, and went into the sewer, too.

"Thanks for the apple," Molly said.

"Yeah, thanks," I added.

"See you," he said.

The questions went around and around in my head again as Molly drove me home. As I was about to get out of the car, she touched my arm. "He's going to need a friend," she said. "More than ever."

I nodded. "From the minute you first saw him, he's had one."

I waved goodby and then hurried to the garage, where Dad was working. I slammed the door behind me and leaned on it, filling my lungs with the smell of sawdust and wood shavings.

"So, you ready for the Olympics yet?" Dad asked.

11 When the alarm rang the next morning, it interrupted a weird dream. One of Dad's totem poles had come to life, eagle, wolf, bear, and was coming after me, one on top of the other, the eagle screaming and flapping its huge wings, and the wolf howling. I had run from it, first across an open field, then down a corridor in school, and finally through the cemetery, all the time going more and more slowly, like a wind-up toy running down. Claws and beak and glittering teeth had been closing in when the eagle's scream intensified and became the shriek of my alarm clock. I opened my eyes to the pale gray light that filled my room and jammed the button down, pulling my tangled covers up over my head. I was so exhausted from running in the dream, I decided not to get up.

I must have fallen asleep again, because suddenly my father was standing over my bed dressed for work, with a mug of coffee in his hand, whistling reveille. "Since when do Olympic runners sleep late?" he said. "Up and at 'em. Juice is poured." With that he was gone, and I groaned and sat up.

When I'd pulled on my sweats and plodded into the kitchen in my sock feet, Dad was mopping egg yolk off his plate with a chunk of bread. The morning paper was open on the table in front of him, and the unmistakable smell of sausage was in the air.

"You had sausage again!" I said, looking under the table for my shoes. "Aren't eggs bad enough? Sausage is terrible for you—fat, cholesterol, nitrites—"

"Next to the fridge," he said, pointing to my shoes. "And lay off. You're the runner, not me." He picked up the section of the paper he'd been reading and waved it at me. "Do you know the kid they're talking about here?"

I got my shoes and sat down across the table from him to put them on. "What kid?"

"They don't give his name. Collier's son. They claim he's living here and going to Ridgewood High. Did you know about this?"

I took the paper he was holding out. The whole top half of the front page was about the Collier case. A picture of Collier was in the middle, an article about the trial was on the left, and an article with the headline "Son Seeks Refuge in Ridgewood" was on the right. "Nobody knew about it till yesterday."

Dad poured himself another cup of coffee. "Biggest story we've had around here since the Vietnam protests. That was the other time the big bad world invaded Ridgewood, New York."

I scanned the article as quickly as I could. There wasn't much to it. It said that Joseph Collier's son had come to Ridgewood to stay with relatives for the duration of the trial and maybe longer, depending on the outcome of the case. He was currently enrolled at Ridgewood High School. Most of the article was background on Collier, all of which I'd seen elsewhere, except that Collier's wife had disappeared fifteen years ago, leaving him to raise their one-year-old son by himself.

One of my questions was answered, then. Bran had lived with his father. And it had been just the two of them. I could feel the goosebumps rising along my arms as I thought about it.

At the end of the article there was a sentence I had to go back and read twice. "There is no evidence that Joseph Collier's son had any involvement with any of the alleged murders." No evidence. Worded that way, it seemed to leave the door open. They weren't saying that Bran had nothing to do with his father's crimes, they were saying only that there was no evidence. The police only had enough evidence to take Collier to trial for four of the murders, even though they'd found nine bodies. Nobody doubted that he'd killed those other kids; it's just that there was no evidence.

"So, how well do you know the kid?" Dad asked.

"Not very well. His name's Bran Slocum. Molly sort of befriended him when he came, because he's had trouble with some of the jocks from the beginning. Bruno and his buddies."

"I suppose she's doing her Saint Francis routine."

I folded the paper so that Joseph Collier's face wasn't looking up at me and nodded.

"You'd better tell her to be careful." Dad tapped the paper. "There's going to be trouble about this." He stared into his cup for a moment, then looked back at me. "Ridgewood isn't just small, David, it's tight—all wrapped around itself like a cocoon." He swallowed the last of his coffee and put the cup down on his greasy plate. "It's a great place to raise kids. An all-American town. And it's scared to death. Molly had better keep her distance."

"You know Molly better than that," I said.

Dad stood up. "Molly thinks she can protect him the way she does her animals. You tell her he's no abandoned puppy. He's dangerous."

"He's not!" I thought about Bran carrying the twins inside to put Band-Aids on their knees. "You should meet him. You'd know if you did."

"It doesn't matter. All that matters is who he is." Dad leaned across the table toward me. "Listen to me, David. There are rotten things in the world that you can't fix. No matter how much you hate them, you can't fix them. You and Molly are young enough that you still think you can, but you'll learn. You tell her to stay away from this kid. There's not a thing in the world she can do for him. And she doesn't want to get mixed up with whatever happens."

"What do you mean, 'whatever happens'?"

Dad took his Windbreaker off the back of his chair and slipped it on. "People aren't going to ignore the fact that the son of a serial killer is going to school with their kids." He headed for the door, and then turned back. "Tell Molly what I said."

When Dad had gone, I sat for a minute, staring at the paper, trying to imagine what could happen. I couldn't. And I couldn't imagine telling Molly to stay away from Bran.

I drank the juice Dad had put out for me and went out to run. I followed my usual route, concentrating on filling my lungs with air and emptying them. When I passed the Ridleys' house, the blinds were all down, and it looked blank and closed. *No sign of life,* I thought, my concentration slipping. It was an ordinary expression that popped into my head naturally. All of a sudden, it had unpleasant associations. *No sign of life,* I thought again, as I ran through the back gate of the cemetery and up the winding road.

A line from the newspaper article came back to me. Bran's mother had "disappeared" when he was a year old. What did that mean? It didn't say she'd left or divorced Bran's father. Was somebody implying that Joseph Collier might have killed his wife and hidden her body somewhere? The idea was too grim to contemplate. I decided to count my steps as I ran. *One, two, three, four. One, two, three, four.* My time on that run was very bad.

When I turned off the shower later, the phone was ringing. I hurried to answer it, a towel wrapped around my waist.

"Have you seen the paper?" Molly never identified herself on the phone; she just started in.

I ran my hand through my wet hair and dripped on the wood floor. "Yeah. So now the whole town knows."

"I'm going by to pick Bran up. Then we'll come by for you."

"Have you talked to him yet?" I asked, stalling for time, trying to figure out what I wanted to say, what I wanted to do.

"Their line's busy. They probably have the phone off the hook."

"How do you know he's even there? How do you know he's going to school today?"

"He said he would."

"That was before the newspaper article. Maybe he's changed his mind. Maybe he's on his way to Australia by now. I would be. Or Mars."

"We'll be there in fifteen minutes."

"Molly, I just got out of the shower. I'm not dressed and I haven't had breakfast. No way I'll be ready. You go ahead, and I'll see you there." There was silence on the other end of the line. I shivered. "Molly? I've got to get some clothes on. I'll see you at school."

"We'll be there in fifteen minutes. If you're not ready we'll wait." Before I could say anything, she'd hung up.

As I got dressed, I kept going over what Dad had said. It had sounded cowardly somehow. But what could we really do to help Bran? It would be different if my hanging around with him could make things any better for him. But I didn't see how it could. Maybe it made sense to think about how it could hurt me. People already associated Molly with him, but not me. Not yet. Maybe Dad was right.

When the doorbell rang, I had just about decided to say I was sick and just stay home from school. But Molly was alone. "He wasn't there," she said as she came in. "Nobody was."

I did my best not to let my relief show, and hoped maybe he'd gone—just left Ridgewood. Then I wouldn't have to think about the whole subject any more. Or face Molly.

On the way to school I half-listened to Molly ranting about the immorality of the newspaper editor. I was remembering the look on Bran's face that first day, when I'd run into him. Like a hunted animal. I understood it now. That's what he was. He'd been tracked down here. If he went someplace else, there'd be someone tracking him again. Whatever his life had been like with his father, it had to have been pretty lousy from the time his father was arrested till now.

Finally, Molly ran down and stopped talking. Her hands were gripping the steering wheel so hard her knuckles were white. She was chewing on her lower lip.

"You can't change his life," I said.

She didn't say anything for a moment. Then, stopped at a red light, she turned to me. "You said last night I was his friend. Well, I am. And don't tell me having a friend doesn't make a difference. It does. I know."

I nodded. We rode the rest of the way to school without talking. There was nothing I could say.

When we turned into the school parking lot, we saw

a television news van in the bus drop-off area. Molly parked in her usual place at the very back of the lot, and we walked toward the school. It was still more than half an hour before classes would start, but already the parking lot was filling up and people were heading up toward the courtyard. Not just students and faculty, but other adults. Lots of them.

"I've got a bad feeling about this," Molly muttered, as a woman pushing a stroller cut in front of us onto the sidewalk. An older couple closed in behind.

There was a crowd around the fountain, growing steadily as more and more people came up from the parking lot and down the sidewalk. A few held signs aloft, lettered in Magic Marker. "Kid Killer Out," said one. Another said, "Expel Collier's Son." A cameraman was filming as a woman reporter interviewed the woman with the Kid Killer sign. A cluster of students stood behind her, waving and mugging at the camera.

"Why isn't Towson out here?" Molly asked, shielding her eyes and scanning the steps. The doors, as usual at that hour, were firmly closed.

"How'd this get organized so fast?" I asked.

"It probably started the minute kids got home yesterday and told their parents. I'll bet there were people on the phone half the night."

Molly shoved her way through the crowd around the camera, using her elbows liberally, even against adults. People grumbled, but moved out of her way. When she'd reached the front, she went straight up to the woman who was facing the microphone. "Who are you calling a kid killer?'' she asked, her voice drowning out the question the reporter was trying to ask.

"Get off camera, kid!" someone yelled.

"Who are you calling a kid killer?" Molly asked again. "The only kid killer I know about is on trial in New Jersey."

The reporter turned to Molly, and the cameraman switched his angle to get her completely into the picture. "Are you a student here?" the reporter asked.

Molly, her eyes snapping, spoke directly into the microphone. "Yes, I am, which is more than I can say for most of these people. And I asked this woman a question. Who is she calling a kid killer?"

"You know perfectly well who," the woman said. A number of voices behind her yelled in agreement. "Joseph Collier's son. We want him out of our school— out of Ridgewood." There was another roar of agreement. "We don't want him here, polluting our town!"

"Joseph Collier's son is not on trial," Molly shouted in order to be heard over the jeers that were growing louder. "He's never hurt anybody and he has as much right to be at this school as any student here."

"We want him out!" someone shouted, and others joined, until it had grown into a chant. "We want him out! We want him out!"

The cameraman backed away, taking in the whole scene, and the reporter angled off, heading for a man with another sign. I saw the reporter from the national tabloid, his notebook out, talking to a woman at the edge of the crowd. I pushed my way forward till I was next to Molly, who was still trying to argue her point against the wall of sound the chant had become.

"Come on, let's get out of here," I shouted into her ear. "Nobody can even hear you. And anyway, they wouldn't listen."

"Collier's the kid killer, not his son," she yelled once more. Then, reluctantly, she let me pull her away. "Towson had better get out here," she said, when we'd broken out into the open by the steps.

"He'll be out when it's time to open the doors," I assured her.

"He'd better have something to say."

I was right about his timing. The doors stayed firmly closed until the first bell rang, and then they opened as they did every day, and Dr. Towson appeared, as if nothing unusual were happening in the courtyard. He stood for a moment, scanning the crowd, and then Mrs. O'Neil hurried out and handed him a bull-horn.

"The school day begins in five minutes," he said, and the horn squealed. He adjusted the volume and raised it again. "Students are to come inside immediately. Others, please go back to your homes and let us get on with educating your children."

The chant, which had stopped when he came outside, began again. "We want him out! We want—"

"I've been on the phone with the superintendent and the president of the school board," Towson said, the bullhorn allowing him to be heard over the chanting. "There will be an emergency school board meeting tomorrow evening at City Hall. You are free to bring your concerns to that meeting. In the meantime, please let us conduct our business in peace.''

The kids started inside, but the chant continued, a little less insistent now as a few of the protesters began wandering away.

''Free speech,'' Molly said bitterly, as we headed for the doors. The television reporter had cornered Dr. Towson now, and the cameraman was coming in for a closeup.

12 Bran wasn't in homeroom, so I was able to keep hoping he'd gone—or at least that he wasn't going to chance coming to school.

Mrs. Campbell had trouble getting people to sit down, let alone shut up. Everybody was talking about the protest and arguing about whether Bran should be allowed to come to Ridgewood High, to live in Ridgewood at all.

Finally, she slammed the classroom door so hard I expected the window to come crashing down on her. "Now that I have your attention," she said, "I want you in your seats and quiet. Since Bran Slocum doesn't seem to be here this morning, all this may be so much wasted breath." She adjusted her glasses. "I fervently hope so," she muttered, as she opened her attendance book.

After the all-school announcements Dr. Towson's voice came over the room speaker, as calm and rea-

sonable as ever through the crackle of the sound system.
"This is your principal speaking. In spite of the unusual
circumstances of the—ah—demonstration going on
outside at this time, students are to carry on as usual.
The citizens outside have a right to make their views
known, but they do not have a right to disrupt the
business of this school. Students who wish to voice their
opinions on any subject concerning Ridgewood High
School may attend tomorrow evening's special school
board meeting along with other citizens. In the meantime, this is a school and no student—I repeat, *no* student—is to be released from class for any reason.
Unexcused absences will be treated according to official
school policy, of which you are all aware."

This statement was greeted with boos and hisses. Mrs.
Campbell slapped her attendance book against the desk.
She was ignored. Some kids said they should have a
right to join the protesters if they wanted. Matt suggested that everybody should boycott the school if Bran
was allowed to stay. Andy Tuttle said he didn't see what
it could hurt to have Bran in school, but several other
kids booed him down. Mrs. Campbell made a few more
ineffectual stabs at restoring order, but when the bell
rang everyone was still talking. I was glad to get out
into the hall.

"Hey, Watson!"

I turned and saw Zach Lewis pushing through the
crush of kids, trying to catch up with me. I waited for
him.

"You know Slocum, don't you?" he asked, falling in
beside me as I started moving again.

I shrugged. "I see him around."

"Come on. You're friends with Molly Pepper, and she's with him all the time. She must have talked to you about him. What's he like? You think he's sort of a serial killer in training?" Zach grinned a ghoulish grin.

"How would I know? How would Molly, for that matter? The guy's only been here a couple of weeks." Some part of me wanted to tell Zach that Bran was okay, to say that everybody was crazy trying to hook him up with what his father did. But I didn't.

"I have this really terrific idea. I want to interview him for the school paper. That would be such a scoop they'd just about have to make me editor next year. What do you say? Can Molly get him to talk to me?"

"I don't think Molly can even get him to talk to her," I said as I went into my American lit class.

"Ask!" he called after me.

To my surprise—and everybody else's—Bran showed up for math. He walked into class, looking the same as ever, his shoulders hunched a little more than usual, maybe, his chin pulled in close. He took his seat, dropping his pack heavily on the floor.

A murmur went through the room like a river rising. Bran ignored it. He pulled his math book out of his pack, opened his notebook and then just sat, staring at the page in front of him. Dana Farmer, whose seat was next to his, picked up her books and papers and moved to an empty place at the back of the next row.

Ms. Caitlin cleared her throat about five times before she asked to have homework papers passed forward. The murmur died down. While Ms. Caitlin was gathering the papers, Molly moved to the seat Dana had just left. Bran glanced over at her, then looked back at his math book. He stayed that way through the class, looking up only when Ms. Caitlin put an equation on the board. She went on as if it were a perfectly normal day, but I doubt that anybody learned any math that period.

When the bell rang, the kids, even the ones who'd been defending Bran's right to be there, were on their feet and moving out into the hall in record time, as if Bran's presence in that confined space had drained all the oxygen out of the air, and they needed to get outside to breathe. Molly took her time getting her stuff together, and I could tell she was going to walk out with Bran, who was waiting, as usual, to be last.

When Molly had first moved to sit next to Bran, I'd almost decided to follow her example, to take a public stand on his side. If she could risk it, why couldn't I? Looking at him there, so quiet, imagining what it must have been like for him to come past that line of protesters, I'd drawn a crusader's sword and shield in my notebook. To heck with what my father thought. But now I scuttled out the door with the others.

"Go back where you came from!" somebody shouted as I hurried away. I glanced back over my shoulder before I turned down toward the gym. The halls were always jammed between classes, bodies practically against bodies so that it was like crossing a busy street even to cut over to your locker. Now there was an empty space moving through the crowd like a bubble, with Molly and Bran in the middle of it. They both had

British lit third period, so they were headed to the same classroom. I couldn't have stayed with them anyway, I assured myself, since I had to change for phys. ed.

All morning the atmosphere got more and more tense. The protesters stayed outside, their chant sounding faintly through the windows on the courtyard side of the building. Arguments inside got louder and more heated, until the anti-Bran side took to shouting down anyone who dared to disagree.

When I came out of chemistry, Kristin was leaning against the wall outside, waiting for me the way she used to after her free period. "If you see your friend," she said, "tell him to get out of here while he still can."

For a minute I thought she was just adding her voice to all the others telling him to get out of Ridgewood. But she grabbed my arm and pulled me down so she could talk into my ear. "Nick and the guys're going to get him at lunch." Before I had a chance to answer, she had hurried off. Partway down the hall, she turned back. "Tell him!" she mouthed.

Molly was coming out of class behind me. "What's up?"

"Kristin says to tell Bran to go home—before lunch. Nick—"

"Bran won't go," Molly said. "I already told him to. But he says he ran once—when he left New Jersey— and he doesn't want to run again. He thinks if he does, he'll have to keep on running forever."

"Not forever. Just today."

"Then what? What about tomorrow? And the day after that?"

"Okay then, till the trial's over. If his father's convicted, everybody'll forget about it."

"Sure. Like people forgot about Charlie Manson and Jack the Ripper. He's going to be Collier's son the rest of his life."

I shrugged. "Tell him to hide out during lunch, at least."

When I got to the cafeteria she and Bran weren't there. Nick, Matt, Gordon, Jerry and a couple of senior friends of theirs were. I hoped she'd gotten Bran to listen to her.

Usually the cafeteria is the loudest place in school, but not that day. From the start of lunch period there was a kind of waiting hush, so the clatter of trays and dishes from the kitchen seemed especially loud. Everybody was looking back and forth from the doors to Nick and his gang, lounging casually against the first table. The faculty monitors, Mr. Girard, the Spanish teacher, and Mrs. Spitelli, typing, were standing together at the very back of the room, as far away as possible. They were talking quietly, apparently unaware that anything unusual was going on.

I'd just begun to relax a little when I saw them coming, Bran and Molly, in the center of that same bubble of space. They walked through the doors and stopped, as Nick and the others came to attention. Kids starting in behind them stopped too, and then stepped back.

"Well, if it isn't the killer's kid and his girlfriend," Nick said. The room was so quiet that his voice carried to the far corners. Even the clatter from the kitchen seemed to have stopped. Girard and Spitelli stayed where they were.

"Lay off, Bruno," Molly said. But she didn't try to move.

Bran looked at Nick and not at him, his right eye, as always, off to the side. He hitched his pack higher onto his shoulder.

Nick looked around the cafeteria. "Didn't I tell you about this creep right from the first? All you had to do was look at him. Sick. Like I said. Him and his daddy."

Bran said nothing. Molly looked around the room and saw the teachers at the back. She started forward and Nick stepped closer, looming over her. She stopped, clutching her backpack strap with both hands. Gordon and Jerry came up next to Nick.

"Sick," Nick said again, and moved toward Bran, till they were almost nose to nose. Even then, it was hard to tell whether Bran was looking directly back at him.

That seemed to enrage Nick. He grabbed Bran by the shoulders and shook him. "Look at me, you pervert!"

Molly started to move and Jerry stepped between her and Nick.

"My father says you must have been your daddy's little helper, mopping up the blood—digging graves in the backyard." Nick let go and Bran stumbled backwards. "He says you ought to get out of here and quit polluting our town. We don't want you here. Isn't that right?" Nick looked around again, and a few people nodded, but still everything was very quiet.

Nick gestured with one hand and the rest of his gang

closed in around them, forming a tight semicircle, edging Molly away and Bran back out through the doors. The kids behind him moved hurriedly out of the way.

I couldn't tell from where I was whether Bran was moving under his own power or whether they were shoving him along. Molly tried to push her way through to Bran, but she was shoved roughly against the wall. She stayed there for a moment and then began fighting her way against the stream of kids who were heading out after them, back toward Spitelli and Girard.

Nearly everyone joined the crowd leaving the cafeteria. I went too, pulled by a force like a magnet, a force that collected kids all along the way. I still don't know why I went along, why I didn't break away, like Molly, to go find someone to stop it. It wasn't as if I didn't know what was going to happen.

Partly, I expected somebody to come and interrupt the procession any minute. Towson. Or Mr. Byrd. Somebody.

But no one did. Mrs. Campbell started out of the faculty lounge, saw the mob heading toward her and ducked back inside. We went on down the back stairs and outside without encountering any other adults. In a few minutes it seemed that about half the student body was gathered behind the school on the grass between the back of the gym and the track.

Once it started, none of us could have done anything to stop it. It was like a shark feeding frenzy, mindless and out of control. They made a circle around Bran and took turns, hitting, pushing him from one to the other, kicking. Bran did his best to shield himself at first, then he began fighting back, using his fists and his feet. But finally he seemed to give up, like a dog being whipped, and just absorbed it all, bleeding from his nose and mouth. Except for sporadic grunts on impact, he made no sound. The spectators began cheering, like the crowd at a boxing match, except that I heard no voices for Bran.

It was Towson who stopped it, charging through the crowd like a linebacker, grabbing Nick and Gordon by the backs of their shirts and pulling them off balance. "That's enough! Stop this instant!" he shouted, shaking them, his face mottled red, his voice crackling with fury.

Bran, already on his knees on the grass, crumpled to the ground, his head on his fists.

"You're suspended," Towson said, and let go of Nick and Gordon. He made a gesture that included the guys who'd been beating Bran and Bran as well. "I don't want to see any of you on school grounds the rest of the week. That goes for everyone who threw a punch. I will not tolerate violence!"

"What about the football game Friday?" Nick asked. "What about practice?"

Towson shook his head. "You should have thought of that before you started this. Get going. Now!" He turned to the rest of us. "Go back inside. This is a school, not a Roman circus."

Molly, who had apparently come out behind Towson, hurried through the crowd, her hands full of wet paper towels. She knelt next to Bran and touched him gently on the cheek. That was the last I saw as Towson herded us all inside.

13

I usually spend my free period after lunch in the library. Sometimes I do homework, sometimes I join the group that hangs out at a table near the magazine racks, talking—quietly, so Mrs. Davidson, the librarian, doesn't kick anybody out.

When I went through the swinging doors that day the usual group was already in place, talking excitedly. Mrs. Davidson was reading at her desk, ignoring the noise.

"—nothing compared to what his father did to those kids," Scott Handleman was saying as I started toward the table. "One of the articles I read said they must have screamed a lot. Collier must have gotten off on that."

Jill Brandon shuddered. "That's so gross! Don't talk about it."

"So if they screamed, how could Collier's own kid not know what was going on? He had to, didn't he?"

"Not necessarily," Evan Lawton said.

"All I know is I don't feel sorry for him because of a bloody nose."

I nodded to them, went past the table and pretended to be looking for a book in the stacks. I didn't want any part of the conversation.

"It was a lot more than a bloody nose," I heard Evan say. "And as bad a beating as he was taking, he didn't holler about it. You have to give him that."

Scott snorted. "I don't have to give him anything. He's spooky, that's what he is. And you don't even have to know about his father to see that."

I didn't want to hear any more. When I got to the rear shelves, I slipped past three more stacks and headed back toward the front desk again, still pretending to check the spines of the books I was passing. Molly materialized next to me so suddenly I jumped.

"Come with me," she said. "I'm taking Bran to the hospital."

"Is he hurt that bad?"

"I don't think so. But the nurse checked him out and said he ought to go to the emergency room just to be sure. Bran didn't want to bother his aunt and uncle at work, and Towson didn't want to call an ambulance and have them come screaming up to the front doors. He doesn't want any more public fuss than he's already got, so when I offered to take Bran, he agreed."

"Why do you want me?"

Molly shrugged. "Bruno and his goons are suspended, too. I don't know where they are, and I don't want to take a chance on running into them, just Bran and me."

"I'm not exactly the bodyguard type."

"You're for moral support. Come on, David. Please!"

It wasn't like Molly to beg. But then, she didn't usually have to. Mostly, I did what she wanted. "I've got history next period."

"If anybody'd understand, Byrd would."

I sighed. "Okay. Where's Bran now?"

"At the car. You go on out. I've got to go get our stuff, but I'll be right there."

We went past the desk together. When Mrs. Davidson looked up from her book, Molly smiled. "Dr. Towson asked me to get David for him. He wants him to run an errand." Mrs. Davidson nodded and told the kids by the magazine racks to keep it down. Then she went back to her reading. When we were out in the hall, Molly headed for her locker and I hurried down the stairs and outside.

A few demonstrators, including the woman with the "Kid Killer Out" sign, were still trudging back and forth by the fountain, but they'd given up the chant.

I went around to the parking lot. Bran was slouched in the passenger seat of Molly's car with his eyes closed. He didn't hear me coming till I opened the driver's side door. He jerked and turned toward me with that look—the one I'd seen the first day. This was a person who expected enemies, not friends. "It's just me," I said. "Molly'll be here in a minute."

I climbed into the back seat, where I had to practically fold myself in half to fit. Getting settled, I bumped Bran's seat back and he groaned. "Sorry."

I could hardly believe the condition of his face. I'd

seen guys with black eyes before, but this was way beyond that. His bad eye was swollen completely shut, his lips were puffed, purple and badly cut up, and his face was bruised and streaked with dried blood. "You look awful! Are you okay?"

After a moment he answered, his voice coming thickly between his swollen lips. "I'll survive."

"I hope it's okay for me to ride along." He nodded gingerly, and neither of us had anything to say after that. The car was warm and stuffy. "Molly said she'd be here soon. I don't know what's taking her so long."

Bran didn't answer. I stared at the faded knees of my jeans, a few inches from my nose. They'd wear through soon, I thought. The cloth was almost white.

"I didn't know," Bran said, his voice so low I could hardly hear.

"What?"

"I didn't know what my father was doing." I couldn't think of anything to say to that. After a moment he went on. "He didn't bring them to the house. Not till afterwards. Till they were dead." He touched his lips with one hand. His knuckles were scraped raw. "I didn't know then, either. That's the truth."

I just nodded.

"He used to dig in the backyard sometimes. Said we had trouble with the septic tank." Bran rubbed the ear with the gold hoop. There was blood there, too. "I *did* help him dig once. We dug a long time but didn't find the tank. Next morning the hole was filled in. He said we'd been digging in the wrong place, that the tank was closer to the house." He swallowed. Clearly, it hurt

him to talk. "He didn't dig closer, though. Said whatever the problem was, it had gone away. I believed
him."

"Why shouldn't you? I would have believed that if
my dad—" I let the sentence hang there unfinished.
My dad wasn't Joseph Collier.

We sat for a while in silence again, and then he went
on. "I built a fort in the corner of the yard when I was
a kid. They found three of them right there. Right where
my fort had been."

I remembered the television pictures of a place with
a tall wooden fence in the background, surrounded by
cop cars with blinking lights, a rescue van pulled up
next to a huge pit, and guys carrying plastic body bags
from the pit to the van. It hadn't looked much like a
backyard where a kid might build a fort.

The jingle of keys announced Molly's arrival. She
opened the door and got in. "David, you keep an eye
out for Bruno and the others. And don't tell me I'm
paranoid." She eyed Bran critically. "You're looking
worse by the minute. I hope the emergency room isn't
busy this time of day."

Bran shook his head and then groaned. "I don't need
the hospital. Just get me home."

"The nurse said—"

"She said she didn't think anything was broken. She
wants me at the hospital to protect the school, that's
all. In case I decide to sue somebody."

Molly shook her head. "You ought to get X-rays. You
can't know for sure—"

"I know."

She still wasn't convinced, but even Molly couldn't force him. "You should at least file a complaint. What they did to you was criminal assault. I could take you to the police. We'll be witnesses." I cleared my throat, to let her know I didn't want her speaking for me, but she didn't take any notice.

"I've had enough police, thanks. Just home."

Molly started the car. "Towson knew who started that fight. I think he suspended you along with the others so you'd have an excuse not to come back for a while."

Bran grunted. "Or so he could get the killer's kid out of his school."

"Towson's okay," she said. "You'll see."

We drove the rest of the way without talking. I was wondering what it must have been like for Bran to find out he'd helped dig one of the graves. Had that been any worse than the rest of it?

"Who's that on your porch?" Molly asked as we neared the house.

"Don't know," Bran said.

I looked between them and recognized the reporter from the tabloid. He was wearing the same jacket. "Don't stop," I told Molly as she started to pull over to the curb behind a red Volkswagen beetle. "That's the reporter who was asking about Bran yesterday. The one who started it all."

Molly started to pull out again, but Bran put his hand on the wheel. "Stay. If he doesn't get me now, he'll just hang around till he does. I'd rather get rid of him before the twins come home."

"You don't have to talk to him," Molly said.

"What I don't say, he'll make up."

"He'll make it up, anyway," I pointed out. "I don't know why he's even bothering to look for you."

We found out soon enough. The moment we started to get out of the car the guy was on his feet, unslinging a camera he was wearing over his shoulder. I hadn't seen the camera. He knew his business with it. I had barely begun untangling myself from the backseat when the clicking started. Molly tried to get around the car and in front of Bran, but at least ten pictures—maybe more—had been taken in the time it took her to get completely between them.

Later, blown up on the front page of the guy's paper, the color picture they used made Bran look as if he'd been hit by a truck. It had been touched up so that the dried blood on his face and shirt was brilliant red. They ran it with a picture of the demonstrators, faces and signs clustered together so that it looked as if there had been hundreds of them. The headline said, "Killer's Son Beaten by Enraged Mob."

Molly shook her fist at the guy. "You've got your pictures, now go back under your rock."

"You got me all wrong, kid," the guy said, snapping the lens cover onto his camera. "I want to help him. I want to tell the world his side of the story."

"You've done a great job of helping already," Molly said. "If it wasn't for you, nobody would know who he is. His blood is on your hands."

The guy just laughed. "Looks to me like it's mostly on his shirt. Doug McKenzie." He held his hand out to

Bran, and when Bran didn't take it, dropped it with a little shrug. "I want to talk, Bran, that's all. Just a little talk. You can have your friends with you to be sure I don't bully you or anything."

Bran, limping, pushed past Molly and went around the reporter as if he weren't there. He stumbled on the steps and caught himself against the iron rail, then pulled open the screen and pushed against the door. It opened. "I guess we ought to learn to lock up," he said, and turned back to the reporter. "I'll talk inside. I need ice for my eye." He looked at us. "Come on in."

I thought about missing history class, then went in with the others. Mr. Byrd was the one teacher I was sure would understand. If Bran was going to talk to this jerk, I wanted to be there. That much I could do.

A few minutes later we were all sitting around the table in the Ridleys' kitchen. Bran had washed and put on a clean T-shirt while Molly made him an ice pack. He held the pack to his face now while McKenzie, like an eager ferret, pulled a miniature tape recorder out of an inside pocket and set it on the table. He looked questioningly at Bran before turning it on. Bran nodded. "I'd just like for you to tell your story," McKenzie said.

Bran sat for a moment, staring down at the red formica table top. Then he looked up. "Which story?"

"You know. Tell me about your father."

Bran moved the ice pack to the other side of his face, groaning a little as it touched. "Not much to tell," he said. "I always thought he was a pretty ordinary father. Of course, he was the only one I had. He did regular father stuff—bought me my first bike. Strapped me for stealing apples from a neighbor's tree. Worked hard. Paid the bills."

I thought of my dad again, a father who didn't always manage to pay the bills.

"Were you aware of what he was doing?"

Bran turned to Molly. "Get some cider, would you? The glasses are over the sink." He didn't say any more until there was a glass of cider in front of each of us. "Did I know his hobby was picking up runaways and murdering them? No. And neither did anybody else. You've talked to the neighbors—" McKenzie nodded. "So you know that everybody thought my father was okay. He always insisted that we mow the grass and paint the house and shovel the sidewalks." Bran lifted his glass and winced as the cider touched his lips. "He had a thing about being a good neighbor. Worked at it. Just like everybody said, he was this nice, plain, normal guy. Except inside his head. And nobody ever saw inside his head. Not the neighbors. Not me."

McKenzie took out a notebook and jotted a couple of notes. I tried to see what he'd written, but from where I was it was only a scrawl. "Do you think your father is crazy?"

"The psychiatrists say he's sane enough for a trial." Bran made a sound in the back of his throat that might have been a laugh. "You want to know something funny? Really funny?"

"Sure."

Bran fingered his earring. "The only real fight I ever

had with my father was about this earring, and my hair and the way I dress. He wanted me to look super respectable. Short hair, sweaters, khakis. Mainstream stuff. He used to talk about how important it was not to stand out. I thought he was dull, you know? The blue suits and striped ties and white shirts. Isn't that funny? I thought my father was dull.'' He shifted the ice pack. A crack at the corner of his mouth had opened up again and was bleeding slightly. The tape recorder hissed on, and I wondered if it was strong enough to pick up the sound of the refrigerator motor coming on that seemed to fill the kitchen.

"I wanted to be different," Bran said. "That's funny, too. I *wanted* to stand out. Be noticed." He looked at the reporter. "Looks like I got my wish."

"Did your father abuse you?" McKenzie asked.

Bran took another sip of his cider, winced again, and set it down. "No. He strapped me a few times, like I said. But that's all. I have all my fingers and toes. And no scars." He pulled at his T-shirt. "You want to see?"

"Not necessary," McKenzie said, and wrote in his notebook.

"My father didn't do *this* to me," Bran said, moving the ice pack again. "Nice, respectable small-town boys did this."

"Why do you think they did that to you?"

Bran shrugged.

McKenzie sighed. "What did you think about the demonstration at the school today?"

"I loved it, what do you think?"

"It doesn't help to play games with me."

Bran snorted. "What does help?"

"Are you going back to school tomorrow?"

"I thought you were such a hot-shot reporter," Molly broke in. "How come you don't know he was suspended?"

"For fighting. I could have hurt those guys real bad," Bran said.

"Then you won't have to go past the protesters again," McKenzie said, and drained his cider glass. He started to put his notebook away and I thought the interview was over. But then he leaned toward Bran. "Do you remember your mother?"

Bran didn't answer. I glanced at my watch. History would be nearly over. The reporter opened his mouth, as if to ask the question again, just as Bran spoke.

"All I know about my mother is her name, Caroline Slocum Collier." He paused for a moment. "That's not true. I also know that she's dead."

McKenzie jotted something in his notebook. "Do you know how she died?"

"Car accident. She was hit by a drunk driver, six months after she left my father. My aunt was with her at the hospital when she died. Sorry if that's a disappointment, but my father didn't kill her and bury her in the basement. Or cut her up and feed her to me for dinner. The truth doesn't make the kind of exciting story your paper likes. Maybe you could stick an alien in somewhere to spice it up."

"With your father," McKenzie said, "we don't need any alien."

Molly smacked her cider glass down on the table and Bran shook his head at her.

McKenzie did put his notebook away then. "Just one more question."

"Shoot."

"You ever hear of a movie called *The Bad Seed?*"

"No," Bran said.

"It's about a very sweet-looking little girl who kills people. Nobody wants to believe it at first. Turns out she's adopted. Turns out her real mother was a homicidal maniac. Bad seed, get it?"

Bran didn't say anything. He just sat there, looking at the reporter.

Molly leaned over the table and switched off the tape recorder. "I think that's about enough." She stood up and when she spoke to McKenzie her voice was venomous. "Don't you think you should be getting back to New Jersey in case something's happening with the *real* story?"

Bran started to get up, and Molly waved him down. "David'll show him out, won't you, David?"

I pushed myself away from the table and stood as McKenzie slipped his recorder back into his inside pocket.

Bran continued to stare at him. "Bad seed," he said, finally. "If that's what I am, I'm a late bloomer. I haven't killed a single person yet. Maybe I'll let you know when I'm ready to start. You can be a witness. Bring your camera. You'd like the blood."

I started out of the kitchen, McKenzie following. He turned back in the doorway. "I'm not your enemy," he said to Bran. "Your father is. He became that the first time he turned a kid as alive as the three of you into a bloody corpse."

"I'm glad you told me that," Bran said. "I might have missed it otherwise."

14 After the reporter left, Molly offered to pick the twins up at day care so Bran could go to bed. He nodded gratefully, started to get up, groaned, and sat back down. I realized, suddenly, how much pain he was fighting, and went to help him up. He'd held himself together so well for the reporter that in spite of how terrible he looked, his face all swollen and discolored, I'd almost forgotten the battering he'd taken.

As I watched him drag himself up the stairs, hanging onto the rail with both hands, I wondered which pain was harder to handle, the outside pain of the cuts and bruises, or the inside pain that must have taken over his life the day he found out about his father. He was sixteen. The same age as me. Just watching him I felt like a little kid. I wished I could go home, crawl into bed and pull the covers up over me while the grown-ups fixed whatever was wrong with the world. I knew, but I hated knowing, that they couldn't do it.

Molly and I went to pick up Kipp and Keith. They had twin fits that Bran wasn't with us. Molly explained that he was very, very tired and was home taking a nap, but the boys said he was too old for naps. Once we got them home, we had to keep them outside, pushing them up and down the block on their go-cart, to keep them from charging upstairs to rout Bran out to play with them.

We were still there when Bran's uncle got home. When he asked where Bran was, and Kipp started in again about how Bran was too old for taking naps, I went inside with Mr. Ridley to explain what had happened. Mr. Ridley was older than Dad, probably in his fifties, but in spite of his sparse hair and paunch, he looked strong—tough. He didn't look like someone I'd want to cross.

As he listened, he closed his eyes and I could see a muscle begin to twitch along his jaw. When I'd finished, he brought one fist down on the back of a chair so hard I jumped. "I'm calling the cops," he said. "We'll give those kids something to think about."

"Excuse me," I said, "but we suggested that already, and Bran says he's had enough of police."

Mr. Ridley shook his head. "Well, he's not me. If we don't get the cops into this now, those punks'll be after him again, as bad and maybe worse. If you don't fight back they run all over you." When I went back outside, he was dialing the phone.

When Angela, the twins' mother, got home from work, Molly and I left. Molly frowned into the growing dark as she drove, her mouth a tight line. I wondered what she was thinking about. As I started to get out at my house, she spoke for the first time. "Do you think there's any such thing as a 'bad seed'?"

I thought about it for a minute, then shook my head. "That's just some Hollywood invention. I don't believe murder's genetic."

"I don't either." She shifted into reverse. "I don't know what Bran thinks, though."

I waved and stood watching as she backed out of our drive. Did Bran believe it? Had that been part of the pain he'd been fighting as he went up those stairs? I shook the thought away.

As usual, Dad was in his workshop. The house was dark, but light poured out of the window in the garage door, cutting through the gray dusk. If a kid could inherit that kind of thing from a parent, wouldn't I go my own way, like Dad, doing my own thing, no matter how weird it was, and thumbing my nose at the rest of the world? And wouldn't my thing be totems? Wouldn't I be in the garage with Dad right now, gouging away at a hunk of wood with a carving knife or banging away with a chisel?

I tried to imagine it. Watson and Son, Carvers Extraordinaires, Makers of Personal and Garden Totems. No way. I pushed the door open and stuck my head inside. "Is supper on yet?"

Dad looked up from the piece he was working on, and wiped his smudged face with one sleeve. He grinned. "Almost. I thawed some chicken."

"I'm impressed," I said, and noticed for the first time in my life how much the dark blue eyes looking at me from my father's face were like the eyes that looked at me from the mirror every morning. The graying hair, the shaggy mustache, the cheekbones, those were different. The eyes were the same.

The next morning there were pictures of the protest on the front page of the paper, along with an interview with a couple of the sign carriers and a short article about the emergency school board meeting that night. Dad read it, shaking his head and grumbling about the small-minded meanness of the citizens of Ridgewood. He reminded me to warn Molly not to get involved. "You can't win against the haters," he said. "There's no sense trying. The only safe thing is to avoid confrontations and hope they leave you alone." I hadn't told him about the beating, or that we'd taken Bran home afterwards. I decided not to.

As I walked to school, I was afraid that things would be even worse than the day before, but when I got there only a few diehard pickets were circling the fountain. Then I saw that there were four new ones clustered at the bottom of the steps, staying well away from the others. One of them carried a sign that said "Land of the Free." Another's said "Public School for All."

"The great placard battle," Zach Lewis said as we went into English class. "Pretty disappointing turnout for both sides, I'd say. Ridgewood can't seem to muster up a really good fight."

"Which side are you on?" I asked him.

"Who, me? Take sides? I'm a floater, Watson, you

know that. Just like you. The minute you take sides you got enemies. I just keep my head down and try to preserve my pudgy self."

The atmosphere at school wasn't bad. Without Bran there for a target or Nick and the guys to keep everybody stirred up, things were almost normal. I stayed out of the conversations about Bran I encountered and did a fair job of convincing myself that Zach was right. Ridgewood wouldn't, or couldn't, muster a major battle over this, and it would eventually blow over. People would get tired of it. Everything would go back to the way it had been before. After all, I told myself, it was only a few people who were upset in the first place.

We had a regional track meet coming up in two weeks, so Coach Morelli worked us hard at practice, giving me an extra hill run because I'd missed practice the day before. I didn't mind. By the time we got back to shower and change, I was so tired that I wasn't thinking about anything except a decent meal and bed. But halfway through supper, Molly called. "Pick you up at quarter of," she said.

"What for?"

"What for? The school board meeting, of course."

"No way. I'm doing my math and hitting the sack. I'm wiped."

"David, we *have* to go. The crazies are trying to have Bran kicked out. We can't let that happen."

"We can't do anything about it," I protested, knowing already that it was no use arguing. Knowing that I would go. It wasn't just that I could never stand against

Molly. I'd sat in the Ridleys' kitchen and listened to Bran talk, watched him fighting a pain I couldn't even imagine. Somehow, I'd become a part of it.

Zach Lewis had been wrong. By the time Molly and I squeezed into the back of the room where they were holding the meeting, it was obvious that Ridgewood could muster not only a real fight, but a big one. We were fifteen minutes early, but already the meeting had had to be moved twice to a bigger room. When we got there only one row of seats was empty. By the time the meeting started the room was jammed with people sitting in the aisles and standing three deep against the walls.

At a long table on a raised platform in front sat Dr. Towson; Dr. Lyons, the superintendent; and a bunch of people who must have been school board members. Off to the side but on the platform with them sat Mayor Mahoney. "I don't know what he's doing here," Molly whispered. "I don't think he's got anything to say about what the school board does."

"He's probably just here to be sure everybody sees him. He's up for election next year."

When the meeting started, Mr. Sullivan, the president of the board, said they'd begin with the public part of the meeting, and when they'd heard people out, they'd go into executive session. He had to pound his gavel a few times to get order after that announcement. Two different television stations were there, with cameras and lights, and they were told they were welcome to stay until the public part of the meeting was over.

Microphones had been set up, one on each side of the room, and people were to make their statements into those. "Please keep your remarks brief," Sullivan said, glancing at the watch he'd laid on the table in front of him. "We would all like to get home at a reasonable hour. As you know, the subject of this meeting is the presence at Ridgewood High School of the son of Joseph Collier, the man currently on trial in New Jersey for the torture slayings of four children.

"I'm sure you're all aware that there are laws governing the eligibility of New York citizens for public education. We cannot change those laws here tonight. We can only decide what is the best way to handle our obligations within them. Before we invite statements from the floor, Dr. John Towson, principal of Ridgewood High, will explain the legal situation."

Towson explained that Bran's aunt, a longtime resident of Ridgewood, had been granted temporary guardianship, so the fundamental legal responsibility was clear. Ridgewood High School had an obligation to educate Bran Slocum. It was not possible to expel Bran solely because his father had been accused of a crime, no matter how terrible, nor because people did not trust Bran himself. "The boy has committed no crime."

There were only two alternatives to allowing Bran to continue to attend Ridgewood High, Towson explained. One was to request that another school district accept him as a student, and the other was to provide him with tutors who would educate him at home.

When Towson had finished, the meeting was opened

to public comment. I could hardly believe what happened then. The first man to speak hadn't said more than ten words before he was shouting into the microphone, accusing Bran of helping his father commit the murders. By the time he quit, Molly had hold of my hand and was digging her nails into me.

Mr. Sullivan reminded everyone that police in New Jersey had never arrested Bran, had never even suggested that he might have been involved in his father's alleged crimes. A second man, with a little more control than the first, said his kids shouldn't have to associate with Bran and that he didn't pay taxes so that the children of out-of-state maniacs could be educated for free.

"Just the children of New York maniacs," I whispered to Molly. She smiled in spite of herself. But joking didn't make the feeling in the room—like static electricity—any less noticeable. Or any less scary.

Then a woman got up and said that our good, honest, innocent kids would be tainted by associating with Joseph Collier's son. That he'd brought the evil and violence of the outside world to our town. "This has always been a safe place to raise kids and we want to keep it that way," she said. A lot of people shouted out their agreement with her, as they had with the man before her. Mr. Sullivan was having to pound his gavel a lot to keep order.

Molly had just told me she was going to get up and say that Bran was a lot less violent than the "good, honest, innocent" Ridgewood kids who'd beaten him bloody, when Mr. Byrd went up to the microphone.

When he began to speak, the audience got quiet so they could hear. "I don't personally know the student in question," he began, "but I'd like to remind everyone here that none of you know this boy either. How do you think he will harm your sons or daughters? As I'm sure you know, the boy was badly beaten on the school grounds yesterday, so it would appear that the harm has gone the other way."

"He got what he deserved!" a woman's voice called out.

"How do you know what he deserved?" Mr. Byrd asked, his voice rising. "What his father may or may not have done has absolutely nothing to do with the boy. Neither does it in any way change his absolute right to receive an education, the same right your own children have, I would remind you."

"Sit down, Byrd," someone shouted from behind me. I turned and saw that it was Nick Bruno's father. "We all know where you stand."

"Yeah, pinko jerk!"

"We know the kind of commie propaganda you been teaching our kids."

"If you deny a person's rights because you don't like him," Mr. Byrd shouted into the mike, "or because you don't like his father, who do you think is going to guarantee *your* rights against someone who doesn't like you—or *your* father?"

Mr. Byrd went on speaking, but even though he was still at the microphone, he was drowned out by the crowd. It was hard to pick out particular words in the noise, but the hostility was absolutely clear. I felt I could reach out and touch it.

The gavel banged and banged, but it wasn't until the policemen who'd been standing at the doorway came to the front of the room that order was restored. Mr. Sullivan called on the man who was standing at the other microphone, and Mr. Byrd, his face white behind his beard, left the room.

I was thinking that even with police in the room, it might not have been safe for Mr. Byrd to stay, and didn't hear what the next speaker was saying till Molly elbowed me. "That's what that reporter—McKenzie—was talking about. 'Bad seed.' "

The man at the microphone was gray haired and distinguished looking in a three-piece suit. He was talking about a teenage boy who had suddenly, without warning, turned on the middle-class family that had adopted him at birth and raised him. Using a hatchet and a butcher knife, he had murdered them all. During the investigation, police had discovered that the boy had been conceived during a rape, that his father was a psychopath and that he'd inherited his father's violent tendencies.

"This is only one case. There are many, many more. We know now that alcoholism can be passed genetically from one generation to the next. We are learning that some kinds of mental illness may also be inherited."

By the time that man had finished, the room was in chaos. One man grabbed the other microphone and said that Bran shouldn't even be allowed to live in Ridgewood. A woman yelled that she wouldn't send her kids to school if Bran was allowed to come back. Other people shouted their agreement, and finally Mayor Mahoney stood up and called to the police to empty the room. Mr. Sullivan, banging away with the gavel, yelled that the public part of the meeting was adjourned.

I grabbed Molly's arm and pulled her out through the crowd of furious people. She was shaking. So was I. I knew nobody could know which side Molly and I were on, but if anybody so much as looked at me, I felt as if it were tattooed on my forehead. "Good thing you didn't get up there," I said in her ear.

She just nodded. Behind us, a man said loudly that if Bran Slocum returned to Ridgewood High School, no one could guarantee his safety. I felt my stomach turn over. This wasn't Nick Bruno. This was an adult. A supposedly rational, civilized adult.

15 The next morning, as I was pulling on my sweats, Dad called me into the kitchen. "Ridgewood's made the national news," he said, waving the newspaper at the TV screen. "And they're doing a feature sometime this half hour. Skip your run this morning. I've already called Paul and told him I'd be late. It isn't every day we make the networks."

While we waited for the segment about Ridgewood, I read the front-page article about the school board meeting. It didn't say what the board had decided, since that part of the meeting had been closed, but it went into gory detail about the meeting itself. Whoever had written it was good. By the time I'd finished reading there were goosebumps all along my arms. It brought back that awful feeling that everyone in the room was an enemy. "It was scary there last night," I told Dad.

He scraped the last of his egg yolk off the plate with

his toast and nodded. "Now you see why I wanted you to warn Molly. I've seen this kind of thing before, and you don't want to be on the wrong side of it. After what Dan Byrd did last night, I wouldn't be surprised if a lot of people started demanding that Towson fire him. He was right in the middle of it the last time, and a lot of people haven't trusted him since." Dad shook his head. "I was surprised when he came back here after college."

"What did he do?"

"Burned his draft card at our graduation. Two other guys did, too. Ridgewood went nuts."

"All he said last night was that Bran had the same rights as their kids."

"That wouldn't cut any ice around here. If the Bill of Rights went up for a referendum tomorrow, it'd be defeated by a landslide."

"After last night, I believe it."

"He should have known better. You can survive in hostile territory, David, so long as you don't confront the natives head on."

"Maybe he thought he had to. What they're doing to Bran—"

"Shush, now, we're on."

"During our news segment a moment ago, you saw a clip from a school board meeting in Ridgewood, New York," the anchorwoman was saying. "Ridgewood is the town where the son of accused serial killer Joseph Collier is trying to attend high school over the strenuous objections of the citizens." She was replaced on the screen by pictures of the man in the three-piece suit, speaking into the microphone last night, as people leaped to their feet around him. Her voice went on over the pictures. "Last night a psychologist told townspeople that Collier may have passed his violence genetically to his son. Is this possible? Is there really such a thing as a 'bad seed'? We've brought in some experts to help us find out."

She introduced two psychiatrists. In the next few minutes one said there was absolutely no evidence for the existence of directly inherited criminal tendencies and the other said that genetically determined psychosis was documented fact. The reporter asked if the two were talking about exactly the same thing, and they argued back and forth about that. By the time they broke for a commercial, nothing was clear except that nobody knew anything for sure. Afterwards, the male anchor said that everyone should remember that Joseph Collier hadn't been convicted yet, so no one should assume his guilt, let alone speculate on his son's mental health.

"So why'd you air the segment, you hypocrite?" Dad said to the TV screen.

"What do you suppose the school board decided?" I asked him.

Dad shrugged. "You can be sure what they want to do is transfer him to another district. But who's going to take him? There's no good answer for them or for the kid." He pushed back his chair. "It's what's known as a dilemma, Davey. Give me wood carving any day. Problems, sure—dilemmas, never."

When he'd left, I turned off the TV set and stacked

the dishes. Maybe that was the real lure of Dad's totems, I thought. They didn't make any demands.

At school the pickets were back in force, their numbers probably more than tripled, and new signs said Bad Seed Out. A police squad car was pulled up near the fountain, probably to protect the couple of pro-Bran pickets who had dared to show up. The chanting seemed to be directed entirely at them, and went on all morning, so loud, now, that it invaded every classroom. In spite of the police, by lunchtime the pro-Bran forces had disappeared, giving the whole show to the other side.

Most teachers did their best to carry on as usual, but the entire day seemed to be about Bran. Should he be allowed to come back? Was he a walking time bomb, a deadly force programmed to go off any minute?

Rumors sprang up like mushrooms. In class, between classes, at lunch, I heard them everywhere. The Ridleys, afraid for their lives, were going to commit Bran to a mental hospital. Collier's neighbors had insisted that Bran move because they were afraid to keep him in their town. New Jersey police had forced Collier to sign over parental rights so New Jersey wouldn't be stuck taking care of Bran. It was all I could do to keep my mouth shut. The worst I heard was that Bran's mother had left because he'd started throwing butcher knives at her as soon as he could walk. There were variations to that one—some said he'd used his father's gun.

Molly hardly looked at me or anyone else all day. She seemed shut inside herself, as if she'd put a wall up against everybody, against the world. But no wall could keep out the sound of the chanting from outside, or the talk that swirled through the halls.

Zach Lewis had quit floating. "Keep him out," he said. "Why take chances?"

"He hasn't done anything wrong," I said.

"Not that we know of. And not yet. Like I said, why take chances?"

"You don't take sides till you're sure which is safe, huh?" I asked him.

"So when do *you* take sides, David?" he asked. I just walked away.

Coach Morelli dismissed us early, partly because it had started to rain—one of those on-again, off-again, blustery October rains that make you feel cold clear to the bone—and partly, I think, because he was as distracted as everybody else.

When I started home afterwards, my hands in my jacket pockets, I found myself heading for the Ridleys' house. I hadn't consciously planned to go there, but when I realized what I was doing, I knew that all day I'd been wondering if Bran was all right, if he'd seen the television show or read the paper.

I saw the huge red letters before I was close enough to see that Bran and his uncle were on the front porch, with rags and paint thinner, trying to rub them out. Someone had written BAD SEED with red spray paint on the side and across the front of the house.

"They did this in broad daylight," Mr. Ridley said as I started up the porch steps. His face was nearly as red as the paint, and he waved his rag at me as he spoke. "Do you believe that?"

I nodded. By now I was ready to believe almost anything.

"The police came to look. Said if we didn't have witnesses or evidence on who did it, there was nothing they could do. Said it was too bad. *Too bad!*" He attacked the aluminum siding as if he were trying to wipe out whoever had done it instead of just the paint.

"I must have been back in the kitchen," Bran said. "Didn't hear a thing." The swelling in his face had gone down some, but the bruises had turned a darker purple. He still looked awful.

"I hope you've started locking your doors," I said.

"Doors, windows, garage," Mr. Ridley assured me. "Not that it'll keep them from doing anything they want outside. The cops won't do nothing but drive by once in a while to check on things. What the hell do we pay taxes for, I want to know. I tried to file charges against that Bruno kid yesterday and they told me Bran had to do it himself. And if he did, he'd need a doctor's report describing his injuries. So that when the bruises went away there'd be evidence of assault."

"I didn't want to file charges," Bran said.

Mr. Ridley turned away from Bran and attacked the siding again. "Right, and even if you did, we don't have no doctor's report. You didn't want to go to the hospital, either. So there's nothing we can do. You got to hide out in the house all day and those punks are out roaming the streets with their paint cans, free as birds." He smacked his left hand against the siding. "And the cops can't touch 'em."

Bran stopped rubbing at the paint. "Getting the police into it would have made it worse. And they'd still be out here."

"Maybe."

"Can I help with that?" I asked.

Mr. Ridley looked from the rag in his hand to the siding, where the red paint was smudged, but still glaringly there, and shook his head. "We're getting nowhere"—he leaned closer to look—"except we're wrecking the finish on the siding." He sighed and held out his hand to Bran for the other rag. "Why don't you go pick up the twins?" He turned to me. "Go with him, would you? We took them over to day care this afternoon so's he could rest, but they might as well come home now. Marie's got the car—you'll have to walk."

"No problem."

He picked up the can of thinner and looked at the red letters again. "I guess we could paint over 'em."

"We'd better wait a while for that," Bran said.

Mr. Ridley nodded. "They better not plan to do any more." He looked up at the blustery clouds. "I'll just bring a chair out and sit awhile. Nice day for sitting on the porch."

I pulled my jacket up around my neck. "Real nice."

"Let's go," Bran said. We started off, taking the short cut through the cemetery.

"You still hurting?" I asked, as we headed up the gravel road. He was still limping a little.

He shook his head. Then shrugged. "Some."

"You look a little better."

"Thanks."

"Did you see that TV show this morning?"

"Yeah." He walked on a few steps, then stopped. "There's one thing wrong with that inherited psychosis theory. The prosecution's doctors say my father isn't crazy." He made that short, sharp sound that passed for a laugh and started walking again. "He's something, they say. But not crazy."

"Do you ever wonder—" I started. But I couldn't go on, couldn't ask him straight out if he was afraid of turning out like his father.

"What?"

"Nothing. Never mind." I left the road that curved around to the right and started across the grass between the gravestones, toward the gate on the other side. After a few steps I realized he hadn't come with me. I turned back to see that he'd stopped where the browning lawn met the gravel.

"I'll stick to the road," he said.

I hurried back to join him, remembering his backyard. Bodies under grass. "Sorry. It's just shorter—"

"I know." He stepped over a puddle. "Did you mean do I ever wonder if I'm a bad seed?"

I nodded.

"I killed a bird once." His voice was so low I had to strain to hear. "It had been chewed by a cat. It was going to die anyway, so I wanted to help it."

Bran stopped walking and stared down at a drift of leaves at the edge of the road. The wind gusted around us and I shoved my hands farther into my pockets.

"I can still feel that body in my hand when I think about it. I don't even know what kind of bird it was. It just had that brown fuzz still, instead of feathers. And blood all down the front. Its heart was beating so fast I couldn't have counted the beats. Like an engine idling too high. It was so little. So light. I thought I'd just twist its neck. You know, a quick twist and it would all be over. Easy."

He fingered his earring and looked up, then, and past me across the stones. "I didn't—couldn't—twist hard enough. I don't know why. Weak as it was, it tried to get away, and I couldn't make myself do it hard enough. It kept looking at me with those bright, black eyes. I tried five times before I felt the bones break under my fingers. By that time I was crying. I probably hurt it a lot more than if I'd just left it to die. It was only this baby bird. But one minute there was that warm, soft body and that heart beating away and those eyes. Then nothing. Just nothing. I threw it in a trash can and went home and washed my hands. And tried not to think about it."

Bran started walking again, his limp more noticeable than before. "Sure, I wonder about him and me. But when I remember that bird, I know nothing in the world could make me kill again. I'm *not* like—" He paused, and when he spoke again the tone of his voice sent a chill through me. "Like that man in the jail cell in New Jersey. I'm not."

Rain started spitting again, so we picked up the pace as much as Bran's limp would let us.

≪   ≫

When we arrived at the day-care center, Kipp and Keith came roaring out onto the porch and dragged us in to look at the art work they'd been doing. "This is you, Bran," Keith said, pointing to a crayoned purple circle with arms and legs and a huge head with a large black circle where one of the eyes should have been. "And those're the jerks that beat you up." Squeezed into one corner were a bunch of smaller people, looking a little like two-ball snowmen. "And here's Super Keith, coming with a big gun to blow 'em all away." Above the purple figure's head there was a smaller head, attached to a big red triangle that must have been a cape. Stick-like hands jutted out from the cape, and perched at the end of the hands was an enormous black banana-looking gun, with black dots and red lines coming out of the barrel. "I'm not done yet," he said. "I gotta put in all the blood."

"Mine's done," Kipp said, pulling Bran over to his paper. "See? Dinosaurs. Them *Tyrannosaurus rex* kind with the teeth. They're to walk to school with you when you go back. If anybody wants to fight, they'll chomp 'em!"

Bran held up Kipp's drawing, looking at the two green ovals with legs and tiny arms, the heads not much more than jaws full of huge red teeth. "We could use these guys to stand next to the house, Kipper."

"They'd be bigger than any old house," Kipp said.

"Big doesn't count," Keith scoffed, "if they don't have guns."

"Maybe if you're big enough, all you need are teeth," Bran said. "Can we take these home with us?"

"We made 'em for you," Keith said. "You can put 'em up on the wall by your bed."

We had just started home, the boys stomping along ahead of us, brandishing their rolled-up pictures like swords at imaginary enemies, when Bran glanced back over his shoulder and put his hand on my arm.

"What?" I asked.

He leaned closer. "Someone's behind us, trying to stay out of sight. I saw him as we came out, standing behind the porch of the next house. He just ducked behind a hedge. Looks like Ritoni."

I looked back, but didn't see anyone. "Where Jerry is, Nick is."

Bran shook his head. "I think he's alone. We can check when we cross Broad."

My throat felt dry all of a sudden, and I swallowed. I was pretty sure Nick was someplace close. The others, too, probably. "We could have trouble."

"Just keep walking."

I tried to think how Jerry had known where we were. "You think he followed us here from your house?"

Bran shrugged. "I didn't see anybody. But I wasn't looking, either."

At the corner of Birch and Broad, we stopped to wait for a break in the fairly steady late-afternoon traffic, and I looked back. It was Ritoni, all right, doing a clumsy job of getting himself behind a tree trunk. I had a feeling he didn't care all that much whether we saw him or not.

"What do you think we should do?" I asked Bran.

"What about?" Keith asked, walking the curb like a balance beam.

"About the space aliens that landed in the cemetery this afternoon," Bran said.

"Ah, there's no such thing as space aliens," Keith said.

"Except on TV," Kipp added. He looked up with an uneasy grin. "You're just teasing us, right?"

Bran pointed down Broad toward the cemetery entrance and I saw what he'd seen. A couple of figures had just stepped back behind the huge brick columns next to the gate. Nick. And Gordon. Matt was probably there too.

"Nope," Bran said to the twins. "This great big flying saucer landed right in the middle of the cemetery this afternoon, and two aliens got out. They're probably in there somewhere right now."

Kipp's eyes got huge, and he gripped his picture with both hands. "You think they're nice ones like Alf?"

Bran shrugged. "What do you think, David? They looked kinda mean to me."

I nodded. "Yeah. Big and mean."

"We'd better go the long way home, guys, what do you say?" They both nodded, their faces solemn.

"Okay," Bran said. "Tell you what. When these next two cars get past, we'll zip across the street as fast as we can, and go that way." He pointed right, away from the cemetery gates. "You two be the front lookouts and we'll be the back lookouts. If you see any aliens, you run for home, okay? As fast as you can. David and I'll keep them busy till you're safe."

"What do they look like?" Keith asked, when we were across the street and walking beside the tall iron fence. He kept glancing into the cemetery and walking faster and faster. Kipp was half skipping, carrying his rolled picture cocked like a bat.

"Purple fur," Bran said. "And four eyes."

I glanced back and saw Jerry, waiting to cross the street, waving to the others. "You think they'll follow us?" I whispered.

"Or cut through and be waiting at the back."

All the way around the cemetery, we pretended to the twins that we were keeping a watch for space aliens. Bran and I each picked up a stick—"To knock 'em on the head with," Keith said with satisfaction. To keep my heart from thumping out of my chest, I thought. Gordon and Jerry were following us, staying well back, but not trying to hide. Nick and Matt were nowhere to be seen.

I wondered if Mr. Ridley had been right, if reporting them to the police would have kept them from going after Bran again. At least in the middle of the day. I was glad we'd seen them in time to stay out of the cemetery. Out here there were cars going by. And houses with people in them.

We had almost made it back to the Ridleys' when I looked over my shoulder and saw that Nick and Matt had joined Gordon and Jerry. They had picked up the pace and were closing in on us fast. I nudged Bran, and he looked back. "Hey, you two!" he called to the twins, who were now fully into the game, dodging from tree to tree and jumping out at each other with wild shrieks.

"Aliens coming in from behind. See how fast you can get home!"

The twins began to run, and I heard the pounding behind us as the guys started running, too. Keith was moving pretty fast, but Kipp had begun to fall behind, when Bran swooped down on him and snatched him up. His picture fell from his hands and Bran kicked it out of the way. "Get Keith," he said to me.

Kipp yelled, "My dinosaurs!" but Bran just kept running. I slowed long enough to pick up the picture then grabbed Keith and carried him, jouncing against my hip, the last hundred yards to the Ridleys' porch.

Mr. Ridley stood up from his chair as Bran and Kipp bounded up the steps, Kipp still hollering about space aliens getting his picture. Keith and I were only seconds behind.

We set the twins down, and I handed Kipp his torn picture. "We're all safe," Bran gasped. "The aliens won't get us here."

Nick had stopped a few houses down and was now sauntering casually toward us, hands in his jacket pockets. The others walked behind him, their eyes fixed on the porch.

"Somebody writed all over our house!" Keith said.

"Go on inside," Mr. Ridley said, his voice stern.

Keith started to object, looked at Mr. Ridley's face, and changed his mind. "Come on, Kipper."

"My picture's all tore up!" Kipp complained as they went inside.

Nick stopped directly in front of the house. Bran was still catching his breath from the run, but he stood very still, looking back at Nick. No one said anything. After a moment, Nick smiled and then turned slowly away and walked on, Gordon and Jerry and Matt at his heels.

"Like a pack of dogs," Mr. Ridley said, and slammed into the house.

Bran stood there, watching, till they turned the corner and disappeared.

16 After dinner that evening I sat on a crate in the garage and watched Dad carve. For a long time we didn't talk. I just watched as he moved the fine knives with a sure, easy touch, defining the individual feathers on an owl that was part of a walnut totem pole. The huge bird was nearly finished, its claws resting on the head of a completed bear cub, its "horns" seeming to grow out of a roughed-in lynx above. This was a commissioned piece, the totem animals chosen by a man who'd ordered it for the entry hall of his new house.

"Did he choose the owl for wisdom?" I asked finally, more to fill the garage with the sound of our voices than because I wanted to know.

Dad laughed. "These three are his family—he's the lynx, his son's the cub, his wife's the owl. And he did mention wisdom. He also pointed out that owls eat vermin—fur, bones and all."

"We could use some owls around here," I said. "Big

ones." I told him about the graffiti on the Ridleys' house. "I gave Molly your advice about staying away from Bran. She didn't."

"That doesn't surprise me."

"The thing is, I didn't either."

Dad didn't say anything. He adjusted the glasses he wears for doing close work, blew on the feather he'd just finished and began another.

"I'm getting to know him. He's okay. What's happening to him—to his family—doesn't make any sense. They haven't done anything. It's crazy."

Dad nodded. "The world *is* crazy, Davey. A lot of the time it is. You can't change that. Best you can do is come to terms with it—one way or another." He put his glasses up on his head and looked at me. "Just be careful. I don't want you to get hurt."

"Depends on what you mean by *hurt.*"

"Right now I'm talking about good old-fashioned physical pain. Be careful."

He put his glasses back down and began working again. *Careful,* I thought. That was pretty much what I'd always been. Not exactly what you'd expect of somebody whose totem was a wolverine. "When did you choose my totem for me?" I asked.

"The day you were born. We had names before, but we wanted to meet you before we chose your totem." Dad laughed. "What an argument that was. They could hear us all over the maternity ward. I was this dove person, and my first kid was a boy. I wanted you to be tougher than me, and I didn't know anything tougher than a wolverine."

"I never thought it was exactly me," I said.

"Maybe we should have chosen a beaver. That's what your mother wanted. You know, worthy and industrious."

I thought about that, and laughed. "You think it would have worked?"

Dad blew on the owl again, sending tiny shavings into the air. "No way to know. Anyway, it's too late now. We chose what we chose. Like it or not, Davey, somewhere inside you is that wolverine."

Friday morning, to the accompaniment of the steady chanting from outside, Dr. Towson announced that because home tutoring at the high school level could not be considered equivalent to class and laboratory work, the school board had decided that Bran should continue attending Ridgewood High. "The suspended students will return to school on Monday, and for the first time in the history of Ridgewood High School, I have asked for a police patrol of our halls. There will be order at this school even if it has to be imposed by outside authority."

This was greeted by a roar of derision, but the voice over the crackling intercom system went on. "For those of you who may be considering a boycott, I should remind you once more of the guidelines governing unexcused absences. Whoever hopes to make passing grades this semester should be present on Monday and every day after that."

I hadn't thought things could get worse, but they did. The few neutral voices had vanished. "Bad seed out," the constant chant from outside, was echoed on walls and blackboards. Someone had written PSYCHO LOVER across Mr. Byrd's classroom door. Teachers who had tried to discuss the situation rationally in their classrooms went suddenly silent. They focused on their regular subjects as if nothing unusual were happening around them.

People were saying that the teachers had sent a petition to Dr. Towson asking to have the suspension extended, "in the interests of maintaining order and promoting education."

The connection between me and Bran had finally been made. No one, not Zach Lewis, not the guys from the track team—no one—was speaking to me. If some kids hadn't made a point of pulling back or turning away when I came near, I'd have thought I'd become invisible. I waved to Kristin once, and she ducked into the nearest classroom. Later, in my locker, I found a note from her, telling me that if I didn't keep away from Bran I was likely to find myself in trouble with "the guys." I supposed it was the best she could do.

Molly had more trouble than I did. People just ignored me. They shoved, bumped and tripped her in the halls. That was the day she called what was happening a plague.

"It's like the Black Death," she said at lunch, as we sat at an otherwise empty table. "It would get into a little village somehow—maybe the germ would come on somebody passing through from a city where it had already started—and in a couple of days the whole village would be sick. That's what's happening here.

First it was only Nick and his cretins who were infected, and now it's everybody. Even the teachers. Did you hear about the petition?" I nodded. "They say it's not to punish Bran. It's to protect him."

"Oh, sure. That's like saying nobody should use the subway in New York because people get mugged there. Have you talked to Bran? Did you hear about their house?"

Molly set her milk carton down so hard the milk splashed into her tray. "Yes, I heard. I mean it, David, it's a plague." She shook her head. "I thought we lived in a civilized world. Not perfect, maybe, but rational. With laws. And rules. Bran's only been here—how long? A couple of weeks?"

I thought back to the day I'd first seen him. A Saturday. Today was Friday. I could hardly believe how short a time it was. "Not quite three weeks." Three weeks to change everything.

"A plague of hatred. Maybe Collier's totally evil, but what about everybody here? Beating Bran up and spraying hate messages on their house are stupid kid things. But look at all the adults in it now. It's as if they're giving permission for whatever anybody wants to do. What's to stop it?"

I didn't have an answer.

"They talk as if Bran could infect Ridgewood," Molly said, mopping up the milk with her napkin. "Seems to me it's the other way around."

During last period Towson announced that he was turning down the teachers' petition to extend the suspension. By the time school was over for the day BAD

SEED OUT was spray painted on the windows of his car and all four tires had been slashed.

I worked from three to nine that night. When I got off, the night was dark, the sky choked with clouds, the wind gusty. As I started home, I kept looking over my shoulder into the moving shadows the trees were throwing across the sidewalk beneath the streetlights, half expecting to see Jerry or Gordon—or Nick. I knew they'd have no reason to follow me, not while I was alone, but I couldn't shake the feeling that they were out there somewhere. They hadn't been able to play in the football game that night. No telling what they'd be doing.

Passing the cemetery, I thought again about the way Nick had stood there in front of the house, looking at Bran. And I decided to go over to the Ridleys' before I went home, to see how he was doing.

Once I was inside the cemetery and away from the streetlights, I could barely see the road. I put on as much speed as I could in the darkness, anxious to be out of there, away from the sound of the wind in the leaves, the darker shapes in the darkness that were the headstones, the idea that someone might be following me.

I heard the chant long before I reached the back gate. "Bad seed out! Bad seed out!" I stopped and listened. There were a lot of voices. More, even, than at school, it seemed. For a moment, I thought of just turning back, going on home. I could check with Bran in the morning. It wasn't as if he was expecting me. But then I thought of the Ridleys, in their own house, surrounded by the kind of hatred I'd felt at the school board meeting. Were the twins there, wishing for dinosaurs and guns? How could anybody explain it to them?

I couldn't stop the craziness, turn back the plague. But if I just chickened out and went home, didn't that make me part of it? If I stayed to the backyards, I could at least get close to the house, see what was going on. And if the chanters were all out front, I could maybe get to the back door.

Sticking close to the bushes that stretched across the backyards, I made my way down the block. When I reached the Ridleys' backyard I crouched against a couple of trash cans. There didn't seem to be anyone back there.

The house was totally dark. There wasn't a light anywhere. Maybe they'd gone, I thought. They could have gotten out the way I'd gotten in, and the mob in front wouldn't have known.

Staying low, I crept up onto the back stoop and knocked on the door. The chant from out front was so loud, I could hardly hear the knock myself, so I knocked louder. Nothing. I pounded with my fist, and this time saw a curtain move in the kitchen window. I backed up a little so whoever was inside could see me in the pale light from the windows of the house next door. The curtain moved again.

"It's me, David," I said, my mouth close to the door. "David Watson."

After a few seconds I heard the lock on the door, and it opened a crack. "You don't want to be here," Bran said.

"You're right about that. But since I am, can I come in?"

He pulled me inside and then slammed and locked the door.

"Have you called the police?" I asked.

"Three times. They came by earlier, when there were just a few people out there, and told them to keep the noise down. The neighbors were complaining. Then they told us—*us*—not to provoke a confrontation. So much for police protection. Molly was here then. She called her folks and they came over and took Aunt Marie and Angela and the twins over to their house. Good thing. A while ago they started throwing bricks."

I followed him through the kitchen and into the small living room. A little light, not much more than a softening of the blackness, came from the streetlights outside. "Can't you turn on a light?"

"Damn near got a brick in the head when we had a light." Mr. Ridley's voice came from my left. Peering into the gloom, I could just make out his figure in the bulk of an overstuffed chair.

"No sense giving them something to aim at," Bran said. Staying behind the curtains, he looked out. "Looks like there are more coming all the time."

"The football game must be over," Mr. Ridley said.

"And we're the after-game party. Terrific town you got here." Bran's voice was bitter. "My aunt got fired today."

"Fired? For what?"

"Nobody'll come into the diner while she's there. They sold one cup of coffee all day."

"Rutkowski claims he's too small to survive a boycott," Mr. Ridley said.

"Yeah. His customers would all be back in twenty-four hours if I left town. So would Aunt Marie."

A shattering crash startled me so that I nearly fell over the footstool in front of Mr. Ridley's chair. Something heavy thudded against the wall behind me and glass clattered to the floor all around.

"You'd better get down," Bran said. "As long as you're in line with a window, they could get you even if they can't see you."

I lowered myself to the floor in front of the couch, aware of the crunch of broken glass under me. The whole thing seemed unreal. This was the town I'd grown up in. We'd never even had to lock our house. "We should call the police again."

Mr. Ridley snorted. He pushed himself to his feet and started for the kitchen. "I'm calling Marie to let her know we're still here."

Bran moved from his place near the window and joined me on the floor. "Before we took the phone off the hook some guy called to say they'd get me out of here any way they had to do it. He mentioned a bomb, specifically, in case I didn't get the general threat. You might not want to hang around very long."

While Bran was talking, I noticed a pale, flickering light reflected on the wall behind Mr. Ridley's chair. "Is that fire?" I asked. My throat was so tight I could hardly get the words out. Bricks I could handle. Fire was something else.

Bran pulled himself up so that he could see out over

the back of the couch. "It's fire, all right," he said. I waited for him to go on, to tell me what they were doing, or suggest we run. Something. He seemed paralyzed there.

"Bran?" I asked, finally. "What is it?"

Still he didn't answer. I moved up next to him and looked out. People lined the sidewalk, chanting. In front of them several figures moved around a bonfire in the middle of the yard. One of them was holding a long pole, from which dangled a dummy, dressed in a shirt and jeans, its head dangling to one side. Pinned to the head was a photograph. The others held burning sticks in the air. The flames and smoke of the fire rose and then flattened as the wind gusted. As the man with the dummy moved through the light, I could see that the picture on the dummy's head was the photo of Joseph Collier from the cover of *Life* magazine.

"It's all wrong," Bran said, his voice flat. "It needs a suit. A suit and tie."

"Burn, psycho, burn!" someone shouted. Almost as if it had been rehearsed, the crowd changed their chant.

"Burn, psycho, burn! Burn, psycho, burn!" they yelled as the two with the torches touched them to the dummy's legs.

"Mr. Ridley, call the fire department," I yelled. "They *have* to come!"

The flames began burning upward while the dummy jounced up and down in a wild, jerking dance.

Where his shoulder touched mine, I could feel Bran start to shake as the flames consumed the dummy, bits of burning cloth and stuffing falling to the grass. The flames were licking the photograph then, crinkling the edges and turning them black. A gust of wind caught the flames. Photo and head seemed to explode, bursting all at once into a mass of fire.

There was a sudden pause in the chant, as if even that mindless crowd had been jolted by the dummy's violent end. In the brief silence I could hear Bran's ragged breathing.

The chant wavered into life again as the leader threw his now empty pole onto the fire. Above the sound of voices the piercing wail of a siren began, growing gradually higher and louder.

17 The hook and ladder truck roared up the street and screeched to a stop in front of the house, its siren drowning out the chant. Firemen in rubber coats and boots dropped off the truck and people backed off, letting them through. Some of them began edging away as the firemen, seeing no sign of a fire except the bonfire on the grass, began, disgustedly, to kick it apart and stamp out the last flames. When another siren announced the arrival of the police, more people moved off. The cops got out of their patrol car, leaving the red lights flashing, and the rest of the crowd suddenly melted away into the darkness.

Mr. Ridley, who had come in from the kitchen when the firemen arrived, went out onto the porch, and Bran and I followed.

"Took the fire department to get you here, huh?" Mr. Ridley said, as a young-looking patrolman with a pale mustache came up the steps, his hand on the gun at his belt.

"We were on our way here when the alarm was turned in," the cop said, his voice defensive. He didn't sound like a public servant who had come to the aid of a citizen. The other one, a little older, came up and stood slightly behind him. "You were told we'd put your house on a regular patrol schedule. What's happened here?"

Maybe it was the tone of his voice. Or maybe it was watching the mob take off, while the police ignored them. Or the memory of Bran's face when that dummy burst into flame. Whatever it was, I pushed past Bran and stood face to face with the first policeman. I had an almost overwhelming impulse to push him backward, into the other, both of them down the steps. "What kind of policemen are you?" I said, my voice shaking. "How come you didn't go after the people who were breaking the law here?"

"David!" Bran said, under his breath.

I felt his hand on my arm, but I shook it off. The policeman's face blurred, and I realized I was crying. "These people are victims, not criminals! They haven't been throwing rocks and bricks and setting fires! Somebody could have been killed here tonight."

"Listen, kid, I don't know who you are or what you're doing here, but—"

"Go inside," Mr. Ridley said. "I'll handle this."

I wiped my face on my sleeve and Bran pulled me back into the dark house. I went, reluctantly.

"Don't start with cops," Bran said, as he pulled the door closed behind us. "It's a fight you can't win."

I followed him into the kitchen, crunching broken

glass on the way. Bran turned on a light. "Sit down," he said, and pulled a chair away from the table for me.

I started to sit, and my knees gave way so that I thudded down onto the padded red plastic seat. "I don't know what happened to me. I was just so mad, all of a sudden. At that mob, at the cops for letting them go—"

Bran leaned on the counter. "Turn it off."

"What?"

"That anger. Turn it off. It'll get you in trouble."

I studied Bran's bruised face. Only his good eye, aimed intently at me, gave away any feeling at all. Behind that eye it was as if a fire were burning, hot and steady, but trimmed down. Under control. I remembered what Scott Handleman had called him— *spooky*. "You can't just turn feelings off," I said.

"Sometimes you have to."

"You can't! Maybe I could have kept my mouth shut, but I couldn't have turned off how I felt, how I still feel." I held my shaking hand out above the table. "See? If that guy with the dummy came into this room right now—"

Bran waved his hand, shooing away my words like flies. "If he was alone, you might have a chance—if you were mad enough. But he wouldn't be. He's like Nick. Those guys are never alone. Besides, it isn't just them, it's the whole town. And don't kid yourself. The police are part of the town. You get so mad you take them all on, and you're done."

"Are you telling me *you* aren't mad? Aren't mad now?"

He took a long, slow breath. "I can't afford to be." With both hands he smoothed back the hair that had come loose from his ponytail. His hands, I saw, were shaking, too. "You're wrong about feelings. You can force them down so deep you don't have to know they're there. Sometimes that's what you have to do." He pushed himself away from the counter. "You want some cider? I think there's a little left."

We sat at the table and drank the rest of the cider. Gradually, my hands stopped shaking and I began to feel like myself again. When Bran had drained his glass he leaned forward. "I won't let this get to me. I can't. When they took my father away I said it was a mistake. They had the wrong guy. He wasn't perfect, but he couldn't have done what they said he did." Bran ran one finger around the top of his glass. "Then the digging started."

After a while he went on. "The police think he probably killed dozens of kids. He'd offer to help them— find them a job, a place to live. He was this nice, respectable guy in a suit. Kids believed him."

I stared at the cider ring on the table next to my glass and thought about all the pictures of him. This nice, respectable guy in a suit. "It must have been awful to find out."

Bran nodded, absently, as if what I'd said was too obvious to notice. "You try to go back and find clues— something that might have told you what he was really like. I guess they were there. Little flashes, sometimes, when he'd get mad about some unimportant thing. But nothing you don't see in anybody else. I've never been able to figure it out, how he could have been the man I thought I knew and that other one, too, who did what he did. I mean, I had to accept it, you know? When they found the bodies in our yard—"

I thought of the body bags being brought from the place where Bran's fort had been, and my stomach twisted.

"I never visited him in jail." Bran sighed and straightened up in his chair. "I thought if I changed to my mother's last name maybe that would keep me from being Joseph Collier's son. I don't look like him, I don't dress like him. But it doesn't change anything. I *am* his son." He held up his arm. "It's my blood in here, but it's his, too."

I looked at my own arm on the table, the veins at my wrist. My blood, just mine.

He looked at me, his right eye nearly focusing with the other. "He did what he did and I can't change that. But I'm here. Me. I have to figure out who that is. I have a life. I have to do something with it."

Bran took our cider glasses to the sink, and I thought about Dad, who'd chosen a dove for his totem and a wolverine for mine. Whose eyes looked out of the mirror at me.

Mr. Ridley came into the kitchen then. In the light from the fluorescent fixture overhead, he looked much older than he'd looked that day I'd first seen him—as if years had been packed into days. I've heard that people's hair can turn white overnight. It wasn't like that. His thinning hair was the same dark brown, just touched with gray. But there were circles under his eyes, and the skin of his face sagged, pulling his eyes and mouth down at the corners. He could have been Bran's grandfather instead of his uncle.

"A squad car'll be out there the rest of the night. Keep people away."

"They should have been there from the start," Bran said.

Mr. Ridley ran a hand over his face. "A lot of things 'should have been.' They were 'too busy.' " He opened a door next to the refrigerator and took out a broom. "The glass needs sweeping up. Can't have it there when the twins get back. And I gotta cover the windows. It's cold—"

"We'll do it," Bran said, taking the broom from him. "You get some sleep. You've got work tomorrow."

Mr. Ridley looked at Bran for a moment, then turned away. "Not tomorrow. Things are slack at the shop. I'll go down and find something for over the windows."

I checked my watch. "I'd better call my father and let him know I'll be late."

"Why not stay?" Bran said. "You don't want to go out there alone."

He was right. Nick and the others were out there somewhere. When I called Dad I didn't say much about what had happened, only that the police were watching the house.

"You sure you're okay?" he asked. "There was something on the news about a fire call."

"False alarm," I said. "The house is intact and the cops are on guard. The world's crazy, but I'm fine."

There was a long silence, and I half expected him to say he'd come get me. But he didn't. "Sleep tight," is what he said.

18 I didn't exactly sleep tight. Every time I closed my eyes, that dummy's head exploded in flames again against my eyelids. I was on the upper bunk in the room the twins shared with Bran, and I probably did get some sleep, but it didn't feel like it. It felt like a whole night, minute by minute, staring into the darkness, turning and twisting and bumping my elbow on the rail that was there to keep Keith from falling out. The cops must have stayed out in front, as they'd promised, because no one came to chant or throw rocks. But the quiet didn't help me sleep.

Finally, when it began to get light, I gave it up. I swung my legs over the side of the bed and saw that Bran, on his bed against the opposite wall, was awake too. He was sitting cross-legged, a blanket wrapped tightly around him. On the wall above his head were the twins' pictures, Kipp's torn and dirty dinosaurs mended with Scotch tape. "What time is it?" I asked.

"Six-thirty."

I yawned and stretched. "I ought to run. You want to go with me?"

"I'd slow you down."

"We can jog. Or just walk if you want." I jumped down and grabbed my jeans off the lower bunk. "I don't always have to go for speed. I just need to get out—blow the cobwebs away."

Bran threw his blanket off and stood up. "Me, too. It was a pretty lousy night."

"What there was of it. The crazies ought to be in bed at this hour, so we'll have the world to ourselves."

Mr. Ridley wasn't up yet. We had some orange juice, then walked through the gloom of the living room, where the cardboard we'd taped over the windows shut out the light. It was a shock to step outside into the crisp brightness, where our breath puffed white in the sun.

The police car was parked against the curb, and the single cop inside looked asleep, his head against the window. "Eternal vigilance," Bran muttered.

The blackened patch on the trampled grass of the front lawn, and the scattered chunks of burnt firewood, reminded me of the night before, and I didn't know whether it was the memory or the chill that made me shiver. The sun, just over the red and yellow trees across the street, blazed against the deep blue sky, but there was no warmth in it. I blew on my hands to warm them. "Chilly."

"So let's get going."

Without thinking about where we should go, I started

my usual way, up the sidewalk toward the cemetery, at an easy jog. Bran, though he was still favoring one leg, managed to keep up so that our feet hit the pavement at nearly the same time.

"You're pretty serious about running," Bran said.

"I don't know about serious. I just like how it feels. And knowing I can keep going longer than most people."

"That night at the quarry—how far'd you go?"

We'd passed through the gates by then, angling uphill as the road curved. I laughed, remembering how the others had dropped behind, one after the other. "I went all the way down to the highway. Bruno didn't make it half that far. No endurance, those guys."

The sun lit the maple leaves over our heads so that the air around us seemed almost to glow. Shadows danced on the headstones. I looked at the names carved into them. Thomas. Heroux. Catallo. Some of the names were old Ridgewood names I knew well. They were people whose children or grandchildren or great-grandchildren still lived in town. Bits of their genes were still living. So did that mean in some way they weren't really dead? I wondered where Bran's father would be buried, if he were executed.

"It's a pretty place," Bran said, startling me out of my thoughts. "But I don't like it."

"I used to. Not so much anymore." It was true. Ridge Lawn had been only a nice place to run. Now that I kept thinking of the bodies under the grass, it would never be that again.

Bran went on a few paces, breathing hard, before he could talk again. "Did you ever think that we're all under a death sentence?"

I slowed down and looked at him. He slowed, too. "That's an ugly way to look at it," I said.

"Not so bad. We're free at least."

"And we've got time." I looked at a bouquet of fading plastic flowers leaning against a headstone. "Lots and lots of time." I speeded up again, anxious to get out the other side.

We alternated jogging and walking, so Bran could get his breath, and went on for another ten minutes. Then, seeing that we were near Molly's, Bran suggested we stop there. "We don't need to worry about waking anybody," he said. "With the twins there, the whole household will be up. I want to see how they're doing this morning. They were scared last night."

"I can imagine."

"Scared and mad." He smiled. "Keith wanted to get his bow and arrow and start shooting people."

"Sounds like a terrific idea to me."

"The arrows have suction cups on the ends. I don't think he would have been satisfied."

We jogged the last couple of blocks to the rolling lawn where both the tall frame house and the single-story veterinary clinic stood, surrounded by shrubs. Molly, in a tattered terrycloth robe, answered the door almost as soon as I rang the bell, her four dogs milling and jumping around her feet, barking and wagging.

"Bran said you'd be up." The smell of coffee and fresh muffins wafted out to us. Molly, frowning, pushed the dogs back. "What's the matter, aren't you glad to see us?" I asked, as Juno, the Rottweiler, put her huge paws on my chest and tried to lick my chin. I shoved her down.

Molly glanced at Bran and smiled, but it wasn't a convincing smile. "Why wouldn't I be glad to see a couple of sweaty guys on my doorstep at the crack of dawn on a Saturday morning? And me all dressed for company. Come on in, you're letting in the cold."

She led us through the front hall, the dogs still milling around us, Muttsy, in spite of her missing leg, holding her own with the others. In the sun-filled kitchen Mrs. Pepper, in a red velour robe, and Mrs. Ridley, dressed, her purse on the floor beside her, sat at the round oak table, coffee mugs in front of them. "Look who's here," Molly said. Her voice was unnaturally cheerful.

"I wanted to see how the twins were doing," Bran said, patting Juno's head as the others circled him, tails waving. Mai-tai, the one-eyed Siamese, came out from under the table to join the group. "They must be crazy about all these animals. Where are they?"

Mrs. Ridley and Mrs. Pepper exchanged glances, and Mrs. Pepper stood up hurriedly. "Why don't you two join us for some breakfast? We just took some muffins out of the oven. And how about coffee? Molly, get mugs for the boys." She took a pan of muffins from the stove and set it on the table. "Sit, sit!"

It wasn't like Mrs. Pepper to play fussy hostess. Mostly, she expected me to help myself, which wasn't a problem, since I'd practically grown up in her kitchen. But I sat, and Bran joined me, while Molly got two mugs from the cupboard and Mrs. Pepper poured coffee.

"Was there any more trouble last night after Frank called me?" Mrs. Ridley asked.

"Just more of the same," Bran said. "They broke most of the rest of the windows. The police came, finally, and the people left. You don't want to see the front of the house." He didn't mention the dummy or the fire.

Mrs. Ridley sighed. "I hope our insurance covers the windows. And the painting." She looked up at Mrs. Pepper. "Frank says we'll have to have the house painted. The spray paint won't come off. Funny, that's why we got the siding, so we wouldn't have to paint anymore."

I took a muffin and buttered it. Mrs. Pepper went to the refrigerator for some jam and set it in front of me. There was a tension in the room I couldn't put my finger on. Something was wrong, something everybody knew but Bran and me. The muffin was warm, but I hardly noticed the flavor of the bite I'd taken.

"So, where are the boys?" Bran asked, taking a sip of his coffee. "They can't still be asleep. And where's Angela?"

No one said anything. I took another bite of muffin and Bran, looking from his aunt to Molly's mother, put his mug down. Molly picked Mai-Tai up and buried her nose in his fur. "What's the matter?" Bran asked, finally.

"They're gone," Mrs. Ridley said, her voice so low it was hard to hear.

"Gone? You mean they went home already?"

Mrs. Ridley shook her head. "Angela took the twins to a friend's place in Utica. She's coming back this evening to pack up their things. We tried to call last night, but the phone—"

"What do you mean, pack up their things?" Bran asked. "You don't mean she's taking them there to stay!"

The only sound in the kitchen was the low rumble of Mai-Tai's purr. Mrs. Ridley carefully wiped up a coffee spill with her paper napkin, then folded the napkin into a tiny square. She nodded.

"But why?"

"She's afraid, Bran."

"But the police came, finally," Bran said. "They left someone there all night. Nobody's going to hurt the boys while the cops're on guard. And I'll watch them the rest of the time. You told her that, didn't you? I'll watch them!"

Mrs. Ridley looked down at her hands. "That's just it. It isn't only the rock throwers." She cleared her throat. "She's afraid to leave them with you, Bran."

Bran stared at her, his bad eye angled wildly off. The color had drained from his face and the bruises looked darker than ever. "She's afraid of *me*?" he said, his voice little more than a whisper.

"I tried to talk to her." Mrs. Ridley looked up at Mrs. Pepper. "We all did. She says she likes you, and if it were just her, she wouldn't even think about it. But you know how much she loves the twins—"

"Yeah—I know," Bran said.

"She says we don't know—even *you* don't know—

whether those doctors are right. Whether you have something of Joseph in you somewhere that might come out someday."

Bran's voice was flat. "She thinks I'd hurt Kipp and Keith?"

"She's afraid you wouldn't be able to help yourself."

Mrs. Pepper put her hand on Bran's shoulder. "She's very young, Bran, and frightened for her children."

"Dad tried to tell her there isn't any evidence for that bad seed thing," Molly said. "She said that didn't mean it couldn't be."

I looked down at the plate in front of me. I had broken my muffin into crumbs. Mrs. Ridley spoke again, and I saw that there were tears on her cheeks. "Try to understand. She loves those boys more than anything in the world."

"So do I." Bran pushed himself away from the table and got up. "Thanks for the coffee," he said to Mrs. Pepper. "I've got to be going."

Molly put down the cat, as if to go to the door with him, but he shook his head at her and hurried out, the dogs following. When she heard the door slam, Mrs. Ridley put her head in her hands.

I stood up to go after Bran. "Let him go, David," Mrs. Pepper said. "Let him go."

19 Not long after Bran left, Molly drove Mrs. Ridley home and I went along, squeezed into the back of the Civic. Bran and his uncle were nailing plywood up over the front windows when we got there. When Molly asked, Bran said he was okay, but he didn't look okay. He was moving in a kind of slow motion, the way he had after the beating. Molly offered to stay and help, but he told her to go on home. I couldn't even offer, because I had to work from eight to four. The police had gone, but Mr. Ridley said they'd promised to come by every couple of hours to check on things.

Molly dropped me off at home, and we agreed to go back to Bran's when I got off work. "Nothing's going to happen during the day," I said, partly to reassure her, partly to reassure myself.

"I just don't think he ought to be alone."

"He isn't alone. His aunt and uncle are there."

"You know what I mean."

≪   ≫

All day as I bagged groceries and dragged carts in from the parking lot, I found myself looking at the people coming and going from the store. Had any of them been there last night?

I watched mothers trying to keep their kids from taking candy bars off the shelves by the checkout line, fathers telling kids to stay away from the gum and prize machines, people with food stamps, guys buying beer and potato chips, elderly couples using handbaskets instead of carts. Just people. The kinds of people I'd seen around Ridgewood all my life and felt comfortable with. Safe with.

By the time my shift was over and Molly had come to pick me up, I'd almost convinced myself that the people who'd cheered the burning of the dummy must have come from someplace else. I wanted to believe that there was something different about people like that, something you could see. Something that could warn you.

"Have you talked to Bran since this morning?" I asked, when I got into the car.

Molly shook her head. "I called a couple hours ago, and Mrs. Ridley said he was taking a nap, but she hoped we'd come over later."

The Ridleys and Bran were in the kitchen when we got there, and all the lights were on. Even the windows that hadn't been broken the night before had been boarded over "just in case," Mr. Ridley explained, so it was like night inside. The blankness of the windows, not being able to see out, gave the house the feeling of a prison. I didn't like it. Didn't like the feeling that if the mob came back, there was no way to see what they were doing.

"I'm glad you're here," Mrs. Ridley said, as Molly and I joined her at the kitchen table. "Maybe you can help us talk some sense into Bran's head. He's talking about going back to New Jersey, letting the state put him back into foster care."

Bran was sitting on a kitchen stool, his hands clasped around his knees, his shoulders hunched. His bad eye wandered as he looked up at us, as if he were too tired to control it at all. "I survived it before."

"It was bad then, and it would be worse now, with the trial."

"It's the only thing that makes any sense. I can't stay here. Look what it's doing to you. Your house is wrecked." He turned to his aunt. "You've lost the job you had for ten years."

"Maybe it was time for a change," Mrs. Ridley said. "You're family." She looked at her husband, who was leaning against the counter, his arms crossed, staring down at the floor. He didn't look up. "I'm not going to send you back to stay with strangers."

"We're leaving Ridgewood," Mr. Ridley said, his voice tight and hard. Bran turned to look at him. "We decided while you were sleeping. We'll sell the house and find someplace else, some other town. Start over."

Bran slammed his fist against the wall and we all jumped. "You can't do that! Your friends are here. Your lives are here!"

Mr. Ridley shook his head. "Looks like that's changed. I didn't see any friends here with us last night."

"Our lives will just have to be wherever *we* are," Mrs. Ridley said.

"But your jobs—" Bran said.

"There are restaurants in other towns. And Frank's been thinking about a garage of his own for years."

Mr. Ridley made a sound in the back of his throat that could have been agreement. Or something else.

"We'll manage." Mrs. Ridley reached out and put a hand on Bran's head. His face, as always, was still, but pale, the bruises turning yellow at the edges. She brushed a strand of hair back. "I had a sister once who got mixed up with the wrong man and before she figured that out, she had you. She was too young to know what to do—younger than Angela—so she ran away. There wasn't a thing I could do then. We lost her, and I thought we'd lost you, too. But you're here now and that's how it should be. I'm not going to lose you again. You're not going to strangers."

"You think it'll be better in any other town?" Bran asked. "As soon as they find out who I am—"

"We'll cross that bridge when we come to it," Mrs. Ridley said.

"I can go some place by myself. Get a job—"

"And end up on the streets?"

Mr. Ridley overrode his wife. "Understand this, Bran. With or without you we're leaving Ridgewood. I won't stay here after this."

Mrs. Ridley turned to us. "You tell him he's got to go with family."

"You can't go back to where the trial is," I said.

"He's right," Molly said. "And you can't live by yourself."

Bran nodded, wearily. "All right. But I'm not making any promises. If we go somewhere else and this starts happening again—"

Mrs. Ridley looked at the clock over the stove. "Angela will be here pretty soon to get the twins' things."

This time Bran's face registered his feelings, and I looked away. It hurt to look. "I don't want to be here," he said.

"Come to my place," I offered. "I'll fix dinner, and you can meet my dad. You come, too," I said to Molly.

Dinner that night was like a time out. A sort of vacation from everything that had been happening. Dad was in the workshop when we got there, and while Molly and I made chili and a salad, he gave Bran the grand tour of the Watson wood sculpture and totem museum, inside and out.

By the time they came in to eat, Dad was talking to Bran as if they were old buddies. He'd been expounding his philosophy of totems, and apparently Bran was hooked. As we ate, Dad talked about the Northwest Indians, who'd invented totem poles, and how some would choose a totem to be their guardian spirit.

He gave us more of his philosophy than I had ever heard before. He had added plenty of his own ideas to the old Indian traditions, and mixed in a little Hinduism, a little Buddhism and a lot of modern art. But basically, he'd kept that guardian spirit idea.

Everybody had a link, he said, with an animal. More than one. Everything people were, an animal was, too. He talked about how a person's totem could influence his life. Whenever he'd stop to concentrate on his chili, Bran would ask him a question and get him started all over again.

Once, when Dad was telling about the woman who had insisted he make her a garden totem that included the grizzly bear she'd chosen as her own symbol, her husband's eagle, and their pet Pekingese—"not a symbol for him, but a carving of the actual dog"—Bran actually laughed. A real laugh. Molly jabbed her elbow into my side, and I nodded. It was as if everything outside our house had disappeared for a while and the four of us were the only people in the world.

"You should have seen the finished piece," Dad said, waving a saltine in the air. "I did it just the way she wanted it. Bear on the bottom, eagle on the bear's head, and on top of the eagle, that Pekingese. Awful little dog, the real one was. Nipped me three times before we were done. So I carved a flea into the base of its tail. Out there in a garden somewhere is a wooden Peke with a wooden flea forever biting its rear end. I like to think for the rest of its life that wretched beast had an itch it couldn't scratch."

For a while Bran was the way he always was with the twins—open and relaxed and easy. When dinner was done, Molly and Bran and I got into a mock battle over who'd wash and who'd dry the dishes. Even Dad got into it, flicking us with a dish towel, dodging around the table when we tried to get him back.

Finally, though, the break was over. The dishes were done, and Dad said goodby and went back to work. Bran looked outside into the darkness. His face seemed to close down again. "If you two don't mind," he said, "I think I'll walk home."

"I'll take you," Molly said. "I don't think it's such a good idea for you to be out there alone."

"I need some time by myself. To think about what to do. Uncle Frank and Aunt Marie are at the house and I'm just not up to facing them right now. Their lives were just fine here till I came along." He looked out the window again. "It's nice out. I thought I'd walk up to the quarry. Climb down to your platform. Just sit and watch the water for a while. It's so quiet up there."

"You can't go there," Molly said. "Nick knows about it now."

"Not the platform. Besides, he won't be up there without his car. If I see it, I'll turn right around and come back."

"Why don't you just stay here," I said. "We'll leave you alone, if that's what you want. You can have my room. Molly and I can watch a movie on TV and you—"

"I'll be okay."

"Then let us take you at least," Molly said.

"I want to walk awhile. You know, blow the cobwebs away."

I didn't want to let him go off alone, but I didn't know how to stop him. We couldn't just force him into Molly's car. And he'd made it clear he didn't want us with him.

"You'll be careful," she said. "Promise."

"I'll be careful."

"If you see Nick and the guys, or if the crazies are hanging around your house, you'll come right back here," I said. "And stay."

"Okay."

We walked outside with him. It was clear and surprisingly warm. "Indian summer's coming," Molly said. "And will you look at that moon!"

Full, or nearly full, the moon rode above the trees, throwing so much light that the bushes cast shadows across the driveway. A plane was moving just above the moon, leaving a silver trail behind it. Bran watched it for a moment. "I wonder where it's headed," he said.

"West," Molly observed. "Chicago maybe. With a stop in Pittsburgh."

"Or Detroit," I said.

"I'd take either of them right now."

Molly shook her head. "It might be Cleveland."

"Even Cleveland." He looked down. "I wonder if it would make a difference."

His voice sounded hollow to me, full of such pain that I threw my arm around his shoulder. Embarrassed, suddenly, I stuck a foot out and pretended to be throwing him to the ground. "We can't let this man go to Cleveland!" I said.

Bran pushed me off. "All right, then. I won't go there." He turned and looked toward the garage, where light spilled from the windows. "I'd like your dad to carve me a totem someday. A bird."

"An eagle?" I asked.

He shook his head. "A raven, I think."

I punched him on the arm. "Nevermore, huh? Okay."

"I couldn't afford it, though."

I laughed. "My dad's not into money, didn't you notice? If you want a raven, he'll carve you a raven."

"If he gets a raven," Molly said, "I get a wolf. I've always wanted a wolf."

"I'd have thought a toad for you," I said.

Molly aimed a kick at me and I jumped out of the way.

"I'd better be going," Bran said. "I don't want to be out too late. Aunt Marie'll get nervous. You running tomorrow?"

"Sure. You want to come along?"

He shook his head. "I'll just wave as you go by. See you." He started down the driveway, his hands in his pockets.

"Is it okay if I come over tomorrow after church?" Molly called.

He turned back. "Could I stop you?"

"Nope."

"See you after church, then." He turned onto the sidewalk and disappeared behind the bushes.

"I'd wish he'd stay," Molly said.

Whenever I think about that night, I hear her say that, over and over, like a stuck record. "I wish he'd stay." But he didn't.

Molly and I went back inside and watched an old movie on television. It was just finishing when the phone rang.

It was Mrs. Ridley. "Is Bran there?" she asked.

"No, Mrs. Ridley, he left." Molly turned off the television and came to stand by the phone. "He said he wanted a little time by himself—to think. He was going up to the quarry for a while."

"There's a crowd outside again. I wanted to warn him. The police are here, but I don't want him to try to get through that mob. Did he say when he was coming home?"

I covered the mouthpiece and whispered to Molly that Bran wasn't home yet. "He didn't want to be out so late you'd get worried about him," I told Mrs. Ridley. It had been two hours, I realized. Had he meant to stay that long?

"Well—" Mrs. Ridley cleared her throat. "I'm sure he's all right, but—" Her voice sort of dwindled away.

"How about if we go up and get him? We can bring him home in Molly's car. Or back here if you think that would be better."

"Would you, David? I hate to put you out—"

"No problem, Mrs. Ridley."

"And have him call me, would you?"

"Sure." Molly was already getting our jackets. "We'll go right now."

20

"I hope he's still up here," I said, as Molly maneuvered the curves of the quarry road. "If we miss him, he'll walk right into whatever's happening at the house."

Molly, concentrating as the car skidded on the gravel, didn't say anything. I remembered how it felt to be inside the Ridleys' house with the windows covered, and was glad we'd be taking Bran back to my house. Maybe we could pretend for just a while longer that things were okay. Just for tonight. And maybe tomorrow. Molly slowed down as we passed the place where the path left the road. "Almost there," I said. "You can pull over just around this curve."

We rounded the curve and she slammed on the brakes as our headlights shone on the back of a rust-spotted Mustang, stopped partway into the weeds. "Nick's car," Molly said. "Damn!"

I felt my hands suddenly go cold, and my stomach tighten. "They must have come after he was already here." Molly pulled up behind Nick's car and set the emergency brake. "Maybe he's still down on the platform, waiting till they go." I shoved open the door, pushing sumac branches out of the way as I scrambled around the car, trying desperately to think of something we could do besides go in there to face Nick. I thought of what Bran had said, that Nick was never alone. They'd all be here.

Molly was out of the car, rummaging under the seat. "Great! The batteries are shot," she said, holding up the flashlight, whose beam was easily outdone by the moonlight that flooded down around us.

"My fault."

She slammed the door. "I'll take it with me anyway. I don't feel like meeting up with those guys barehanded."

We hurried down the road and stopped where the path led into the darkness under the trees. "We should get help," I said.

"No time," Molly answered.

I nodded, trying to swallow the lump in my throat that seemed to be cutting off my breathing. "Hang on," I whispered, as she started down the path. "Let me find a stick or something." I scrabbled around in the tall grass, and found nothing.

Molly stopped and kicked at something. "Take this," she said, her voice low. "By my right foot."

I found what she meant, a rock she had loosened. I dug it out, and it filled my hand with a rough, heavy sense of security. "Okay, let's go."

Moving as fast as we could in the thick darkness, we

stumbled along the trail, slipping in the leaves, ducking low branches. As we got close to the shack I thought I heard voices. I put out my hand to stop Molly. "I think I hear them," I whispered.

We stood for a moment, listening. The only sounds were the wind in the trees and the slow murmur of late crickets.

"We don't want them to hear us coming," Molly whispered.

We went on, picking our way carefully, until the path opened out onto the quarry rim. The clearing was empty in the moonlight. But a gleam shone through the bullet holes in the old shack briefly, as if someone had moved a light inside. A dark form came out of the trees across the clearing and toward the shack. "Nobody up here," it called. It was Gordon.

"Here either." Nick's voice came from the shack, and the light gleamed again. "Might as well bring the beer."

"It's in the car," Gordon said, pushing through the sumacs toward the shack entrance.

"You want to go check the platform, or should I?" Molly asked, her lips against my ear, when Gordon was out of sight.

"If he's there, he'll know better than to come up now," I whispered. "We could go back to the car."

"We can't leave till they do, in case they find him. You stay here," she said, "and I'll go around to the other side. Get hidden, and wait."

Before I could say anything, Molly slipped off, weaving her way through the underbrush. I could hear Nick's and Gordon's voices now, but I couldn't make out what they were saying. I hoped they'd stay in the shack till Molly got herself hidden, because I was terribly aware of the sounds she was making.

I moved off the path and stood against a tree trunk, the rock heavy in my hand. Bran had probably started for home already, I thought, and we ought to be out trying to find him, to warn him about the mob he was walking into.

The dark bulk of Gordon came back through the sumacs. "Why don't you get something yourself, once in your life," he was saying. He came out into the full moonlight, started toward me and then stopped. "Is that you, Matt?" He turned toward where Molly had to be by now. There was no sound for a moment. "Matt?" Gordon called again.

"Wh-wh-what?"

I jumped and nearly dropped my rock. The voice had come from behind me, between the path and the quarry rim. I turned and could see Matt, a heavy form with a pale face, moving among the shadows. I froze. Then, as slowly as possible, turned back toward the tree trunk, to cover my face.

By that time, Gordon had gone to find out what was making the noises he'd heard. "Nick!" he shouted. "Come see what I found. The Goblin Girl!"

"Get your hands off me, you slime!" Molly's voice rang across the clearing.

Matt was still coming toward the path. If I moved, he'd see me for sure. I decided to wait till he got past me. If Molly was going to need help, I figured I'd have a better chance if I could surprise them.

Around the edge of the tree trunk, I could see the beam of Nick's light as he joined Gordon. "Well, looky here," he said. "You all alone up here, or you got that psycho friend of yours with you?"

"He's no psycho," she said.

"That's not what I hear. I'm surprised you aren't scared of him. Don't you think she ought to be scared, Gordo? Bad seed and all."

"I'm not scared of Bran Slocum," Molly said.

Matt had gone on past me now, and joined the others. I looked around the tree and saw the four of them clearly in the moonlight and the light of the big, fluorescent lantern Nick was holding. Gordon had a tight grip on both Molly's arms, and she was squirming to get loose. "Looks like *you* are, though, Bruno," Molly said, aiming a kick sideways at Gordon.

"What makes you think I'm scared of him?" Nick asked.

"You must be, the way you gang up on him. You haven't the guts to take him on by yourself."

"Nick Bruno's not scared of anything or anybody," Nick said.

The sumacs behind him moved and Bran stepped out from beside the shack. He was breathing hard, but his voice was steady. "Good," he said. "Because here I am."

Nick spun around and dropped his lantern. It tipped over onto its side, spreading its light at their feet. Bran stood, his legs braced, his arms loose at his sides. His earring glinted in the moonlight.

"Get him, guys!" Nick yelled, and Gordon flung

Molly to the ground to go after Bran. Matt, too, started moving toward him, a little more warily. As Molly went down, I started forward. By the time I reached the clearing, Gordon had pinned one of Bran's arms and Nick had punched him in the stomach. Bran's foot flashed up and caught Nick in the groin. As Nick staggered backward, doubled over, Molly threw herself onto his back, hitting at him with the flashlight. Matt was grabbing at Bran's other arm. I yelled and ran at Matt, the rock balanced in my right hand. He turned toward me and I was on him, my hand and the rock slamming against the side of his head.

I gasped, partly with the pain in my hand, partly with horror at the sound the blow had made. Matt took a step toward me, crashed into my chest and sank to his knees. He grabbed at me and caught my right leg as he went down, throwing me off balance. I fell sideways, the two of us tangled together, and my ankle twisted under me as I hit the ground, sending a searing pain up through my leg. Matt lay against me, curled on his side, both hands to his head.

I struggled to my feet, but my ankle gave way under me, and I fell again. Nick had thrown Molly off onto the ground and was kicking at her as she scrambled backward to get out of his way.

He turned to where Gordon and Bran were locked together, each trying to wrestle the other to the ground. Gordon had Bran by the hair, and Bran's arms were wrapped around Gordon's chest. They careened into the side of the shack and nearly went down together.

But they managed to stay on their feet, their struggle taking them crashing into the low sumacs. They were hardly more than gray shapes, locked together in the moonlight.

Molly threw herself at Nick again, but he shoved her roughly to the ground and picked up a heavy stick. "Get off him now, Gordon," he yelled. "He's mine!"

He advanced on Gordon and Bran, the stick cocked like a baseball bat, and when he got close enough, he swung. There was a crack that seemed to echo off the quarry walls and Gordon screamed. "My arm, Bruno! You broke my arm!" He seemed to be trying to pull away, but Bran held on.

Nick didn't stop. He didn't even pause. On top of them now, he swung the stick again and again. The three of them disappeared around the shack toward the quarry's rim. I could hear the stick landing and Gordon screaming at Nick to stop, to let him get out of the way.

Molly was in front of me suddenly, pulling at my arm. "Get up! We've got to stop him. Get up!" She dragged me to my knees, and while I was still trying to stand, she took off, limping, around the other side of the shack, the broken flashlight still clutched in her hand. I touched my right foot to the ground and found that I could put a little weight on it. Hopping and limping, my teeth gritted against the pain, I headed for the sounds of the struggle. As I came around the shack, Bran, Gordon and Nick were dark figures, silhouetted against the silvery water of the quarry below them. Gordon, one arm dangling, was hitting at Bran's face with the other, trying to get him to loosen his grip, but Bran held on. Nick raised the stick again, tinged with moonlight, and brought it down on Bran's head.

For a moment everything seemed to stop, as if a camera had frozen the frame. Then slowly, almost gently, Bran and Gordon tipped toward the water. There was a sound of rocks sliding as they went over and Nick, too, began to slip toward the edge. Molly screamed. Nick dropped the stick and, as his feet slid from under him, grabbed at the plants that grew along the rim. Whatever he touched gave way, and suddenly he, too, was gone. There was a tremendous splash, followed by a thud as something—someone—hit the platform. Then a heavy, sliding sound and another splash.

I sank to my knees and pulled myself over to the edge, where I could look down into the water. In the moonlight, the ledge was empty. The little tree that had held the door in place dangled, leaves downward, its trunk broken. Ripples and splashes broke the silvery sheen of the moonlight on the water, and I thought I could make out dark heads bobbing against the confusion of reflected light. One, two—only two of them.

Carefully, I turned and lowered myself over the edge, feeling with my left foot for the cracks and steps that led down to the ledge and from there to the rocks below. My eyes blurred and burned, and I rubbed my sleeve across my face. When I reached the broken tree I strained to see those dark heads against the water again. There was a flurry of splashing.

"Gordo," came Nick's voice from beneath the ledge I was on, "is that you? Are you okay?"

There was a gulping, choking sound, and then, unmistakably, it was Bran who answered. "It's me—I'm not—okay." I wanted to shout with relief, knowing he was still there.

The splashes grew louder now, and I could make out Nick, heading in a flurry of awkward arm movements toward the center of the quarry, where Bran was a dark spot, his hands splashing sporadically among the silver ripples.

For a moment, I couldn't believe it. I considered jumping in after Nick, dragging him back so Bran could make it to the side. But he was already too far from me, too close to Bran. Desperately, I felt around the ledge for something to throw, but came up with nothing but small sticks and bits of rock. I threw them as hard as I could, but they only pitted the water with little circles.

"Stop it, Nick!" Molly screamed and something bigger splashed into the water behind him. I looked up to see Molly crouched on the rim. "Let him alone. That's enough!"

But Nick didn't stop. He'd closed with Bran now, and another struggle was going on, punctuated with splutters, coughs and splashes. I was on my feet, getting ready to jump in, when the sounds of the struggle slowed and then stopped. There was a silence, and I became aware of crickets from above and all around, their chirring filling the quarry.

I peered out into the water and saw one head. Just one. There was a feeble splash and the head went down, then came up again, gagging. A hand moved. Then again.

I lowered myself over the ledge and climbed down to the rocks at the water's edge, my ankle sending flashes of pain with every step. Above me, I heard Molly starting down. Another splash, still far from the rocks.

"Help me." It was Nick's voice, choked and desperate. "I can't make it. Help."

I heard a sob and realized it had come from me. My chest burned and I felt tears hot on my cheeks. I crouched by the water and looked out into the shimmering silver. Molly jumped down from the ledge and landed next to me. I rubbed my eyes, clearing them, and saw Nick, about ten feet out. He was moving only a little, his head barely breaking the surface of the water. I took a deep breath that seemed to tear at my throat, and sat down. After a moment, Molly sat down next to me.

Neither of us moved. I'm not sure I even breathed until Nick, too, had disappeared beneath the water, and the moon's reflection, unbroken now, smoothed itself into a silver trail. We sat there, not touching, not talking, for what seemed like a long time. And then we climbed up and went for the police.

21 Molly and I told the police our story, though I only said that during the fight all three of them fell into the water. I didn't say anything about what happened after that. I guess Molly didn't either. I don't know what Matt told them. But in all the news about it afterward, nobody ever said anything about a fight.

Mayor Mahoney gave a speech the following Monday to "lament the tragedy at the quarry." He said Ridgewood had lost young men with great promise who had been important to its future. He didn't mention mobs or special school board meetings, or the house on Larch Street with the boarded-up windows and BAD SEED in red paint all over it. By the end of the week new No Swimming signs had been put up around the quarry—as if what had happened was a swimming party that got out of hand.

That's how the newspaper handled it, too. They printed yearbook pictures of Nick and Gordon, the "top student athletes" who had drowned, and said that no pictures were available of Bran Slocum, the other victim, who had been new to Ridgewood High School. They never even mentioned Joseph Collier. It was almost as if the month of October had never happened, except for that "accident" at the end.

Bran didn't have much of a funeral—just the Ridleys and Molly and her parents and Dad and me. Mr. and Mrs. Ridley didn't want to bury Bran in Ridgewood, but there wasn't anyplace else. He was off in a far corner, sort of by himself, a mound of dirt and a little metal marker with his name and dates. Dad and I promised to take care of his grave after they moved. They couldn't afford a headstone, so Dad said he'd carve the raven Bran had wanted.

Nick's and Gordon's parents held a memorial service at the high school, and most of the town turned out. I felt like I had to go. Matt was there, his head bandaged, and Jerry Ritoni, who'd missed being up at the quarry because he'd gone to join the mob at Bran's house instead. Mr. Byrd and Dr. Towson and the woman who'd carried the Kid Killer sign were there. All the kids came, and there was a lot of crying.

A minister read from the Bible and said that in our grief at the loss of "our children" we should be happy that God had taken them to be with him. That was when I walked away. I couldn't think about heaven right then, and who did or who didn't deserve to be there.

The national paper that had started it all reported Bran's death inside, with no picture and just a tiny headline, "Collier's Son Drowns." I guess dead, Bran wasn't a very exciting story.

The Collier trial lasted through November and all the way into January. In late January it ended. Collier was convicted and sentenced to death. There was a front page headline in the paper, but that was all. Nobody in Ridgewood seemed to care much about Joseph Collier anymore. He was just another story that was happening outside somewhere.

I quit the track team and just about everything else, except work. But I kept running every morning when the weather wasn't too bad. It felt good and clean and alone, the one thing I could almost enjoy. I didn't run through the cemetery very often, but when I did, I always stopped at Bran's grave.

Dad carved Bran's raven out of oak so that it would last. He carved it with its wings spread, looking up. I think Bran would have liked it. Among all those gravestones, the Bibles and the angels, it was certainly different. Bran would have liked that, too.

Molly and I never talked about that night, about Nick and what we did. What we didn't do. I don't think we ever will. We don't talk much at all anymore.

I thought about the plague a lot all winter, whatever it was that had infected us all and turned ordinary people into people who could kill someone—or let someone die. It had all started because Joseph Collier was a murderer, and I tried to figure out what the difference was between him and the rest of us. There was one, I knew that. But I couldn't seem to draw a clear line.

Finally, I decided Molly was wrong to call what happened here a plague. This wasn't something that came from outside. It was inside all of us, whatever it was. And it still is.

Bran ran out of time before he could do anything with his life. But maybe all he needed to do was just be who he was. He never forgot that bird's heart, beating in his hand. I think he would have helped Nick if he'd been where I was that night.

Molly will go on taking in strays and standing up for someone everybody else is against because that's who she is. That part of her is stronger than the part that sat with me on the rock that night. Most times it'll win.

That leaves me. When the new grass was up and the tulips Molly planted on Bran's grave were blooming, I went out there. The sun was shining on the raven, and it seemed almost ready to take off. Bran was just another body under the grass, but I kept seeing him with a twin under each arm, taking them to get Band-Aids for their knees.

And I made a decision. I'm going to save all the money I earn at the store so I can go to college. I'm going to do something with my life, like Bran wanted to—something to stand against what I found inside myself that night. I don't know what I'll do, yet. I just know it will be something that makes a difference.